HSDPA/ HSUPA
HANDBOOK

T0172521

INTERNET and COMMUNICATIONS

This new book series presents the latest research and technological developments in the field of Internet and multimedia systems and applications. We remain committed to publishing high-quality reference and technical books written by experts in the field.

If you are interested in writing, editing, or contributing to a volume in this series, or if you have suggestions for needed books, please contact Dr. Borko Furht at the following address:

Borko Furht, Ph.D.
Department Chairman and Professor
Computer Science and Engineering
Florida Atlantic University
777 Glades Road
Boca Raton, FL 33431 U.S.A.

E-mail: borko@cse.fau.edu

HSDPA/ HSUPA
HANDBOOK

Edited by
Borko Furht
Syed A. Ahson

CRC Press
Taylor & Francis Group
Boca Raton   London   New York

CRC Press is an imprint of the
Taylor & Francis Group, an **informa** business

CRC Press
Taylor & Francis Group
6000 Broken Sound Parkway NW, Suite 300
Boca Raton, FL 33487-2742

Printed in the United States of America on acid-free paper
10 9 8 7 6 5 4 3 2 1

International Standard Book Number: 978-1-4200-7863-3 (Hardback)

Library of Congress Cataloging-in-Publication Data

HSDPA/HSUPA handbook / editors, Borko Furht and Syed A. Ahson.
    p. cm. -- (Internet and communications ; 12)
  "A CRC title."
  Includes bibliographical references and index.
  ISBN 978-1-4200-7863-3 (alk. paper)
  1. Packet switching (Data transmission)--Handbooks, manuals, etc. 2. Digital communications--Handbooks, manuals, etc. 3. Universal Mobile Telecommunications System--Handbooks, manuals, etc. I. Furht, Borivoje. II. Ahson, Syed. III. Title. IV. Series.

TK5105.3.H724 2010
621.382--dc22
                                           2009052320

Visit the Taylor & Francis Web site at
http://www.taylorandfrancis.com

and the CRC Press Web site at
http://www.crcpress.com

Contents

Preface

Mobile users are demanding higher data rates and higher-quality mobile communication services. The 3rd Generation Mobile Communication System is an outstanding success. The conflict of rapidly growing numbers of users and limited bandwidth resources requires that the spectrum efficiency of mobile communication systems be improved by adopting some advanced technologies. It has been proven, both in theory and in practice, that some novel key technologies such as MIMO (multi-input, multi-output) and OFDM (orthogonal frequency division multiplexing) improve the performance of current mobile communication systems. Many countries and organizations are researching next-generation mobile communication system, including the ITU (International Telecommunication Union), European Commission FP (Framework Programme), WWRF (Wireless World Research Forum), Korean NGMC (Next Generation Mobile Committee), Japanese MITF (Mobile IT Forum), and China Communication Standardization Association (CCSA). International standards organizations are working for standardization of the E3G (Enhanced 3G) and 4G (4th Generation Mobile Communication System), such as the LTE (Long Term Evolution) plan of the 3GPP (3rd Generation Partnership Project) and the AIE (Air Interface of Evolution)/UMB (Ultra Mobile Broadband) plan of 3GPP2.

The HSDPA (High-Speed Downlink Packet Access) standard was introduced in Release 5 in 2002, followed by the introduction of HSUPA (High-Speed Uplink Packet Access) in Release 6 in 2004. The HSUPA and HSDPA are combined under the same standard and known as the HSPA standard. HSDPA is an enhancement of UMTS (Universal Mobile Telecommunications System) networks that supports data rates of several megabits per second (Mbps), making it suitable for data applications ranging from file transfer to multimedia streaming. The introduction of High-Speed Packet Access (HSPA) greatly improves the achievable bit rate. HSDPA has been standardized as an extension of the UMTS as a part of the 3GPP Release 5. It is spectrally the most efficient WCDMA (Wideband Code Division Multiple Access) system commercially available at the moment.

UMTS networks that are currently offering both legacy and HSDPA/HSPA services have upgraded their UTRAN (UMTS Terrestnial Radio Access Network) functionalities based on the Release 5/6 or higher 3GPP standard. The new standard supports both legacy services as well as advanced packet-based HSPA services. Introduction of HSDPA and HSUPA services has increased the packet-switched traffic volume in the UTRAN and in the core network (CN). The UTRAN architecture is currently evolving toward a high data rate and high QoS (Quality of Service) network. Recently, the E-UTRAN (Evolved UTRAN) architecture has been introduced and was designed to support advanced packet-switched services using a flat network architecture to accommodate new services as well as to offer high QoS to all services.

This book provides technical information about all aspects of HSPA technology. The areas covered range from basic concepts to research-grade material, including future directions. This book captures the current state of HSPA technology and serves as a source of comprehensive reference material on this subject. It has a total of 13 chapters authored by 30 experts from around the world. The targeted audience for this Handbook includes professionals who are designers and/or planners for HSPA systems, researchers (faculty members and graduate students), and those who would like to learn about this field.

The book is expected to have the following specific salient features:

- To serve as a single comprehensive source of information and as reference material on HSPA technology
- To deal with an important and timely topic of emerging technology of today, tomorrow, and beyond
- To present accurate, up-to-date information on a broad range of topics related to HSPA technology
- To present the material authored by the experts in the field
- To present the information in an organized and well-structured manner

Although the book is not precisely a textbook, it can certainly be used as a textbook for graduate courses and research-oriented courses that deal with HSPA. Any comments from the readers will be highly appreciated.

Many people have contributed to this Handbook in their unique ways. The first and foremost group that deserves immense gratitude is the group of highly talented and skilled researchers who have contributed 13 chapters to this Handbook. All of them have been extremely cooperative and professional. It has also been a pleasure to work with Rich O' Hanley, Amy

Blalock, and Kari Budyk of CRC Press, and we are extremely grateful for their support and professionalism. Our families have extended their unconditional love and strong support throughout this project, and they all deserve very special thanks.

Borko Furht

Syed Ahson

About the Editors

Borko Furht is a professor and chairman of the Department of Computer Science and Engineering at Florida Atlantic University (FAU) in Boca Raton, Florida. He is also Director of the NSF-sponsored Industry/University Cooperative Research Center at FAU. Before joining FAU, he was a vice president of research and a senior director of development at Modcomp (Fort Lauderdale, Florida), a computer company of Daimler Benz, Germany; a professor at the University of Miami in Coral Gables, Florida; and a senior researcher in the Institute Boris Kidric-Vinca, Yugoslavia. Borko received a Ph.D. degree in Electrical and Computer Engineering from the University of Belgrade. His current research is in multimedia systems, video coding and compression, 3D video and image systems, video databases, wireless multimedia, and Internet computing. He has been Principal Investigator and Co-Principal Investigator of several multiyear, multimillion dollar projects—on Coastline Security Technologies, funded by the Department of Navy, One Pass to Production, funded by Motorola, and NSF PIRE project on Global Living Laboratory for Cyber Infrastructure Application Enablement, and NSF High-Performance Computing Project. He is the author of numerous books and articles in the areas of multimedia, computer architecture, real-time computing, and operating systems. He is a founder and editor-in-chief of *the Journal of Multimedia Tools and Applications* (Springer). He has received several technical and publishing awards, has consulted for many high-tech companies including IBM, Hewlett-Packard, Xerox, General Electric, JPL, NASA, Honeywell, and RCA, and has been an expert wetness for Cisco and Qualcomm. He has also served as a consultant to various colleges and universities. He has given many invited talks, keynote lectures, seminars, and tutorials, and also has served on the board of directors of several high-tech companies.

Syed Ahson is a Senior Software Design Engineer with Microsoft Corporation (Redmond, Washington). As part of the Mobile Voice and Partner Services group, he is busy creating new and exciting end-to-end mobile services and applications. Prior to Microsoft, Syed was a Senior Staff Software Engineer with Motorola, where he contributed significantly in leading roles toward the creation of several iDEN, CDMA, and GSM cellular phones. Syed has extensive experience with wireless data protocols, wireless data applications, and cellular telephony protocols. Prior to joining Motorola, Syed was a Senior Software Design Engineer with NetSpeak Corporation (now part of Net2Phone), a pioneer in VoIP telephony software.

Syed has published more than ten books on emerging technologies such as WiMAX, RFID, Mobile Broadcasting, and IP Multimedia Subsystem. His recent books include *IP Multimedia Subsystem Handbook* and *Handbook of Mobile Broadcasting: DVB-H, DMB, ISDB-T and MediaFLO*. Syed has authored several research articles and teaches computer engineering courses as an adjunct faculty member at Florida Atlantic University (Boca Raton, Florida) where he introduced a course on Smartphone technology and applications. Syed received his M.S. degree in Computer Engineering in 1998 from Florida Atlantic University, and his B.Sc. degree in Electrical Engineering from Aligarh University, India, in 1995.

List of Contributors

Belal AbuHaija
University of Glamorgan,
Pontypridd, Wales, United Kingdom

Syed Ahson
Microsoft Corporation
Redmond, Washington, United States
and
Department of Computer Science
Florida Atlantic University
Boca Raton, Florida, United States

Ahmet Baştuğ
Vestek Electronics
Istanbul, Turkey

Khalid Al-Begain
University of Glamorgan
Pontypridd, Wales, United Kingdom

Sebastian Caban
Institute of Communications and
    Radio-Frequency Engineering
Vienna University of Technology
Vienna, Austria

Shuping Chen
Key Lab of Universal Wireless
    Comunications
Beijing University of Posts
    and Telecommunications
Beijing, China

Salah E. Elayoubi
Orange Labs
Issy-les-Moulineaux, France

Borko Furht
Department of Computer Science and
    Engineering
Florida Atlantic University
Boca Raton, Florida, United States

Irfan Ghauri
Infineon Technologies
Sophia Antipolis, France

Fredrik Gunnarsson
Ericsson Research
Ericsson AB
Linköping, Sweden

Tero Isotalo
Department of Communications
  Engineering
Tampere University of Technology
Tampere, Finland

Jamil Yusuf Khan
School of Electrical Engineering and
  Computer Science
The University of Newcastle
Callaghan, New South Wales,
  Australia

Junsu Kim
Department of Electrical and
  Computer Engineering
University of British Columbia
Vancouver, Canada

Panu Lähdekorpi
Department of Communications
  Engineering
Tampere University of Technology
Tampere, Finland

Jukka Lempiäinen
Department of Communications
  Engineering
Tampere University of Technology
Tampere, Finland

Christian Mehlführer
Institute of Communications and
  Radio-Frequency Engineering
Vienna University of Technology
Vienna, Austria

Szilveszter Nádas
Traffic Analysis and Network
  Performance Laboratory
Ericsson Research, Ericsson Hungary
Budapest, Hungary

Pál L. Pályi
Department of Telecommunications
  and Media Informatics
Budapest University of Technology
  and Economics
Budapest, Hungary

Sándor Rácz
Traffic Analysis and Network
  Performance Laboratory
Ericsson Research, Ericsson Hungary
Budapest, Hungary

Gerardo Rubino
DIONYSOS Research Group
IRISA/Université de Rennes I
Campus Universitaire de Beaulieu
Rennes, France

Markus Rupp
Institute of Communications and
  Radio-Frequency Engineering
Vienna University of Technology
Vienna, Austria

Shakti Prasad Shenoy
Infineon Technologies
Sophia Antipolis, France

Kamal Deep Singh
DIONYSOS Research Group
IRISA/Université de Rennes I
Campus Universitaire de Beaulieu
Rennes, France

Iana Siomina
Ericsson Research
Ericsson AB
Stockholm, Sweden

Dirk T.M. Slock
EURECOM
Sophia Antipolis, France

Dan Keu Sung
Department of Electrical Engineering
Korea Advanced Institute of Science
  and Technology
Daejeon, Korea

Cesar Viho
DIONYSOS Research Group
IRISA/Université de Rennes I
Campus Universitaire de Beaulieu
Rennes, France

Wenbo Wang
School of Telecommunication
  Engineering
Beijing University of Posts and
  Telecommunications
Beijing, China

Martin Wrulich
Institute of Communications and
  Radio-Frequency Engineering
Vienna University of Technology
Vienna, Austria

Xinzhi Yan
School of Electrical Engineering and
  Computer Science
The University of Newcastle
Callaghan, New South Wales,
  Australia

Suleiman Y. Yerima
University of Glamorgan
Pontypridd, Wales, United Kingdom

Di Yuan
Department of Science and
  Technology
Linköping University
Linköping, Sweden

Dong Zhao
China Industry Environment
Nokia Siemens Networks
Beijing, China

Chapter 1

TD-SCDMA HSDPA/HSUPA: Principles, Technologies, and Performance

Shuping Chen, Wenbo Wang, and Dong Zhao

Contents

1.1 Overview

Jointly developed by the China Academy of Telecommunications Technology (CATT) and Siemens Ltd., Time Division Duplex-Synchronous Code Division Multiple Access (TD-SCDMA) is one of the IMT-2000 standards that was approved by the International Telecommunication Union (ITU). At the time when the 29th Olympic Games were held in Beijing, the performance of the China Communications Standards Association (CCSA) N-frequency TD-SCDMA DCH (Dedicated Channel)-based services was well tested in the commercial network operated by the China Mobile Communication Corporation (CMCC). Nowadays, with the ever-increasing demand for the higher data rates and aiming at providing various multimedia services, TD-SCDMA networks are evolving toward an enhanced version, TD-SCDMA HSDPA/HSUPA (High Speed Downlink/Uplink Packet Access).

    This chapter aims to provide an overview of the evolution of TD-SCDMA networks to the HSPA version, including key techniques, channel processing, and operation principles. Performance evaluations are also provided to aid in evaluating the capability of TD-SCDMA HSPA.

1.2 Introduction

As the basic 3G (the 3rd Generation mobile communication system) choice in China [1, 2], TD-SCDMA has been widely accepted and adopted. The performance of the CCSA N-frequency TD-SCDMA DCH-based services has been well tested in both trials and commercial networks, especially during the time of the 29th Olympic Games held in Beijing in 2008. It was shown that DCH-based TD-SCDMA is able to reliably provide both voice service and packet services.

Nowadays, with the ever-increasing demand on the data transmission rates and the various multimedia services, TD-SCDMA networks are evolving toward TD-SCDMA HSDPA/HSUPA. Single-frequency TD-SCDMA HSDPA was introduced in the 3rd Generation Partnership Project (3GPP) in the R5 (Release 5) version as the downlink evolution of TD-SCDMA networks. Multi-frequency TD-SCDMA HSDPA, introduced by the CCSA, is the enhanced version of the 3GPP single-frequency TD-SCDMA HSDPA; it adopts multi-frequency to improve system performance. It offers backward-compatible upgrades to both former N-frequency TD-SCDMA networks and single-frequency TD-SCDMA HSDPA systems. In 2006, the 3GPP made an effort to standardize the uplink evolution of TD-SCDMA networks. Released as the 3GPP R6 version, TD-SCDMA HSUPA is believed to be able to significantly enhance the system uplink capacity. To offer the backward-compatible upgrade to both the N-frequency TD-SCDMA and multi-frequency TD-SCDMA HSDPA, the CCSA started the standardization work of multi-frequency TD-SCDMA HSUPA in August 2007. Now, in both the CCSA and 3GPP, multi-frequency TD-SCDMA HSDPA/HSUPA have been specified. This chapter introduces the key concepts of TD-SCDMA HSDPA/HSUPA networks, including key techniques, protocols, channel processing, and principles of operation. Performance evaluations are also provided to aid in evaluating the key aspects of TD-SCDMA HSPA.

1.3 What Is TD-SCDMA?

Before delving into TD-SCDMA HSDPA/HSUPA, we first provide a short review of TD-SCDMA systems. Jointly developed by CATT and Siemens Ltd., TD-SCDMA is one of the IMT-2000 standards approved by the ITU. Let us begin with TD-SCDMA. The "TD" part of the term has two meanings:

1. It is based on TDD modes, which brings many advantages, such as easily accommodating asymmetrical traffic and high correlation between the downlink (DL) and uplink (UL) channel.
2. TDD operation also makes full use of the asymmetrical frequency resource and makes it very flexible in frequency band occupation. It uses combined TDMA (Time Division Multiple Access) and CDMA (Code Division Multiple Access) for multiple access. The signal is separated in both the time domain and the code domain [3].

The "S" denotes synchronous operation. Such a characteristic brings not only the relative low interference operation, but also the cost-efficient operation of the system.

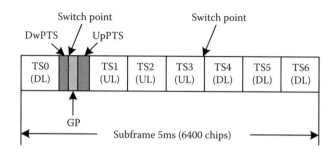

Figure 1.1 Sub-frame structure of TD-SCDMA (symmetric configuration).

Figure 1.1 shows the TD-SCDMA frame structure. The TDMA frame has a duration of 10 ms and is divided into two subframes of 5 ms each. The frame structure for each subframe in the 10-ms frame length is the same. Time slots #0 through 6 make up the traffic time slot, and each contains 864 chips. The DwPTS and UpPTS are the downlink and uplink pilot time slots, containing 96 and 160 chips, respectively, and are designed for downlink and uplink synchronization purposes. GP is the guard period for TDD operation, containing 96 chips. The entire 5-ms sub-frame contains 6,400 chips, indicating that the chip rate of TD-SCDMA is 1.28 Mchips/s. The operation band for TD-SCDMA is 1.6 MHz. Compared to wideband CDMA (WCDMA) [4], TD-SCDMA is a narrowband system.

Among these seven traffic time slots, time slot #0 is always allocated for DL (downlink) while time slot #1 is always allocated for UL (uplink). The time slots for the UL and the DL are separated by switching points. Between the downlink time slots and the uplink time slots, the special period is the switching point to separate the uplink and downlink. In each subframe of 5 ms, there are two switching points (UL to DL and vice versa). Using the above frame structure, TD-SCDMA can operate on both symmetric and asymmetric mode by properly configuring the number of DL and UL time slots. In any configuration, at least one time slot (time slot #0) must be allocated for DL, and at least one time slot must be allocated for UL (time slot #1). In a multi-frequency cell, the traffic time slots allocated for UL and DL pair(s) for one UE should be on the same carrier.

Due to the above characteristics, TD-SCDMA is able to adopt various advanced techniques to enhance system performance and boost system capacity. The TDD mode makes the UL channel and the DL channel highly correlated. Thus, channel estimation based on the UL channel can be used for the DL if the interval between UL reception and DL transmission is less than the channel coherent time. This provides a good condition for the implementation of a smart antenna [5–8], which can boost the signal strength while compressing or nulling the interference. For low chip rate and narrowband operation, advanced receivers can be implemented. TD-SCDMA adopts multi-user detection [9] to combat the various interferences that the CDMA system holds, that is, inter-symbol interference (ISI) and multiple access interference (MAI). The synchronous operation facilitates the operation of baton handover, which smoothes the interruption experience when users hand over from one cell to another. The combination of TDMA and CDMA enables the adoption of dynamic channel allocation (DCA) [10], which can adjust the load of different time slots, achieving load balancing at any time slot. Due to space limitations, we end here with the introduction of TD-SCDMA systems. Interested readers can refer to [2, 11–19] for the details of TD-SCDMA systems, including operation protocols [11, 13–19], key techniques (see [2] and the reference therein), resource management [12], etc.

1.4 TD-SCDMA HSDPA

TD-SCDMA HSDPA, which can be recognized as the 3.5G evolution of the TD-SCDMA system, aims to improve the downlink capacity and throughput of the TD-SCDMA system. By adopting advanced techniques, such as adaptive modulation and coding (AMC), Hybrid Automatic Repeat reQuest (Hybrid ARQ, HARQ) and fast packet scheduling, TD-SCDMA HSDPA can accommodate both real-time and non-real-time services with quality-of-service (QoS) guarantee. The peak data rate for TD-SCDMA HSDPA is 2.8 Mbps adopting five time slots with full code utilization, 16QAM (quadrature amplitude modulation), and non-multiple-input, multiple-output (MIMO) operation. In this section we introduce the TD-SCDMA HSDPA system, including its concepts and principles, key techniques, transport and control channels, and the related operation principles.

1.4.1 General Concepts and System Architecture

Figure 1.2 shows the system architecture of TD-SCDMA HSDPA. The downlink data originates from a multimedia service server, such as a Web server or streaming media server, and then passes through the core networks (CNs) and arrives at the radio network controller (RNC). After the data is processed by each layer residing at the RNC—namely, the PDCP/RLC/MAC-d (medium access control-dedicated) layer—the data is encapsulated into MAC-d PDUs (Protocol Data Units), which are further routed to the specific base station to which the target user belongs. The MAC-d PDUs are further packed into MAC-hs (high speed) PDUs at the base station and then sent via the TD-SCDMA HSDPA air interface. When the data correctly arrives at the user terminal, each layer at the user terminal does opposite operations. The new entity introduced into TD-SCDMA HSDPA is

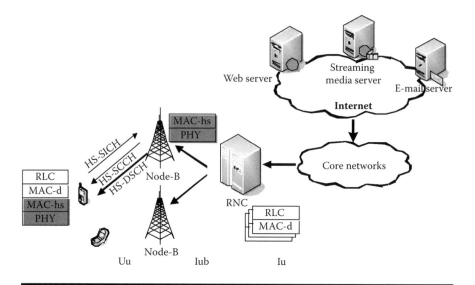

Figure 1.2 TD-SCDMA HSDPA system architecture.

the MAC-hs sublayer, which is in charge of user/packet/PDU scheduling, transmit format selection, and HARQ-related processing. Due to the HARQ retransmission operation, packets arriving at the MAC layer of Node-B (a call for base station in 3G enviroments) are typically out of sequence. Another key function of MAC-hs is the in-sequence delivery of the MAC-d PDU to the MAC-d layer. There are three HSDPA-related channels: (1) High-Speed Downlink Shared Channel (HS-DSCH), (2) High-Speed Shared Control Channel (HS-SCCH), and (3) High-Speed Shared Information Channel (HS-SICH). The HS-DSCH is the transport channel; it is used to carry the HSDPA-related traffic data. HS-SCCH and HS-SICH are physical control channels designed for transmitting HSDPA signaling.

1.4.2 Key Techniques

1.4.2.1 Link Adaptation

The benefits of adapting the transmission parameters in a wireless system to the changing channel conditions are well known. In fact, fast power control is an example of a technique implemented to enable reliable communications while simultaneously improving the system capacity. The process of modifying the transmission parameters to compensate for variations in channel conditions is known as *link adaptation.*

One technique that falls into the category of link adaptation is AMC [20]. The principle of AMC is to change the modulation and coding format in accordance with variations in the channel conditions. The channel conditions can be estimated, for example, based on feedback from the receiver.

In a system with AMC, users in favorable positions (e.g., users close to the cell site) are typically assigned higher-order modulation with higher code rates (e.g., 64QAM with $R = 3/4$ Turbo codes), while users in unfavorable positions (e.g., users close to the cell boundary) are assigned lower-order modulation with lower code rates [e.g., QPSK (quadrature phase-shift keying) with $R = 1/2$ Turbo codes]. The main benefits of AMC are (1) higher data rates being available for users in favorable positions, which, in turn, increases the average throughput of the cell, and (2) reduced interference variation due to link adaptation, based on variations in the modulation/coding scheme instead of variations in transmit power.

In TD-SCDMA HSDPA, the user measures the downlink channel quality and sends the channel quality indicator (CQI) on HS-SICH. After decoding the CQI information, the AMC entity at the base station decides which modulation and coding scheme (MCS) should be used when it performs the next transmission to this user. The chosen MCS information, together with the allocated resource information, is notified on HS-SCCH. Due to the interference fluctuation brought by smart antennas, the MCS selection accuracy is much lower in TD-SCDMA HSDPA than in WCDMA HSDPA [21, 22]. It has a negative impact on the performance of TD-SCDMA HSDPA.

ARQ is another type of link adaptation technique. ARQ compensates for channel variations in an implicit way via retransmissions. Hybrid ARQ with soft combining allows the rapid retransmission of erroneous transmitted packets at the MAC layer. It is able to reduce the requirement of RLC layer retransmission and the overall delay; thus, it can improve the QoS of various services. Prior to decoding, the base station combines information from the initial transmission with that of later retransmissions. This is generally known as *soft combining* and is able to increase the successful decoding probability. Incremental redundancy (IR) is used as the basis for the Hybrid ARQ operation, and either the same or different versions of parity bits can be sent in the possible retransmissions. If the same parity bits are sent, the well-known Chase Combining is used to combine different transmissions, and thus the energy gain can be obtained. Otherwise, if retransmissions contain different parity bits, additional coding gain can also be obtained. If the code rate of initial transmission is high, the coding gain provided by retransmitted parity bits is important; otherwise, if initial code rate is already low, the energy gain is more obvious. With the fast retransmission ability and combing gain brought by Hybrid ARQ, the initial transmission can be performed in relatively higher data rate and target at higher error ratio but with the same power as that of lower data rate targeting lower error ratio. The required lower error ratio can be achieved by subsequent retransmissions. Such operation can obtain the early termination gain (ET gain). Such gain comes from the fact that retransmissions can bring time diversity gain. The shorter transmission time interval (TTI) and the larger allowed transmissions will result in larger gain exploitation. Therefore, the

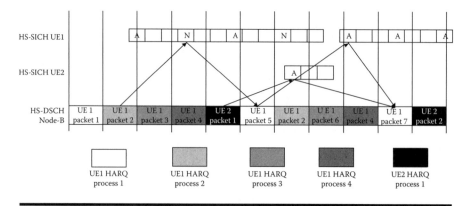

Figure 1.3 N parallel stop-and-wait Hybrid ARQ processes.

delay-tolerant services, such as file-uploading and e-mail, can be operated in such a way.

In TD-SCDMA HSDPA, N parallel stop-and-wait Hybrid ARQ processes are adopted to facilitate ARQ management and allow continuous transmitting. In N parallel stop-and-wait Hybrid ARQ, data transmitted over HS-DSCH is associated with one Hybrid ARQ process. Initial transmission and possible retransmission(s) are restricted to perform on the same process. Figure 1.3 illustrates the operation principle of $N = 4$ parallel stop-and-wait in TD-SCDMA HSDPA.

1.4.2.2 Fast Packet and User Scheduling

When serving packet services, resource sharing among users is believed to be more efficient than dedicating to a certain user. Resource sharing is enabled in TD-SCDMA HSDPA via the shared channel HS-DSCH. Here, scheduling is an important function in determining when, at what resource, and at what rate the packets transmit to a certain user. Combined with AMC, scheduling in TD-SCDMA HSDPA takes advantage of fat-pipe multiplexing, that is, transmitting as many packets as possible to a certain user when it has relatively good channel quality. Different from TD-SCDMA, in TD-SCDMA HSDPA, the scheduling entity resides in the base station, which enables the scheduler to quickly adapt to both the interference and user channel variations.

The scheduling policy is an implementation issue and is flexible based on the system requirements. In general, greedy methods can improve the overall system throughput. Also, other scheduling methods may take the user fairness, traffic priority, service QoS, and operator's operating strategy into consideration [23]. Examples are the PF (Proportional Fair) scheduling [24] method in providing non-real-time services, while EXP (Exponential Rule) and M-LWDF (Modified Largest Weighted Delay First Rule)

[25] are believed to be suitable for serving the real-time services. Under the mixed services scenario (e.g., simultaneously serving VoIP (Voice over Internet Protocol) and other background services), in order to guarantee different QoS requirements of different services, a differential scheduling mechanism is required [26].

1.4.3 Multi-frequency TD-SCDMA HSDPA

Because TD-SCDMA is a narrowband system, the peak data rate provided by TD-SCDMA HSDPA is limited. To overcome this drawback, multi-frequency operation for TD-SCDMA HSDPA is adopted [27]. For multi-frequency operation, multiple 1.6-MHz frequency bands are combined and operated together. As shown in Figure 1.4, in multi-frequency TD-SCDMA

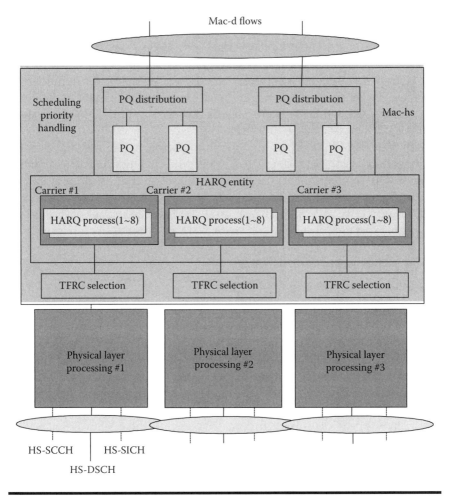

Figure 1.4   Data processing in multi-frequency TD-SCDMA HSDPA.

HSDPA, there are multiple sets of HSDPA-related channels—that is, multiple {HS-DSCH, HS-SCCH, HS-SICH} sets, one set for each carrier. Scheduling and resource allocation should be performed over all carriers simultaneously based on the CQI information on each carrier. One UE (user equipment) may or may not receive multiple data streams from multiple carriers. Carrier data distribution is performed at the MAC layer right above the Hybrid ARQ entity. Below the Hybrid ARQ entity, there are multiple data flows and one for each carrier. TFRC (transmit format and resource combination) selection should be performed for each carrier separately.

1.4.4 TD-SCDMA HSDPA Channels

As discussed above, for the operation of TD-SCDMA HSDPA, three additional channels are introduced: one transport channel (HS-DSCH) and two control channels (HS-SICH and HS-SCCH). HS-DSCH is used for carrying the TD-SCDMA HSDPA-related data information. HS-SCCH is used for downlink signaling purposes, indicating the resource position, conveying the Hybrid ARQ-related parameters, etc. HS-SICH is used for the feedback of channel quality information and the information of correctly decoding or not (ACK/NACK). In this section, detailed processing procedures for these channels are introduced. Using these processing procedures, the key techniques introduced above are incorporated. Before going into the detailed channel processing procedure, we first introduce the timing and association relationship between the TD-SCDMA HSDPA-related transport and control channels, that is, the timing and association between HS-DSCH and HS-SCCH, and between HS-SCCH and HS-SICH. These associations and timing determine the right operation of TD-SCDMA HSDPA.

1.4.4.1 Association and Timing

1.4.4.1.1 HS-DSCH and HS-SCCH

The transport channel HS-DSCH can be associated with a number of downlink control channels HS-SCCHs. In a multi-frequency HS-DSCH cell, HS-DSCH may be mapped on HS-PDSCHs (the physical layer channel conveying the HS-DSCH data information) on one or more carriers for UE supporting multi-carrier HS-DSCH reception. HS-DSCH transmission on each carrier is associated with an HS-SCCH subset, and the number of HS-SCCHs in one HS-SCCH subset can range from a minimum of one to a maximum of four. All the HS-SCCH subsets for one UE constitute an HS-SCCH set. For UE not supporting multi-carrier HS-DSCH reception, only one HS-SCCH subset is allocated. All relevant Layer-1 control information is transmitted in the associated HS-SCCH; that is, the HS-PDSCH does not carry any Layer-1 control information, which is used for conveying Layer-2 (MAC) information only.

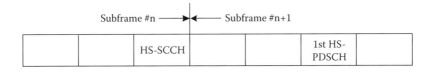

Figure 1.5   Timing for HS-SCCH and HS-DSCH (DwPTS and UpPTS not included).

The HS-DSCH-related time slot information that is carried on the HS-SCCH refers to the next valid HS-PDSCH allocation, which is given by the following limitations: The indicated HS-PDSCH is on the subframe next to the HS-SCCH carrying the HS-DSCH related information. The HS-DSCH-related time slot information does not refer to two subsequent subframes, but always refers to the following subframe, as illustrated in Figure 1.5. In case of multi-carrier HS-DSCH reception, the timing for HS-DSCH transmission on each carrier and its associated HS-SCCH applies the same rule.

1.4.4.1.2 HS-SCCH and HS-SICH

The HS-SCCH is always associated with one HS-SICH, carrying the ACK/NACK and CQI. The association between the HS-SCCH in DL and HS-SICH in UL is predefined by higher layers and is common for all UEs.

The UE with a dedicated UE identity transmits the HS-DSCH related ACK/NACK on the next available associated HS-SICH with the following limitation: There shall be an offset of nine time slots between the last allocated HS-PDSCH and the HS-SICH for the given UE. DwPTS and UpPTS are not taken into account in this limitation. Hence, the HS-SICH transmission is always made in the next but one subframe, following the HS-PDSCH transmission, as illustrated in Figure 1.6. In the case of multi-carrier HS-DSCH reception, the timing for HS-DSCH transmission on each carrier and its related HS-SICH applies the same rule.

The timing between the HS-SCCH and the HS-SICH for the given UE is illustrated in Figure 1.7. The UE transmits the HS-SCCH-related ACK on the next available, associated HS-SICH with the following limitation: There shall be an offset of 14 time slots between the HS-SCCH and the HS-SICH

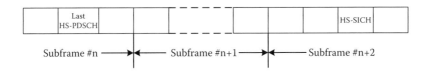

Figure 1.6   Timing for HS-DSCH and HS-SICH (DwPTS and UpPTS not included).

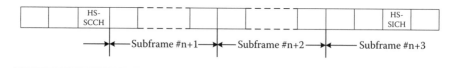

Figure 1.7   Timing for HS-SCCH and HS-SICH (DwPTS and UpPTS not included).

for the given UE. DwPTS and UpPTS shall not be taken into account in this limitation.

1.4.4.2 Channel Processing

1.4.4.2.1 HS-DSCH

Figure 1.8 illustrates the overall procedure of processing for HS-DSCH. Data arrives at the coding unit in the form of one transport block once every TTI, which is 5 ms for TD-SCDMA HSDPA. The entire processing procedure includes the following steps:

1. A CRC (cyclic redundancy check) of 24 bits to each transport block
2. Code block segmentation
3. Channel coding
4. Hybrid ARQ processing

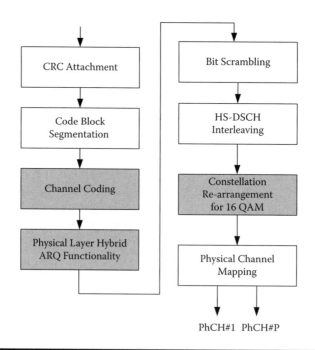

Figure 1.8   Physical layer processing for HS-DSCH.

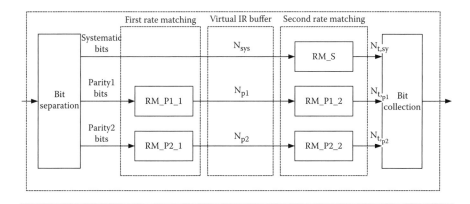

Figure 1.9 HS-DSCH Hybrid ARQ functionality.

5. Bit scrambling
6. Interleaving
7. Constellation re-arrangement for 16QAM
8. Mapping to physical channels

As WCDMA HSDPA, only Turbo code with rate 1/3 is used for the HS-DSCH channel. This was motivated by the fact that Turbo coding outperforms convolution coding otherwise expected with the very small data rates. Additional new issues for HS-DSCH processing include the handling of 16QAM constellation rearrangement and Hybrid ARQ processing on the physical layer for the HS-DSCH.

Figure 1.9 shows the functionality of Hybrid ARQ. This functionality matches the number of bits at the output of the channel coder to the total number of bits of the HS-PDSCH set to which the HS-DSCH is mapped. The Hybrid ARQ functionality is controlled by the redundancy version (RV) parameters. The exact set of bits at the output of the Hybrid ARQ functionality depends on the number of input bits, the number of output bits, and the RV parameters. The Hybrid ARQ functionality consists of two rate-matching stages and a virtual buffer, as shown in Figure 1.9. The first rate-matching stage matches the number of input bits to the virtual IR buffer. Note that if the number of input bits does not exceed the virtual IR buffering capability, the first rate-matching stage is transparent. The second rate-matching stage matches the number of bits after the first rate-matching stage to the number of physical channel bits available in the HS-PDSCH set in the TTI.

For 16QAM modulation, there is the specific function of constellation rearrangement, which maps the bits to different symbols depending on the transmission numbers. This is beneficial because, as with 16QAM, all the symbols do not have equal error probability in the constellation. The reason is that different symbols have different numbers of neighboring symbols,

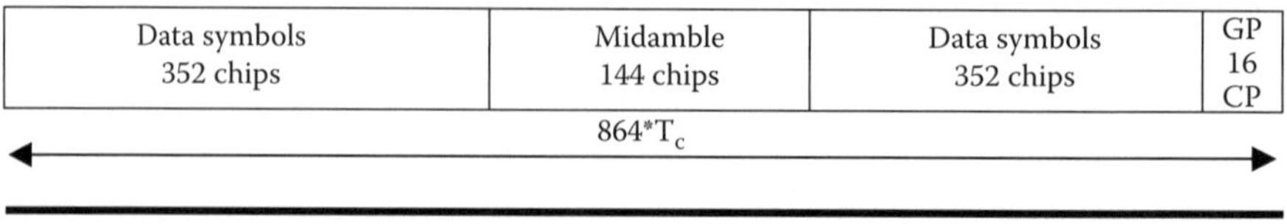

Data symbols 352 chips	Midamble 144 chips	Data symbols 352 chips	GP 16 CP

$864 * T_c$

Figure 1.10 Burst type for HS-SICH (T_C is chip duration, midamble is training sequence).

which places the symbols closer to the axis, with a greater number of neighboring symbols more likely to be decoded incorrectly than the other symbols further away from the axis.

1.4.4.2.2 HS-SICH

HS-SICH is used to carry the CQI and ACK/NACK information. The following burst type is used for HS-SICH. HS-SICH will carry the transmit power control (TPC) and Synchronozation Shift (SS) bits for power control of HS-SCCH and for synchronous purposes, respectively. The spreading factor is 16 for HS-SICH. Thus, HS-SICH will carry 44 information bits in the first data field and 40 information bits plus 2 bits TPC and 2 bits SS in the second data field (see Figure 1.10).

The physical layer processing for HS-SICH is shown in Figure 1.11. The following information is transmitted by means of the HS-SICH:

1. Recommended modulation format (RMF) (1 bit), which is used by UE to recommend its favorable modulation format, namely, QPSK (0 indicates) or 16QAM (1 indicates). The RMF is repetition coded to 16 bits.
2. Recommended transport-block size (RTBS) (6 bits). UE uses this field to recommend the data amount that is preferred to be received by UE in the next TTI. The 6 bits of the RTBS field are coded to 32 bits using a (32, 6) first-order Reed-Muller code.
3. Hybrid ARQ information ACK/NACK (1 bit), with the value 0 indicating NACK and 1 indicating ACK. For the coding of this field, the repetition code is adopted. The one indication bit is repeated to 36 bits.

All these bits (84 bits) are then multiplexed and interleaved before mapping and being transmitted on the code channel.

1.4.4.2.3 HS-SCCH

HS-SCCH is used for the transmission of HS-DSCH-related control information. The following burst type is used for HS-SCCH. HS-SCCH contains two code channels (HS-SCCH1 and HS-SCCH2). HS-SCCH1 will carry the TPC and SS bits for power control of HS-SICH and synchronization purposes, respectively. The spreading factor is 16 for HS-SCCH. Thus, HS-SCCH1 will

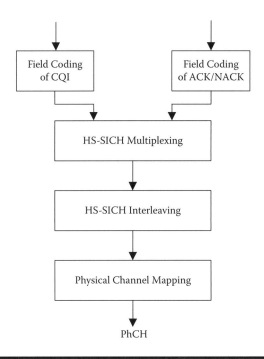

Figure 1.11   Physical layer processing for HS-SICH.

carry 44 information bits in the first data field and 40 information bits plus 2 bits TPC and 2 bits SS in the second data field, while HS-SCCH2 will be used to convey 88 information bits only (see Figure 1.12). The physical layer processing of HS-SCCH is shown in Figure 1.13.

The following information is transmitted on HS-SCCH:

1. Channelization code set information (8 bits). HS-PDSCH channelization codes are allocated contiguously from a signaled start code to a signaled stop code, and the allocation includes both the start and stop code. The start code is signaled by the first 4 bits (the code length is 16 chips) and the stop code by the remaining 4 bits.
2. Time slot information (5 bits). The time slots used for HS-PDSCH resources are signaled by the bits $x_1, x_2, \ldots, x_5$, where bit x_n carries

Figure 1.12   Burst type of HS-SCCH (T_c is the chip duration).

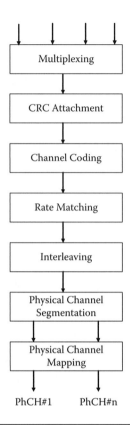

Figure 1.13   Physical layer processing for HS-SCCH.

the information for timeslot $n + 1$. Timeslots 0 (conveys common control channels, such as P-CCPCH) and 1 (always used for uplink) cannot be used for HS-DSCH resources. If the signaling bit is set (i.e., equal to 1), then the corresponding time slot is used for HS-PDSCH resources. Otherwise, the time slot is not used. All used time slots employ the same channelization code set, as signaled by channelization code set information bits.

3. Modulation scheme information (1 bit). The modulation scheme used by the HS-PDSCH resources is signaled by this bit, with the value 0 indicating QPSK and the value 1 indicating 16QAM.

4. Transport block size information (6 bits). The transport block size information is an unsigned binary representation of the transport block size index from 0 to 63.

5. Hybrid ARQ process information (3 bits). The hybrid-ARQ process information is an unsigned binary representation of the HARQ process identifier from 0 to 7.

Table 1.1   RV Mapping

	QPSK		16QAM		
X_{rv}	s	r	s	r	b
0	1	0	1	0	0
1	0	0	0	0	0
2	1	1	1	1	1
3	0	1	0	1	1
4	1	2	1	0	1
5	0	2	1	0	2
6	1	3	1	0	3
7	0	3	1	1	0

6. RV information (3 bits). The RV parameters r, s and the constellation version parameter b are mapped jointly to produce the value X_{rv}. This is done according to Table 1.1 according to the modulation mode used.

The other information conveyed on HS-SCCH includes new data indicator (1 bit), indicating whether or not the related HS-DSCH transmission is a new transmission; HS-SCCH cyclic sequence number (3 bits); and UE identity (16 bits).

1.4.5 HS-DSCH Operation Procedure

This section overviews the basic HS-DSCH operation procedures, including the link adaptation procedure, the channel quality reporting procedure, the HS-SCCH monitoring procedure, and the Iub interface flow control procedure. From these procedures, one obtains a general understanding of how the TD-SCDMA HSDPA system operates.

1.4.5.1 Link Adaptation

For HS-DSCH, the modulation scheme and effective code rate are selected by higher layers located within Node-B. This is achieved by appropriate selection of an HS-DSCH transport block size, modulation format, and resources by higher layers. If UE supports multi-carrier HS-DSCH reception, higher layers may select multiple carriers to transfer the data. Carrier selection may be based on CQI reports from the UE. If the UE supports multi-carrier HS-DSCH transmission, the UE will report the CQI information of every carrier via HS-SICH. The overall HS-DSCH link adaptation procedure consists of the following two parts:

1. *Node-B procedure.* Node-B transmits HS-SCCH carrying a UE identity indicating the UE to which HS-DSCH TTI is to be granted. In the case of HS-DSCH transmissions in consecutive TTIs to the same UE, the same HS-SCCH is used for associated signaling. If UE supports multi-carrier HS-DSCH reception, the above HS-SCCH detection procedure is applied on each independent carrier. Node-B transmits HS-DSCH to the UE using the grant indicated in the HS-SCCH. Upon receiving the HS-SICH from the respective UE, the status report (ACK/NACK and CQI) is passed to higher layers.

2. *UE procedure.* When indicated by higher layers, the UE starts monitoring all HS-SCCHs that are in its HS-SCCH set. In the case that an HS-SCCH is identified as correct by its CRC, the UE reads the HS-PDSCHs indicated by the HS-SCCH. If UE supports multi-carrier HS-DSCH reception, UE may acquire HS-PDSCH resource allocation information of each carrier according to the associated HS-SCCHs. In the case that an HS-SCCH is identified to be incorrect, the UE will discard the data on the HS-SCCH and return to monitoring. After reading the HS-PDSCHs, the UE generates an ACK/NACK message and transmits this to Node-B in the associated HS-SICH, along with the most recently derived CQI. If UE supports multi-carrier HS-DSCH reception, the CQI and ACK/NACK of every carrier are transferred via individual HS-SICHs.

1.4.5.2 HS-DSCH Channel Quality Indication

The channel quality indicator (CQI) provides Node-B with an estimate of the code rate that would have maximized the single-transmission throughput of the previous HS-DSCH transmission if decoded in isolation. The CQI report must be referenced to a given set of HS-PDSCH resources by Node-B. The reference resources for a CQI report is a set of HS-PDSCH resources that were received by the UE in a single TTI and contain a complete transport block. These resources will be known to Node-B from the relative timings of the HS-SICH carrying the CQI and previous HS-DSCH transmissions to the UE.

As described above, the CQI consists of two fields: RTBS and RMF. The UE uses the same mapping table for these fields as is being used for the time slot information and modulation scheme information fields, respectively, of the HS-SCCH. The detailed reporting procedure is as follows:

1. The UE receives a message on an HS-SCCH telling it which resources have been allocated to it for the next associated HS-DSCH transmission.

2. The UE reads the associated HS-DSCH transmission and makes the necessary measurements to derive a CQI that it estimates would have given it the highest single-transmission throughput for the allocated resources while achieving a BLER (block error ratio) of no more than 10%. BLER is defined as the probability that a transport block transmitted using the RTBS and RMF is received in error if decoded in isolation. For the purposes of this calculation, it assumes that the transport block that would be transmitted with these parameters would use redundancy version parameters $s = 1$ and $r = 0$. Using this definition of BLER, single-transmission throughput can be defined as single-transmission throughput $= (1 - \text{BLER}) \times \text{RTBS}$.

3. The CQI report derived from a given HS-DSCH transmission is reported to Node-B in the next HS-SICH available to the UE following that HS-DSCH transmission, unless that HS-SICH immediately follows the last allocated HS-DSCH time slot, in which case the subsequent available HS-SICH is used by the UE. This HS-SICH may not necessarily be the same HS-SICH that carries the ACK/NACK information for that HS-DSCH transmission. The UE will always transmit the most recently derived CQI in any given HS-SICH.

1.4.6 HS-SCCH Monitoring

In a multi-frequency HS-DSCH cell, a UE divides its HS-SCCH set into one or more HS-SCCH subsets; in each HS-SCCH subset, all HS-SCCHs are associated with the same frequency's HS-PDSCH. When indicated by higher layers, the UE will start monitoring all HS-SCCHs in all HS-SCCH subsets to acquire the configuration information of HS-PDSCHs. In the case that one HS-SCCH is detected carrying its UE identity, the UE skips monitoring the remaining HS-SCCHs in this HS-SCCH subset, and restricts its monitoring to only previously detected HS-SCCH in the following TTIs. The UE sets all HS-SCCHs carrying its UE identity in all HS-SCCH subsets into an active set, and sets all HS-SCCH subsets in which no HS-SCCH carries its UE identity into a remaining set.

In the case that the multi-carrier number is not configured by higher layers, a UE will always monitor all HS-SCCH subsets. Otherwise, the UE may skip monitoring the remaining HS-SCCH subsets when the number of HS-SCCHs carrying its UE identity; that is, the number of HS-SCCHs in the active set is equal to the configured value.

During the following TTIs, the UE updates and maintains the active set and the remaining set. If one or more HS-SCCHs in the active set do not carry its UE identity, the UE removes them from the active set and sets their corresponding HS-SCCH subsets into the remaining set. Meanwhile, if one or more HS-SCCHs in the remaining sets are detected carrying its UE identity, the UE sets these found HS-SCCHs into the active set and removes their corresponding HS-SCCH subsets from the remaining set.

1.4.6.1 Iub Flow Control Procedure

The HSDPA architecture splits the MAC layer between the RNC and Node-B. MAC PDUs generated by the RNC, called MAC-d PDUs, are aggregated and sent to Node-B over the Iub interface in HS-DSCH DATA FRAME. Node-B buffers the PDUs until they are scheduled and successfully transmitted over the air interface to a UE. The delivery of PDUs over the Iub is managed by a flow control protocol, which can act independently for each CmCHPI (common transport channel priority) of each UE.

Node-B is the master of the flow control. The whole procedure includes the transmissions of two control frames (HS-DSCH CAPACITY REQUEST FRAME and HS-DSCH CAPACITY ALLOCATION FRAME) and one data frame (HS-DSCH DATA FRAME). The HS-DSCH CAPACITY REQUEST FRAME is used for the RNC to request HS-DSCH capacity by indicating the user buffer size in the RNC for a given priority level. The RNC is allowed to reissue the HS-DSCH Capacity Request if no CAPACITY ALLOCATION has been received within an appropriate time threshold. The HS-DSCH CAPACITY ALLOCATION FRAME is used by Node-B to control the user data flow. In the CAPACITY ALLOCATION FRAME, HS-DSCH Credits IE (information entity) indicates the number of MAC-d PDUs that the RNC is allowed to transmit for the MAC-d flow and the associated priority level indicated by the CmCHPI Indicator IE, and the Maximum MAC-d PDU length, HS-DSCH Credits, HS-DSCH Interval, and HS-DSCH Repetition Period IEs indicate the total amount of capacity granted. Any capacity previously granted is replaced.

When the RNC has been granted capacity by Node-B via the HS-DSCH CAPACITY ALLOCATION FRAME or via the HS-DSCH initial capacity allocation and the RNC has data waiting to be sent, then the HS-DSCH DATA FRAME is used to transfer the data. If the RNC has been granted capacity by Node-B via the HS-DSCH initial capacity allocation, this capacity is valid for only the first HS-DSCH DATA FRAME transmission. When data is waiting to be transferred, and a CAPACITY ALLOCATION is received, a DATA FRAME will be transmitted immediately according to allocation received. Multiple MAC-d PDUs of the same length and same priority level (CmCHPI) may be transmitted in one MAC-d flow in the same HS-DSCH DATA FRAME. The HS-DSCH capacity allocation procedure is generated within Node-B. It may be generated either in response to an HS-DSCH capacity request or at any other time. Node-B can use this message to modify the capacity at any time, irrespective of the reported user buffer status. Figure 1.14 illustrates one possible use of the flow control messages. In the example, the first two allocation messages are unsolicited and are generated despite the user buffer size being zero.

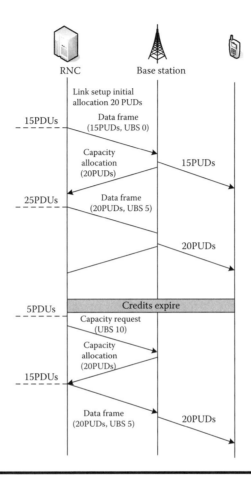

RNC   Base station

Link setup initial
allocation 20 PUDs

15PDUs   Data frame
(15PUDs, UBS 0)

Capacity
allocation   15PUDs
(20PUDs)

25PDUs   Data frame
(20PUDs, UBS 5)

20PUDs

5PDUs   Credits expire

Capacity request
(UBS 10)

Capacity
allocation
(20PUDs)

15PDUs

Data frame
(20PUDs, UBS 5)   20PUDs

Figure 1.14   **Example of Iub flow control procedure (UBS, user buffer size).**

1.5 TD-SCDMA HSUPA

1.5.1 Concept and Principles

TD-SCDMA HSDPA focuses mainly on downlink improvement of TD-SCDMA systems. Limited uplink capacity is becoming a bottleneck. In 2007, the 3GPP finished the low chip rate TDD-HSUPA (i.e., TD-SCDMA HSUPA) specification, which is believed to be able to enhance the uplink TD-SCDMA networks significantly.

Due to the TDD nature, TD-SCDMA HSUPA is quite different from WCDMA HSUPA [22, 31]. WCDMA HSUPA is based on enhanced-dedicated channel (E-DCH). Hybrid ARQ and fast rate scheduling, which are located in the base station are used to enhance DCH performance. Rate scheduling is responsible for uplink interference resource (RoT [rise over thermal noise] resource) scheduling. In TD-SCDMA HSUPA, the concept is mainly on the basis of shared channel. The scheduler, located in the base station, is in charge of both resource and rate scheduling, which is more like that done in TD-SCDMA HSDPA. Another difference is that higher modulation was adopted in TD-SCDMA HSUPA (i.e., 16QAM), while WCDMA HSUPA uses only QPSK modulation. Due to these essential differences, TD-SCDMA HSUPA has its specific characteristics in both the technique aspect and the protocol aspect. This section aims to dig out such essential differences and tries to present a "real" TD-SCDMA HSUPA from both the technique and protocol aspects. Also, this section presents the distinct resource management characteristic of TD-SCDMA HSUPA, and some useful conclusions are drawn.

TD-SCDMA HSUPA aims to enhance the uplink performance of TD-SCDMA networks. Due to the limited uplink channelization code resource, the scheduler, which is located in base station, is designed to manage not only the uplink interference resource (i.e., RoT resource), but also the code resource. In the performance evaluation section of this chapter, we show that due to the effect of interference suppression by smart antennas, the RoT resource control may not be as urgent as that in a WCDMA HSUPA system. Hybrid ARQ is adopted in TD-SCDMA HSUPA to allow for error correction in the physical layer, and less RLC layer ARQ is required in order to meet a certain quality, thus improving the overall end-to-end latency performance. Due to the gain brought by smart antennas and the power-saving nature inherent in TDD operation, high modulation has potential gain in TD-SCDMA HSUPA system. Close loop power control is used to overcome the near–far problem and facilitate the operation of transmission format combination (TFC) control. In the following of this section, we will introduce the general system architecture and channels that facilitate the operation of these features.

Figure 1.15 shows the system architecture of TD-SCDMA HSUPA on both the UE side and the UTRAN [Universal Mobile Telecommunications System (UMTS) Terrestrial Radio Access Network] side. The main modification of the TD-SCDMA HSUPA protocol stack with respect to a traditional TD-SCDMA system concerns the physical and MAC layers. For the UTRAN side, Hybrid ARQ soft combining is introduced in the physical layer to combine the information of different transmissions, which results in a higher successful decoding probability. MAC-e (enhanced) and MAC-es are newly added MAC sublayers. MAC-e, which is located in the base station, is in charge of uplink scheduling, rate control, transmission of scheduling grants, and the Hybrid ARQ related operation. Due to the Hybrid ARQ operation, the PDU arriving in the RLC (radio link control) layer may not be in sequence. MAC-es, which is located in the RNC, is responsible for MAC-es PDUs reordering. On that UE side, the MAC-e/es sublayer performs TFC selection and handles the Hybrid ARQ protocol-related functions.

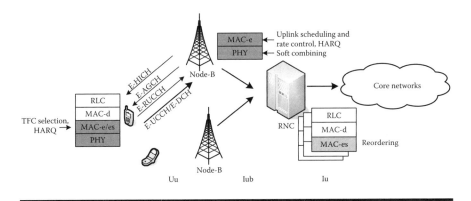

Figure 1.15   TD-SCDMA HSUPA system architecture.

TD-SCDMA HSUPA introduces one transport channel and four physical channels. The newly added transport channel is the E-DCH, which is used to transmit traffic data. Its physical layer channel is the E-DCH physical uplink channel (E-PUCH). The four physical channels are E-DCH random access uplink control channel (E-RUCCH), E-DCH absolute grant channel (E-AGCH), E-DCH uplink control channel (E-UCCH), and E-DCH Hybrid ARQ indicator channel (E-HICH). These four physical channels are used for control purposes. The E-RUCCH is used to carry Scheduling Information, such as the UE buffer status and power headroom, when E-PUCH resources are not available. The E-AGCH carries the UE-specific resource grant, which includes both code and interference resources. The grant information indicates the specific UE to transmit data using what physical resource and at what maximum allowable transmit power. The E-UCCH is the E-DCH associated channel, the function of which is the same as HS-SCCH in HSDPA. It indicates the TFC used in E-DCH and carries Hybrid ARQ process ID information The E-HICH is used to carry Hybrid ARQ acknowledgment sent from the base station. The transmission of Scheduling Information is necessary in the case of separate operation of the scheduling entity (i.e., base station in HSUPA) and data transmission entity (i.e., UE in HSUPA). The scheduling entity needs such information to make a reasonable judgment. When the UE has no resource to send traffic (i.e., no E-PUCH resource), it can initiate E-RUCCH transmission to inform the base station of such information. When the UE has an E-PUCH resource, the scheduling information is transmitted as a MAC-e header.

1.5.2 Key Techniques

1.5.2.1 Uplink Hybrid ARQ

The Hybrid ARQ in TD-SCDMA HSUPA utilizes the N-process ARQ protocol, which is the same as that in HSDPA. The ARQ used here operates in an asynchronous way. The time between data transmission and its feedback is fixed, while the time between different transmissions is flexible and is determined by scheduling policy.

The Hybrid ARQ profile specified in TD-SCDMA HSUPA can provide MAC-layer QoS differential. The Hybrid ARQ profile includes the power offset and the maximum allowed transmissions. The power offset allows certain traffic to pump more power than what is typically needed. Higher power means lower probability of needing a retransmission and thus, low latency. These two attributes allow for flexible Hybrid ARQ operation. For example, delay-sensitive services can use a relative high power offset and low retransmission probability, while delay-tolerant traffic can have more transmissions and obtain more ET gain.

1.5.2.2 Power Control and TFC Selection

The near–far problem is inherited in CDMA uplink operation. Closed-loop power control is a well-known solution to settle such a problem. Different from WCDMA HSUPA, which is based on always-on-DPCCH for closed-loop power control, in TD-SCDMA HSUPA, the E-AGCH and E-PUCH is a closed-loop power control pair. The transmitting power of E-PUCH is determined according to the following formula:

$$P = P_{e\text{-}base} + L + \beta_e + K_{E\text{-}PUCH} \tag{1.1}$$

where L is the path-loss between the base station and UE, β_e is the gain factor and is specific for individual TFCs, $K_{E\text{-}PUCH}$ is the Hybrid ARQ power offset, and $P_{e\text{-}base}$ is a closed-loop control component that is adjusted according to the TPC command carried on E-AGCH

$$P_{e\text{-}base} = PRX_{des\_base} + \eta \cdot \sum_i TPC_i \tag{1.2}$$

where $PRX_{des\text{-}base}$ is the required received reference power and η is the power adjusting step. Because the value of $P_{e\text{-}base}$ is known at both the base station and UE, the base station can effectively control the transmit power of certain UE in such a way that no extra code channel is required for maintaining the closed-loop power control, which is especially important for TD-SCDMA HSUPA for its relative lower chip rate.

TFC selection is performed at the UE MAC-e layer based on the transmitting power each TFC needs and the maximum transmitting power allowed by the network. TFC selection performs the same way as that in R99 DCH. In brief, the TFC requires the largest power, but not higher than the maximum power allowed by network. Because TD-SCDMA HSUPA adopts a higher modulation level (i.e., 16QAM), a separate gain factor list should be provided. In TD-SCDMA HSUPA, the base station controls the maximum allowed power that a certain UE can assume, and thus the network can effectively control the data rate at which UE may transmit. As mentioned above, because power saving is inherited in TDD mode and smart antenna can provide potential high gain, a higher modulation level has potential gain in TD-SCDMA HSUPA, which will be shown later in the performance evaluation section.

1.5.2.3 Scheduling, Rate, and Resource Control

Scheduling, rate, and resource control are the main radio resource management functions in a packet services-oriented system. In TD-SCDMA HSUPA, the corresponding functional entity is the MAC-e located in the base station; it allows rapid resource allocation and exploiting the burstiness in packet data transmissions. It enables the system to admit a larger number of high data rate users and rapidly adapt to both interference and user channel variations, thereby leading to an increase in capacity as well as an increase in the likelihood that a user will experience high data rates.

The TD-SCDMA HSUPA uplink resource includes not only the tolerable interference (i.e., the maximum allowed received power at base station), but also the uplink code channel. For TD-SCDMA HSUPA, due to the use of smart antennas, things can be different. Uplink code channel scheduling should receive more attention than the interference resource control. As commonly known, smart antennas not only can boost the absolute signal strength but also have a positive effect on interference suppression. Besides, the joint detection used in uplink can eliminate a large part of the multiple access interference. The possible scenario that may results large interference between users is that the interfering users (UE2) are in the same direction as that of the victim user (UE1), which is shown in Figure 1.16.

Because TD-SCDMA HSUPA is based on sharing mechanism, the possibility of such a large interference scenario is somewhat low. Even when such a scenario occurs, Hybrid ARQ can further recover the former interference-corrupted packet, and the probability that the same scenario will also occur in the retransmission is very small.

TD-SCDMA HSUPA is code-limited in the uplink for its relative lower chip rate and adopting common scrambling code for all uplink transmissions. [30] Code channels (i.e., orthogonal variable spreading factor [OVSF] codes) should be carefully managed, and one should ensure that these limited codes can be efficiently utilized.

Just as for the downlink user and packet scheduling, the scheduling policy is an implementation issue and is flexible based on system requirements. The scheduling method may take user fairness, traffic priority, service QoS, and the operator's operating strategy into consideration. The fast responding ability to the interference and channel variation of scheduling enables the system to accommodate larger numbers of packet traffic users and can fully exploit the multi-user diversity gain.

The scheduling, rate, and resource control related framework includes two channels (E-RUCCH and E-AGCH) and the user status information,

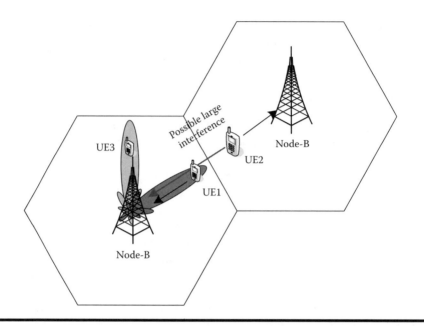

Figure 1.16   Interference scenario under smart antenna for TD-SCDMA HSUPA.

which is Scheduling Information. When a certain user has no uplink resource to transmit data, it initiates E-RUCCH transmission to report its
Scheduling Information to networks. After the scheduling, rate, and resource control procedure, a judgment is made whether to allocate a resource
to that user. If true, E-AGCH is transmitted to inform the user the uplink
resource information, which includes time slot, code channel, and power
resource and may be the duration of such allocation depend on whether
this allocation is valid more than one TTI or not. If the allocation is valid
more than one TTI, E-AGCH will carry the duration of the allocation. If
not, T-AGCH will not contain the information. After decoding E-AGCH, the
user will transmit data using the resource allocated by the network. Unlike
WCDMA HSUPA, TD-SCDMA HSUPA does not employ soft handover. No
other grant channel is transmitted besides E-AGCH. E-RGCH (E-DCH Relative Grant Channel) is another grant channel in WCDMA HSUPA besides
E-AGCH, which is used mainly for interference control. Because interference control is not as urgent as in a WCDMA system, it is not necessary to
adopt another channel for such a purpose in TD-SCDMA HSUPA.

In addition to the dynamic scheduling policy, TD-SCDMA HSUPA also
allows non-scheduling transmission, which is especially suitable for serving
the guaranteed bit rate services, such as VoIP. Such a mechanism allows
users to transmit urgent data, such as delay-sensitive traffic data or signaling,
without waiting for the base station's scheduling grant. It enables the QoS
guarantee for some special traffic.

1.5.2.4 Multi-frequency Operation

For the relative low chip rate that TD-SCDMA HSUPA deploys, the peak data rate is 2.2 Mbps. It is relatively low compared to that of WCDMA HSUPA, which is 5.76 Mbps. Multi-frequency TD-SCDMA HSUPA is also specified. The basic operation principle is the same as that of multi-frequency TD-SCDMA HSDPA. The carrier data distribution is performed at the MAC layer below the MAC-es layer. Multi-frequency operation has potential gain inherited in frequency-selective gain.

1.5.3 TD-SCDMA HSUPA Channels

Similar to the TD-SCDMA HSDPA section, in what follows we first introduce the association and timing between the transport channel and control channel because it encompasses the basic operation requirements. Then we go into the details of channel processing for both transport channels and the related control channels.

1.5.3.1 Association and Timing

1.5.3.1.1 E-DCH and E-AGCH

The E-DCH is always associated with a number of E-AGCHs and up to four E-HICHs. A grant of E-DCH transmission resources may be transmitted to the UE on any one of the associated E-AGCHs. All relevant Layer-1 control information related in an E-DCH TTI is transmitted in the associated E-AGCH and E-HICH.

The E-DCH-related time slot information that is carried on the E-AGCH refers to the next valid E-PUCH allocation, which is given by the following limitation: There will be an offset of seven time slots between the E-AGCH carrying the E-DCH-related information and the first indicated E-PUCH (in time) for a given UE. DwPTS and UpPTS is not be taken into account in this limitation, as illustrated in Figure 1.17.

For semi-persistent E-DCH resources, the timing between E-AGCH and the first E-PUCH can be indicated by the information conveyed on E-AGCH. Once the semi-persistent resources are assigned to UE, the UE can use those resources continuously until the semi-persistent resources have been released or reconfigured by Node-B or RNC.

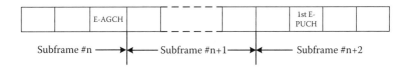

Figure 1.17 Timing for E-AGCH and E-PUCH (UpPTS and DwPTS not included).

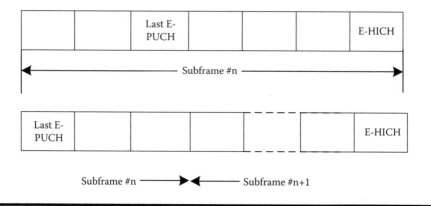

Figure 1.18   Timing for E-DCH and E-HICH (UpPTS and DwPTS not included).

1.5.3.1.2 E-DCH and E-HICH

For a given UE, an E-HICH is synchronously linked with the E-DCH TTI transmission to which it relates. The associated E-HICH resides on the first E-HICH instance of the E-HICH channelization code to occur after $n_{E\text{-}HICH}$ time slots have elapsed since the start of the last E-PUCH of the corresponding E-DCH TTI (see examples in Figure 1.18). The value of $n_{E\text{-}HICH}$ is configurable by higher layers within the range of four to fifteen timeslots, and DwPTS and UpPTS is not taken into account in this limitation.

1.5.3.2 Channel Processing

1.5.3.2.1 E-DCH

Figure 1.19 shows the processing structure for the E-DCH transport channel mapped onto a separate coded composite transport channel (CCTrCH). Data arrives at the coding unit in the form of a maximum of one transport block once every TTI. A TTI of 5 ms will be used, which is the same as the DL. As for HS-DSCH, a CRC length of 24 bits is attached to each transport block, and the rate 1/3 Turbo coding is used as channel coding for E-DCH. The Hybrid ARQ functionality matches the number of bits at the output of the channel coder to the total number of bits of the E-PUCH set to which the E-DCH transport channel is mapped. The Hybrid ARQ functionality is controlled by the RV parameters. Figure 1.20 shows the detailed processing of E-DCH Hybrid ARQ functionality. The parameters of the rate matching stage depend on the value of the RV parameters s and r. Similar to HS-DSCH, constellation rearrangement is performed in the case of 16QAM in accordance with the general method described for HS-DSCH. For QPSK, this function is transparent.

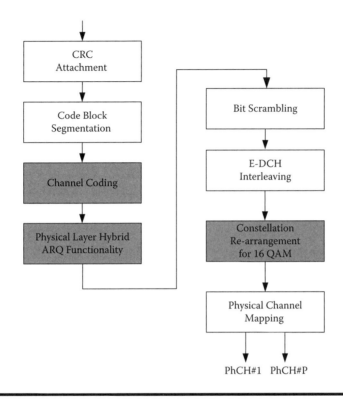

Figure 1.19   Physical layer processing for E-DCH.

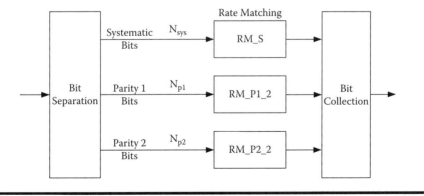

Figure 1.20   E-DCH Hybrid ARQ functionality.

Data symbol
E-RUCCH part1
... ...
E-RUCCH part1
Midamble 144 chips
TPC
E-RUCCH part2
... ...
TPC
E-RUCCH part2
Data symbol
GP
16*S bits
(16+2)*S bits

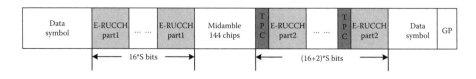

Figure 1.21 Multiplexing of E-DCH and E-UCCH (S is an indication of repetition level).

1.5.3.2.2 E-UCCH

The E-UCCH carries uplink control information associated with the E-DCH and is mapped to E-PUCH, which means that E-UCCH is multiplexed with E-DCH and mapped to the same physical channel. Depending on the configuration of the number of E-UCCH instances and the number of E-PUCH time slots, an E-PUCH burst may or may not contain E-UCCH and TPC. When E-PUCH does contain E-UCCH, TPC is also transmitted. When E-PUCH does not contain E-UCCH, TPC is not transmitted. For one E-UCCH instance, we have 32 physical channel bits, modulated using QPSK modulation. There is at least one E-UCCH and TPC in every E-DCH TTI. Multiple instances of the same E-UCCH information and TPC can be transmitted within an E-DCH TTI. The detailed number of instances can be set by the Node-B MAC-e layer for scheduled transmissions and signaled by higher layers for nonscheduled transmissions. When an E-DCH data block is transmitted on multiple time slots in one TTI, there will be multiple E-PUCH time slots. All repetitions of E-UCCH and TPC are evenly distributed on multiple E-PUCH time slots. The burst composition of the E-UCCH information and the E-DCH data is shown in Figure 1.21.

Figure 1.22 shows the E-PUCH data burst with and without the E-UCCH/TPC fields.

The E-UCCH is used to convey the following information:

- The modulation type of the selected E-TFC (0 bits)
- The transport block size of the selected E-TFC (6 bits)
- The retransmission sequence number (RSN) (2 bits)
- The HARQ process ID (2 bits)

Different from HS-DSCH, the occupied modulation type is not explicitly signaled (0 bits for this information), which is inferred from the transport

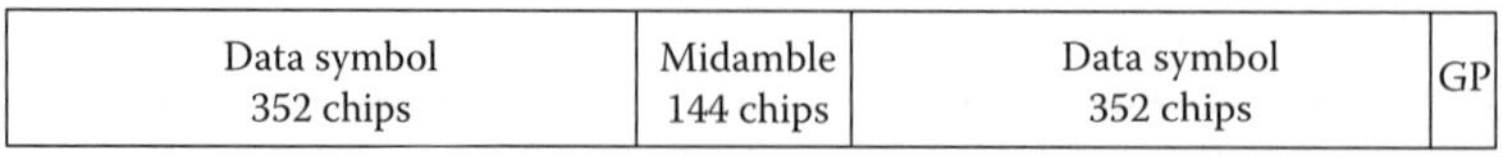

Data symbol 352 chips
Midamble 144 chips
Data symbol 352 chips
GP

Figure 1.22 E-PUCH data burst without E-UCCH/TPC

Data symbol 352 chips	Midamble 144 chips	SS	TPC	Data symbol 352 chips	GP

Figure 1.23   E-AGCH1 burst structure.

block size. The same as HS-DSCH, there are total 64 kinds of transport block size (6 bits). Because it is the UE's responsibility to choose the transmitted TFC, which, combined with the granted resource, determines the block size, the UE informs Node-B of such information via these 6 bits.

To indicate the RV of each Hybrid ARQ transmission, a 2-bit RSN is signaled. Node-B can avoid soft buffer corruption by flushing the soft buffer associated to one Hybrid ARQ process in case the last received RSN for that Hybrid ARQ process is incompatible with the current one. For a given Hybrid ARQ process, once the maximum RSN value of 3 is reached, the RSN alternates between the values of 2 and 3 for any further retransmissions. The used RV and the constellation rearrangement parameter are implicitly linked to the transmitted RSN [14]; as such, the Node-B is always able to determine the correct RV and constellation rearrangement parameter if the RSN information is correctly obtained.

These 10 bits are then multiplexed and coded using a (32, 10) subcode of the second-order Reed-Muller code, which results in 32 bits for each E-UCCH instance.

1.5.3.2.3 E-AGCH

The E-AGCH is a downlink physical channel carrying the uplink E-DCH absolute grant control information. The E-AGCH uses two separate physical channels (E-AGCH1 and E-AGCH2). The spreading factor used for E-AGCH is 16. The burst structures for E-AGCH1 and E-AGCH2 are shown in Figure 1.23 and Figure 1.24, respectively. E-AGCH1 will carry the TPC and SS bits.

The E-AGCH carries the following fields multiplexed into 23 to 26 bits:

■ *Absolute grant (power) value (5 bits).* This field indicates the granted power by the base station to specific UE for E-DCH transmission. This information can be used for uplink interference and load control purposes. There are a total of 32 different kinds of power grant levels.
■ *Code resource related information (5 bits).* This field indicates the specific code resource grant for the E-DCH transmission.

Data symbol 352 chips	Midamble 144 chips	Data symbol 352 chips	GP

Figure 1.24   E-AGCH2 burst structure.

- *Timeslot resource related information (5 bits).* This field indicates the specific time slot resource grant for the E-DCH transmission, indicating the allocation for E-DCH resources from TS1 to TS5. If the bit is set (i.e., equal to 1), then the corresponding timeslot will be used for E-DCH resources.
- *E-AGCH cyclic sequence number (ECSN) (3 bits).*
- *Resource duration indicator (3 bits, if present).* This is used for semi-persistent scheduling, indicating the valid resource allocation duration.
- *E-HICH indicator (2 bits).* This is used to indicate the UE that E-HICH will use to convey the acknowledgment indicator in the following schedule period.
- *E-UCCH number indicator (3 bits).* This is used to calculate the number of E-UCCH instances.

Figure 1.25 illustrates the overall coding chain for the E-AGCH. After multiplexing the above information, a 16-bit CRC is calculated and attached

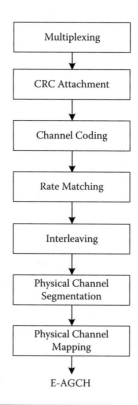

Figure 1.25   Physical layer processing for E-AGCH.

| Data symbol | Spare bits | Midamble 144 chips | Spare bits | Data symbol | GP |

Figure 1.26   E-HICH structure.

to the sequence, and then a 1/3 rate convolutional channel coding is applied. After rate matching and interleaving, the sequence is segmented and mapped to the two physical code channels (E-AGCH1 and E-AGCH2).

1.5.3.2.4 E-HICH

The E-HICH is defined in terms of an SF16 downlink physical channel and a signature sequence. The E-HICH carries one or multiple users' acknowledgment indicator. Figure 1.26 illustrates the structure of the E-HICH. The spare bit values are undefined. The power of each user's acknowledgment indicator may be set independently by Node-B. The number of E-HICHs in a cell is configured by the system. Scheduled traffic's and non-scheduled traffic's acknowledgment indicators are transmitted on different E-HICHs. The acknowledgment indicators for the E-PUCH semi-persistent scheduling operation can be transmitted on the same E-HICH carrying indicators for scheduled traffic or on the E-HICH carrying indicators for non-scheduled traffic.

For scheduled transmissions, at most four E-HICHs can be configured for one user's scheduled transmission. Which E-HICH is used to convey the Hybrid ARQ acknowledgment indicator is indicated by the 2-bits E-HICH indicator on the E-AGCH. A single E-HICH can carry one or multiple Hybrid ARQ acknowledgment indicator(s), which are decided by the Node-B.

For non-scheduled transmissions, E-HICHs carry not only the HARQ acknowledgment indicators, but also TPC and SS commands. The 80 signature sequences are divided into 20 groups, and each group includes four sequences. Every non-scheduled user is assigned only one group. Among the four sequences, the first one is used to indicate ACK/NACK, and the other three are used to indicate the TPC/SS commands.

For E-DCH semi-persistent scheduling, E-HICHs carry not only the HARQ acknowledgment indicators, but also the TPC and SS commands. Each user is also assigned one signature sequence group, including four sequences whose usage is completely compliant with the definition in non-scheduled transmissions.

Detailed coding and multiplexing of E-HICH is not introduced here due to space limitations. One can refer to [14] for more detailed information. In general, because TD-SCDMA HSUPA supports various uplink transmission modes (i.e., scheduled, non-scheduled, and semi-persistent scheduled), the coding and multiplexing of E-HICH is different for different transmission modes.

1.5.3.2.5 E-RUCCH

The E-RUCCH is used to carry E-DCH associated uplink control signaling when E-PUCH resources are not available, for example, at the beginning of the data transmission. It is mapped to the same random access physical resources defined by UTRAN. It uses spreading factor SF = 16 or SF = 8. The set of admissible spreading codes used on the E-RUCCH is based on the spreading codes of PRACH. The time slot format depends on the spreading factor of the E-RUCCH. In general, no TPC and SS bits are transmitted.

1.6 Performance Evaluations

From the above sections, one can obtain a general understanding of the principles, technologies, and concepts of TD-SCDMA HSDPA/HSUPA. In this section, we provide performance evaluations for certain key aspects of TD-SCDMA HSDPA/HSUPA, including the performance benefits of the key techniques, the system average throughput in a typical macro-environment, etc. The performance results given here are intended to provide readers with further understanding of the capabilities of TD-SCDMA HSDPA/HSUPA. Readers can refer to [28–30] for more performance evaluations results: for example, the performance of different beamforming algorithms in [28], the performance of different detection methods in [29], and the performance of VoIP [30].

For the performance evaluation, a dynamic system level simulator was developed [32]. This simulator is based on OPNET Modeler® software with the assistance of link level simulation by the actual value interface (AVI). The simulation scenario is a typical Macro 19 cells with 3 sectors in each cell. The wrap-around technique is used to eliminate the boundary effect. The inter-site distance is 1 km. The propagation model, including path loss and shadow fading, is based on the UMTS 30.03 specification. The channel model is Pedestrian-A and the user mobility is 3 km/h.

In the simulator, key techniques of TD-SCDMA HSDPA/HSUPA, such as adaptive smart antenna, power control, HARQ for UL and DL, UL and DL scheduling and resource control, are all implemented. For UL, closed-loop power control is implemented according to Equations (1.1) and (1.2), and the power control step is 0.5 dB. In the simulation, in order to get enough position samples, 25 users are randomly distributed in each sector. A full buffer traffic model is used, and Proportional Fairness (PF) scheduling is used to calculate user priority. If not explicitly stated, Hybrid ARQ with a maximum of four transmissions is enabled. Note that valuations are separately performed for DL and UL, that is, concurrent UL and DL traffic transmissions are not simulated. The results provided mainly focus on the UL because for TD-SCDMA, UL is the bottleneck link.

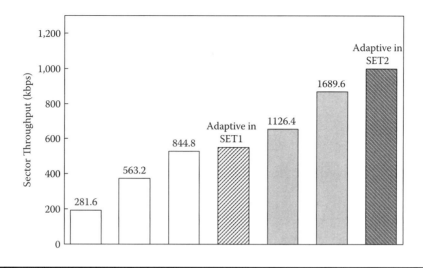

Figure 1.27   Sector throughput comparison of different TFCs.

1.6.1 Higher Modulation Level Gain

Figure 1.27 shows the sector throughput comparison of different TFC sets for UL. TFC SET1 means the data rate of {70.4, 140.8, 211.2} kbps using one time slot, while SET2 is {70.4, 140.8, 211.2, 281.6, 422.4} kbps, and adaptive in SET1 and SET2 means that TFC selection can be performed adaptively among such sets. The other bars mean only use the corresponding TFCs. The latter two TFCs in SET2 (i.e., 281.6 and 422.4 kbps) use 16QAM to pump more bits. In the simulation, four uplink traffic time slots are assumed. This boosts the TFC rate value to four times, which is shown above each bar in the figure. We can see that a higher modulation level has potential gain. From the throughput point of view, adaptive selection in SET2 gains 80.6% over that only in SET1. Although, in real network scenarios, the gain may not be as high due to the channel estimation error and beamforming error, the potential gain still can be obtained. It can also be seen that allowing for the selection of TFC freely among all TFCs is always better than sticking to just one. From these results, one can see the performance of high modulation and also the gain brought about by freely choosing among multiple modulation and coding sets.

1.6.1.1 Hybrid ARQ

To observe Hybrid ARQ related performance, Dynamic TFC selection is disabled in the simulation and a rate of 140.8 kbps is assumed (for a total of four time slots, the rate is 563.2 kbps). Figure 1.28 shows the normalized sector throughput with different maximum allowed transmissions.

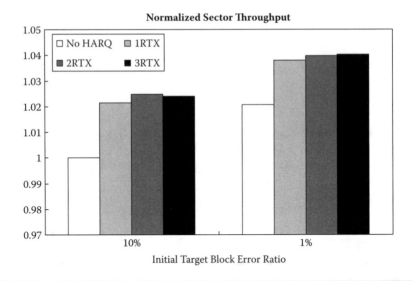

Figure 1.28   Sector throughput under different Hybrid ARQ operation scenarios.

The target block error ratio of initial transmission is set to 10% and 1%. It can be seen that Hybrid ARQ can bring significant gain concerning the system throughput while the contribution of the first retransmission is the most. The succeeding retransmissions can reduce the system's overall error ratio with little improvement from a throughput point of view. Good trade-off should be made between the overall system error ratio requirement and the air interface resource usage efficiency. Targeting a lower error ratio improves the overall system throughput but more transmit power is required.

Figures 1.29 and 1.30 show the ET gain of Hybrid ARQ operation. Two operating methods are adopted and compared. In the first method, TFC with the rate of 422.4 kbps is used and we aiming at three times of transmissions to achieve the 10% BLER target. In a second method, TFC with the rate of 140.8 kbps is adopted but only one transmission is allowed to achieve the same BLER target. Consequently, the overall target rate is 140.8 kbps. The first method aims at three transmissions, while the second aims at only one transmission to achieve such a rate. Figure 1.29 shows the sector throughput for the two methods, while Figure 1.30 shows the successful decoding probability of each transmission and also the residual BLER. It can be seen that the first method, which has a higher target error ratio for each transmission, also has more chance to retransmit to achieve the same overall error ratio and gets much better performance than the second method.

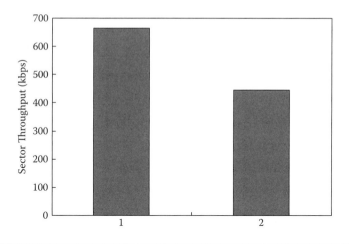

Figure 1.29   Throughput of the different schemes.

1.6.2 Interference Distribution and Control

Figure 1.31 and Figure 1.32 show the PDF (probability density function) and CDF (cumulative distribution function) curves of RoT value in UL under smart antenna and directional antenna, respectively. The smart antenna used here is 60° sector smart antenna, and the pattern is collected from trial networks. The directional antenna is a typical 120° sector directional antenna. In the simulation, no power grant is sent; that is, the maximum allowed transmitting power the user can adopt is the maximum power that the user can support. It can be seen that most of the RoT value under a

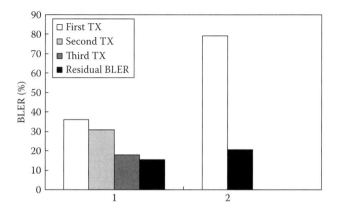

Figure 1.30   Successful transmission probability of each transmission and the residual BLER.

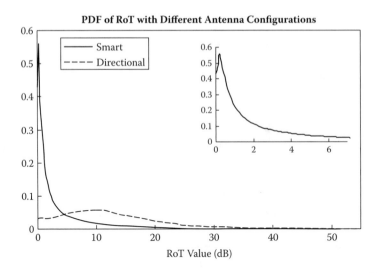

Figure 1.31 Probability density function (PDF) of RoT.

smart antenna is very small compared to that under a directional antenna. The mean RoT value under smart antennas is 3.65 dB while it is 11.93 dB under directional antennas. The overshot probability, which is defined as the RoT value higher than 7 dB, is 16%. This is mainly due to the scenario mentioned in the above section. Such overshot probability does have a negative impact on the user's data, but it is relatively difficult to avoid for the independent scheduling procedure performed in each cell. Fortunately,

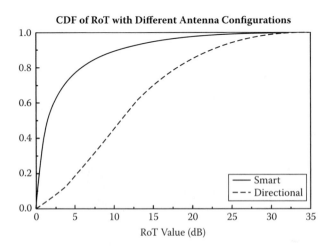

Figure 1.32 Cumulative density function (CDF) of RoT.

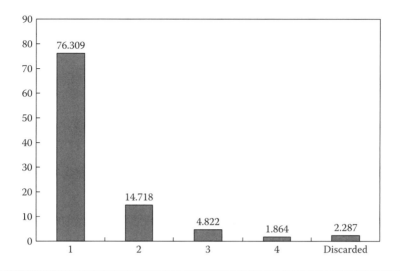

Figure 1.33   Successful transmission probability of each transmission and residual BLER.

however, we have Hybrid ARQ to help us cope with such a bad situation. The residual BLER after certain retransmissions can be very low. Figure 1.33 shows the successful transmission probability of each transmission and also the residual BLER. The initial transmission target BLER is 10%. That means that we aim to get about 90% of the packets transmitted successfully at the initial transmission. But due to the severe interference scenario mentioned above, such a target is somewhat hard to achieve. Only 76.3% of the packets are successfully transmitted at the initial transmission. About 24% of the packets encounter such a scenario and also heavy fading during the initial transmission. Here, Hybrid ARQ is an attractive way to solve the problem. After the first retransmission, another 14.72% of the packets are successfully decoded. That is to say, after the first retransmission, the overall percentage of incorrect packets is only 8.97% $(1 - 76.3\% - 14.72\%)$. If we take the overshot probability by RoT higher than 7 dB as judgment for the heavy interference scenario, most of the 16% of the packets which interference by such heavy interference scenario is successfully recovered by Hybrid ARQ first retransmission $(24\% - 8.97\% = 15.03\%)$.

1.6.3 Multi-frequency Operation

One unique characteristic for TD-SCDMA HSPA is its multi-frequency operation. In what follows, the potential gain brought by multi-frequency is demonstrated. As we know, the basic benefit brought about by multi-frequency operation is the enhancement of transmission capability. The single-link peak data rate can be boosted to K (the number of frequencies

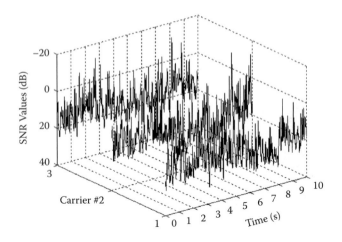

Figure 1.34 Received SNR plot.

adopted) times compared to single-frequency operation. Because the occupied frequency band is also K times, no improvement in the spectral
efficiency is observed (there is even loss due to the guard band). However, for multi-user operation, the spectral efficiency can be improved due
to the multi-user diversity gain (MUD). MUD in single-carrier TD-SCDMA
HSPA can be exploited by serving users at or near their channel peaks,
while multi-frequency TD-SCDMA HSPA provides additional multi-user diversity gain in the frequency domain. This indicates the potential higher
spectral efficiency over single carrier for multi-user operation. In the frequency domain, multi-user diversity gain comes from the different fading
characteristics across carriers. It is the function of inter-carrier channel correlation. High channel correlation reduces the independence of signal paths,
thus limiting the diversity gain. The frequency domain channel correlation
can be described by coherence bandwidth f_c, which is the reciprocal of
the delay spread. Typically, f_c is hundreds of kilohertz (kHz), which is far
less than the single-carrier bandwidth of multi-frequency TD-SCDMA HSPA.
Thus, sufficient channel de-correlation can be achieved, which results in
the possible exploitation of multi-user diversity gain in the frequency domain. Figure 1.34 plots the received SNR of one user observed from three
carriers simultaneously in the DL. It can be seen that there exists adequate
fading in both the time and the frequency domains that can be used to
exploit multi-user diversity gain in both domains.

For exploiting such gain, the design of user and packet scheduling methods is of great importance. The following three scheduling methods are
considered in the evaluations. These methods are the possible modifications of the well-known PF scheduler to multi-frequency operation.

1. *Multi-carrier integrated scheduling (MC-IS).* In MC-IS, all carriers' resources are combined into one resource block. Scheduling and resource allocation are performed on this integrated block. At each scheduling time t, MC-IS allocates all carriers' resources to UE j if

$$j = \arg \max \left. r_i(t) \middle/ \bar{R}_i(t) \right. \tag{1.3}$$

where $r_i(t)$ is the sum of instantaneous supporting rate of user i on all K carriers at time t:

$$r_i(t) = \sum_{k=0}^{K-1} r_{ik}(t) \tag{1.4}$$

$r_{ik}(t)$ is the instantaneous supporting rate of user i on carrier k. $\bar{R}_i(t)$ is the window averaged rate of user i achieved on all carriers:

$$\bar{R}_i(t) = (1 - 1/t_c) \cdot \bar{R}_i(t-1) + 1/t_c \cdot \sum_{k=0}^{K-1} r_{ik}(t) \tag{1.5}$$

where t_c is the time filter constant. If the user has not been served on carrier k at time t, then, we have $r_{ik}(t) = 0$.

2. *Multi-carrier separately scheduling (MC-SS).* MC-SS schedules carriers one by one. The scheduler allocates the resource of carrier k to UE j_k if

$$j_k = \arg \max \left. r_{ik}(t) \middle/ \bar{R}_{ik}(t) \right. \tag{1.6}$$

$\bar{R}_{ik}(t)$ is the averaged rate of user i got on carrier k:

$$\bar{R}_{ik}(t) = (1 - 1/t_c) \cdot \bar{R}_{ik}(t-1) + 1/t_c \cdot r_{ik}(t) \tag{1.7}$$

3. *Multi-carrier independence scheduling (MC-IDS).* At time t, MC-IDS allocates the carrier k resource to the UE j_k if

$$j_k = \arg \max \left. r_{ik}(t) \middle/ \bar{R}_i(t) \right. \tag{1.8}$$

Figures 1.35 through 1.37 show the sector DL throughput and residual BLER performance of both single-frequency and multi-frequency systems under the considered scheduling schemes. It can be seen that the throughput of the multi-frequency system is much higher than that than of the single frequency system for the additional frequency resources. System throughput under MC-SS is exactly three times that of single-frequency with PF scheduling. Because of the additional multi-user diversity gain exploited by MC-IDS in the frequency domain, system throughput under

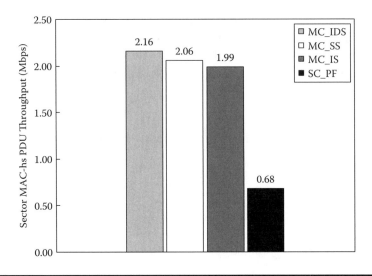

Figure 1.35   Sector MAC-hs PDU throughput for different schedulers.

MC-IDS is higher than that of MC-SS. When it comes to normalized through-put, MC-IDS gains 5.88% (0.228 dB) over MC-SS and also single-frequency with PF scheduling. Due to the impairment to multi-user diversity gain in both the time and frequency domains mentioned above, systems with MC-IS scheduling achieve the lowest throughput performance. The normalized throughput under MC-IS is even lower than single frequency with

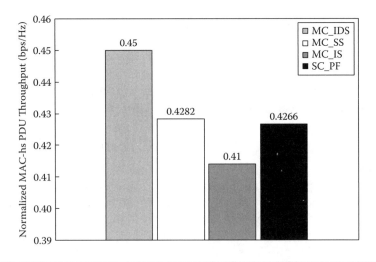

Figure 1.36   Normalized MAC-hs PDU throughput for different schedulers.

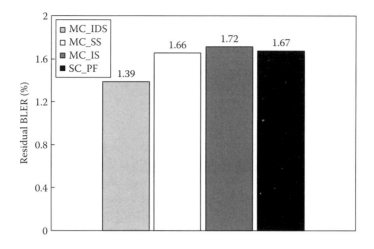

Figure 1.37   Residual block error ratio for different schedulers.

PF scheduling. Performance of system residual BLER is of the same trend as system throughput. Systems with MC-IDS achieve the lowest residual BLER performance and then the MC-SS, which is almost the same as SC-PF; the system residual BLER under MC-IS is the highest.

The higher system spectral efficiency under MC-IDS should also bring benefit to user experience. Figure 1.38 shows the complementary cumulative distribution function (CCDF) of user throughput. The curve labeled

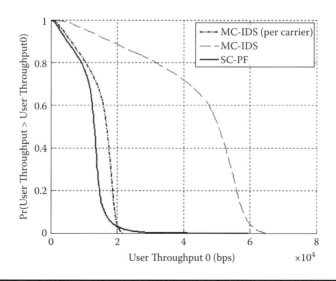

Figure 1.38   CCDF of user throughput.

"MC-IDS (per carrier)" is the CCDF of user throughput averaged per carrier in a multi-frequency system. It can be seen that users achieve more than three times the data rate in a three-frequency system as in a single-frequency system due to the additional diversity gain. The benefit for very high data rate users is limited and it increases for lower data rate users.

1.7 Summary

As the basic 3G choice in China, TD-SCDMA has been well developed and has been operated by the CMCC, one of the biggest operators in China. With the ever-increasing demand for higher data rates and the aim of providing various multimedia services, TD-SCDMA networks have evolved toward their enhanced version, TD-SCDMA HSDPA/HSUPA. This chapter provided an overview of this enhanced version, including the key techniques, channel processing, operation principles, and also some performance evaluations. The characteristics introduced here were based mainly on the 3GPP R5 and R6 versions. After these, some advanced techniques and operation options have been further absorbed into TD-SCDMA HSDPA/HSUPA, including MIMO operation, VoIP service, and 64QAM in DL, etc. With these features, TD-SCDMA HSDPA/HSUPA is believed to be able to provide various multimedia services, including file downloading and uploading, Web-surfing, e-mail, games, and VoIP, etc.

Acknowledgment

The work of Shuping Chen is supported by BUPT Excellent Ph.D. Students Foundation (CX200903).

Links

3GPP homepage. http://www.3gpp.org/
CCSA homepage. http://www.ccsa.org.cn/english/
TD-SCDMA forum. http://www.td-forum.org/en/

References

[1] H.-H. Chen, C.-X. Fan, and W.-W. Lu, China's perspectives on 3G mobile communications and beyond: TD-SCDMA technology, *IEEE Wireless Commun.,* 9(2): 48–59, 2002.
[2] M. Peng, S. Chen, and W. Wang, 3G commercial deployment, *Encyclopedia of Mobile Computing and Commerce,* IGI Global, 2007, pp. 940–945.

[3] S.G. Klein, M.J. Irwin, and P. Roberto, On the capacity of a cellular CDMA system, *IEEE Trans. Vehicular Technol.*, 40(2): 303–312, 1991.

[4] H. Holma and A. Toskala, *WCDMA for UMTS*, 3rd edition, John Wiley & Sons, 2004.

[5] J.C. Liberti and T.S. Rappaport, *Smart Antennas for Wireless Communications*, Prentice Hall PTR, 1996.

[6] J.H. Winterw, Smart antennas for wireless systems, *IEEE Personal Commun.*, 5(1): 23–27, 1998.

[7] L.C. Godara, Application of antenna arrays to mobile communications. I. Performance improvement, feasibility, and system considerations, *Proc. IEEE*, 85(7): 1029–1030, 1997.

[8] L.C. Godara, Application of antenna arrays to mobile communications. II. Beamforming and direction-of-arrival considerations, *Proc. IEEE*, 85(8): 1195–1245, August 1997.

[9] S. Verdu, *Multiuser Decetion*, Cambridge Unversity Press, 1998.

[10] H. Jiang, CBWL: A new channel assignment and sharing method for cellular communication system, *IEEE Trans. Vehicular Technol.*, 43: 313–321, 1994.

[11] S. Li, TD-SCDMA: Standardization and prototype, in *Proc. APCC/OECC'99*, Beijing, pp. 1715–1716.

[12] M. Peng and W. Wang, A framework for investigating radio resource management algorithms in TD-SCDMA systems, *IEEE Commun. Mag.*, 43(6): 12–18, June 2005 .

[13] 3GPP, TS 25.211, Physical Channels and Mapping of Transport Channels onto Physical Channels (TDD).

[14] 3GPP TS 25.222, Multiplexing and Channel Coding (TDD).

[15] 3GPP TS 25.223, Spreading and Modulation (TDD).

[16] 3GPP TS 25.224, Physical Layer Procedures (TDD).

[17] 3GPP TS 25.225, Physical Layer Measurements (TDD).

[18] 3GPP 25.321, Medium Access Control (MAC) Protocol Specification.

[19] 3GPP TS 25.331, Radio RRC Protocol Specification.

[20] A.J. Goldsmith and S.-G. Chua, Adaptive coded modulation for fading channels, *IEEE Trans. Commun.*, 46(5): 595–602, 1998.

[21] S. Chen, J. Du, M. Peng, and W. Wang, Performance analysis and improvement of link adaptation techniques in TDD-HSDPA/SA system, in *Proc. 13th Int. Conf. Tele.*, 1: 102–106, 2006.

[22] H. Holma and A. Toskala, *HSDPA/HSUPA for UMTS*, John Wiley & Sons, 2006.

[23] S. Chen, H. Lin, J. Han, D. Zhao, and K. Asimakis, Generalized scheduler providing multimedia services over HSDPA, in *Proc. IEEE ICME 2007*, Beijing, pp. 927–930.

[24] A. Jalali, R. Padovani, and R. Pankaj, Data throughput of CDMA-HDR: a high efficiency-high data rate personal communication wireless system, in *Proc. IEEE VTC*, 3: 1854–1858, May 2000.

[25] M. Andrews, K. Kumaran, and K. Ramanan, Providing quality of service over a shared wireless link, *IEEE Commun. Mag.*, 39(2): 150–154, 2001.

[26] J. Fan and T. Chen, Packet scheduling algorithms for mixed traffic flow over HSDPA, in *Proc. IEEE PIMRC*, September 2007, pp. 1–5.

[27] H. Lin, S. Chen, and K. Asimakis, Scheduling and performance of multi-carrier TD-SCDMA HSDPA, in *Proc. of ChinaCom2007*, August 2007, pp. 1042–1046.

[28] H. Fu and Y. Han, The application research about the smart antenna in the TD-SCDMA standard, in *Proc. China-Japan Joint Microwave Conf.,* September 2008, pp. 427–431.

[29] Y. Wang Y.-C. Liang, and W.S. Leon, Signal detection and interference cancellation for TD-SCDMA downlink in fast time-varying environment, in *Proc. IEEE WCNC,* March 2005, 1: 291–296, May 2005.

[30] S. Chen, W. Wang, J. Han, G. Liu, and N. Li, Providing VoIP service over TD-SCDMA HSDPA, in *Proc. IEEE VTC - Spring,* May 2008, pp. 2066–2070.

[31] S. Parkvall, J. Peisa, and J. Torsner, et al., WCDMA enhanced uplink—Principles and basic operation, in *Proc. of IEEE VTC—Spring,* 3: 1411–1415, May 2005.

[32] J. Du, S. Chen, Y. Li, and W. Wang, Modeling HSDPA in TDD-CDMA/SA systems, in *Proc. IEEE MAPE,* 2: 1501–1505, 2005.

Chapter 2

Receiver Designs and Multi-User Extensions to MIMO HSDPA

Shakti Prasad Shenoy, Irfan Ghauri, and Dirk T.M. Slock

Contents

2.1 Introduction

Any wireless communication system that leverages the use of multiple antennas both at the transmitter and the receiver qualifies as a multiple-input, multiple-output (MIMO) wireless system. Multiple antennas at the transmitter and receiver add an additional spatial dimension to the communication channel. By taking advantage of this fact and by exploiting the spatial properties of the MIMO channel, it is possible to provide the following features to the communication system.

1. *Making the communication link resilient/robust to channel fades.*
   Diversity techniques have long been considered effective means to combat channel fading. In simple terms, diversity is achieved by combining multiple copies of the same transmit signal. If the fading characteristics of each copy are statistically independent of the rest, the combined signal is more robust to channel fading. In the context of MIMO systems, using the concept of *spatial diversity*, it is possible to show that the probability of losing the signal due to deep fades reduces exponentially with the number of decorrelated transmit-receive antenna pairs (spatial links) between the transmitter and receiver [19].

2. *Increasing the link capacity.* Instead of using the multiple spatial channels to provide diversity, it is possible to use these channels for multiplexing in the spatial domain. A high data rate stream is first split into multiple substreams of lower data rates. Subject to certain channel conditions [6], $\min(N_{tx}, N_{rx})$ streams can be transmitted over the MIMO channel. Here, N_{tx}, N_{rx} refer to the number of antennas at the transmitter and receiver, respectively. Because this requires no extra spectral resources, the total data rate (bits per second, b/s) transmitted over the communication link is increased.

3. *Increasing coverage area.* Transmit beamforming is a technique in which signals transmitted from multiple antennas are multiplied by a complex weighting factor (different for each antenna) such that the transmitted signal power is concentrated in certain spatial directions

(or spatial signatures). The resultant signal can now travel over a larger distance in that direction, thus increasing the coverage area of the base station. A similar type of processing can be employed at the receiver, whereby the received signal power is increased by combining the signals at each receive antenna after application of suitable weights (receive beamforming).

4. *Improving spectral efficiency.* By reusing the multiple access resources [for instance, spreading codes in Code Division Multiple Access (CDMA)] over the spatial dimension, MIMO systems can increase spectral efficiency (b/s/Hz) of the communication system.

However, not all these features can be provided simultaneously. For instance, there exists a trade-off between the coverage range and the link quality in any MIMO system [15]. Similarly, using multiple transmit antennas for spatial multiplexing reduces the available spatial degrees of freedom for spatial reuse.

MIMO systems first attracted attention due mainly to the tremendous increase in channel capacity that is promised [5,19]. While there has been sustained academic interest in MIMO over the past decade, as witnessed by the huge number of research publications in this topic, true MIMO systems are only recently being standardized. This has been mainly due to the increased system complexity of MIMO systems. While MIMO can potentially provide huge gains at no extra cost in terms of spectral resources, these gains can only be realized at the cost of increased system and hardware complexity. Moreover, until recently, multiple antennas at the user equipment (UE) were not considered desirable due to space, battery, and cost constraints of mobile terminals. As a result, standardization bodies have to date concentrated more on the subclass of MIMO systems [Multiple-input, single-output (MISO)/single-input, multiple-output (SIMO)], whereby some kind of antenna diversity at the base station (BS) is used to exploit transmit and/or receive diversity in the interest of enhancing link quality or increasing the total system capacity. With the emergence of Internet-centric applications and an increased demand for high-data-rate applications in cellular systems, this trend is changing very quickly. The present generation of smart phones and Internet-enabled devices have both the form factor as well as the computational powers to support multiple antennas at the receiver. Foreseeing these developments 2×2 MIMO has been standardized [1]. In fact, the world's first HSPA+ (High Speed Packet Access) or evolved HSPA network with support for 2×2 MIMO was launched in early 2009 [10]. Along with enabling technologies such as adaptive modulation and coding (AMC), fast hybrid automatic repeat request (HARQ), and user feedback based scheduling, MIMO in HSDPA (HS Downlink PA) can lead to peak data rates of 42 Mbps in downlink (DL). However, in the present form, MIMO in HSDPA can support only single-user (SU) scenarios in DL.

While shifting from a single user to a multi-user (MU) paradigm mandates a whole new level of increased system complexity [7], the associated gains are significant. For instance, MU-MIMO opens up the possibility of code-multiplexing, which can lead to increased system capacity. As mentioned earlier, the gains promised by MIMO can only be realized at the cost of increased computational complexity. MIMO has largely been discussed in the context of the frequency non-selective (OFDM) case, where optimal joint-stream maximum a-posteriori (MAP) detection can be employed. Spatio-temporal receivers based on ordered successive interference cancellation (OSIC) in frequency-selective environments were considered in [13] while [17] proposed a class of maximum likelihood (ML) receivers for multipath channels. For MIMO WCDMA (wideband code division multiple access) transmission in frequency-selective channels, where the multipath mixes up signals in space and time, proposals for receiver (RX) solutions include chip-level equalization and despreading, followed by joint detection of the data streams at symbol level [14]. More generally, a two-stage approach is considered, where the first stage is the chip-equalizer correlator followed by some kind of joint processing or decision-feedback approach [21]. In this chapter we explore a class of receivers based on the two-step processing strategy. We consider receivers for HSDPA downlink that use MIMO-linear MMSE chip equalizer at the chip level, followed by further processing at the symbol level after despreading. To fully harness the potential of this approach, we will find it necessary to do away with the customary assumption that the scrambler used at the transmitter can be modeled as a random *i.i.d.* sequence and instead treat it as a deterministic, known sequence. This results in enhanced performance at the cost of increased complexity at the receiver.

This chapter is divided into three sections. First we provide a brief review of MIMO support in HSDPA. Then, in the first section we discuss various receiver designs for HSDPA when the UE is configured in MIMO mode. Throughout all analysis in this section we make the customary assumption that the scrambler can be considered a random *i.i.d.* sequence. We also address here the issue of optimal choice of precoding matrix for the receiver. Because the receiver is required to choose the precoding matrix that maximizes its aggregate transport block size, we derive analytical expressions for the choice of the optimum precoding matrix that maximizes the sum-capacity of the receiver when it is based on MMSE designs. In the second section we do away with the assumption that the scrambler is random and treat it as deterministic. We introduce the time-varying model of the resulting symbol-level spatial channel and show that the deterministic point of view leads to a set of reduced dimension linear receivers and interference cancelers with increased achievable capacity. In the third section we extend the current single-user MIMO scenarios in HSDPA to the multiuser case. These extensions require minimal changes to existing standards.

When multiple UEs must be simultaneously serviced in the downlink, we suggest practical multi-user scheduling strategies that can be employed at the BS so as to maximize the downlink capacity. Finally we wrap up with some concluding remarks.

2.1.1 MIMO in HSDPA

3GPP (3rd Generation Partnership Project) has introduced a variant of Per-Antenna Rate Control (PARC), namely, dual stream transmit antenna array (D-TxAA) for transmit adaptive array transmissions [1] in UMTS WCDMA. Code reuse is made across the two streams and the scrambling sequence is also common to both transmit (TX) streams. All (15) spreading codes are allocated to the same user in the HSDPA MIMO context. In general, all UEs served by a BS feed an SINR (signal-to-interference-plus-noise ratio)-based (or based on some other appropriate measure) channel quality indicator (CQI) back to the BS. In addition, the UE also computes (and feeds back) the weighting vector(s) that would ideally provide the best instantaneous rate for the next time slot. Together, these feedbacks translate into a specific transport block size and a specific modulation and coding scheme (MCS) for each UE. Based on this information, the BS is capable of maximizing the downlink throughput for each transmission time interval.

Both transmit diversity and spatial multiplexing have been incorporated by 3GPP as standard in the form of TxAA (Figure 2.1) and its dual stream counterpart D-TxAA (Figure 2.2) for MIMO HSDPA. HSDPA supports a closed loop transmit diversity technique called (TxAA). In the 2 transmit–1 receive (2 × 1) antenna configuration of TxAA, the UE feeds back optimum beamforming weights that the BS uses while transmitting data to UE. D-TxAA is the extension of TxAA when UEs are configured in MIMO mode. Here, two separately encoded, interleaved, and spread transport blocks are

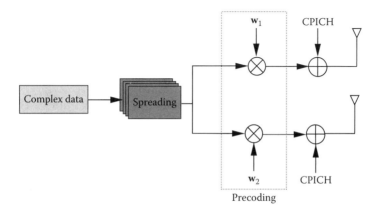

Figure 2.1 **Simplified block diagram of transmitter processing for TxAA.**

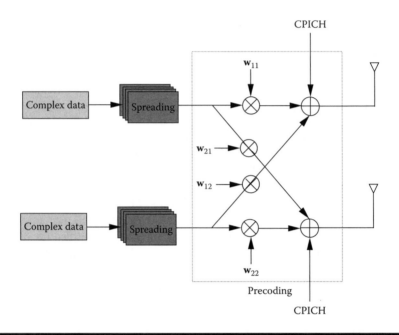

Figure 2.2 Simplified block diagram of transmitter processing for D-TxAA.

transmitted in parallel. In this case, the UE decides the precoding matrix that the BS must use when transmitting data to the UE. Let us now look at the beamforming/precoding aspect in more detail.

2.1.2 Precoding and CQI Feedback

In HSDPA, the UE is required to submit regular CQI and precoding control indicator (PCI) reports to the BS. The CQI can be mapped to a particular MCS. The data packet size associated with a particular MCS can then be mapped to obtain the supported throughput for each stream for a certain predefined packet error rate (PER). The mapping strategy has been subject to significant simulation study (see, e.g., [12]) and the SINR $\rightarrow$ CQI $\leftrightarrow$ PER $\leftrightarrow$ throughput relationship has been agreed to, appearing as CQI to MCS tables in the 3GPP standard document [1]. In addition to this, for each TTI (transmission time interval) over which the UE computes the CQI, PCI is computed using the Common Pilot Channel [CPICH(s)] transmitted from both transmit antennas to decide the beamforming vector. When applied by the BS, this beam forming vector maximizes the aggregate transport block size that the UE can support given the present channel conditions. To this end, when UE is not configured in MIMO mode, or when it requests transmission of a single transport block, the UE is required to choose one of four beamforming weights that control the antenna phase at the BS. This then constitutes its PCI feedback. The UE indicates the number of transport blocks to be transmitted to it as part the CQI report. The BS fixes the phase of its primary (reference) antenna and alters the phase of the secondary antenna accordingly. Because the precoding weight applied to the reference antenna is a constant $(1/\sqrt{2})$, the feedback consists of the weight for antenna-2 and is one of the following weights $w \in \{\frac{1+\sqrt{-1}}{2}, \frac{1-\sqrt{-1}}{2}, \frac{-1+\sqrt{-1}}{2}, \frac{-1-\sqrt{-1}}{2}\}$. One choice of beamforming weight vector, let us call it $\mathbf{w}$, might be one that maximizes the received signal power (or equivalently the receive SNR). For frequency-flat channels, this corresponds to the beamforming vector that is "closest" to the maximum right singular eigenvector of the 2×2 channel matrix $\mathbf{H}$. However, for frequency-selective channels with a delay spread L, there are L such MIMO channel taps. In general, it is not possible to choose $\mathbf{w}$ to match all channel taps; precoding gain in such conditions is in practice very low.

When the UE is configured in MIMO mode and requests two transport blocks to be transmitted, a precoding matrix must be used in place of a single beamforming weight vector. 2×2 unitary precoding based on receiver feedback is applied alongside spatial multiplexing at the base station in HSDPA [1] in D-TxAA. To keep feedback overhead low, both columns of the precoding matrix have exactly the same structure as the beamforming weight vector in TxAA. Moreover, the second column of this matrix is a unique function of the first. This severely restricts possible gains due to precoding. In fact, of the four precoding matrices, two of them are related to the remaining as follows. Let $w_1 = \beta$; then by design, $w_3 = \beta$, $w_4 = -w_2$, and

$$w_2 \in \left\{ \frac{1 + \sqrt{-1}}{2}, \frac{1 - \sqrt{-1}}{2}, \frac{-1 + \sqrt{-1}}{2}, \frac{-1 - \sqrt{-1}}{2} \right\} \rightarrow \in \{\gamma, \theta, -\theta, -\gamma\}$$

(2.1)

Therefore,

$$\mathbf{W} = \begin{bmatrix} w_1 & w_3 \\ w_2 & w_4 \end{bmatrix},$$

$$\mathbf{W}_1 = \begin{bmatrix} \beta & \beta \\ \gamma & -\gamma \end{bmatrix}, \quad \mathbf{W}_2 = \begin{bmatrix} \beta & \beta \\ \theta & -\theta \end{bmatrix}$$

The other two matrices are formed by interchanging the first and second columns of $\mathbf{W}_1$ and $\mathbf{W}_2$. Because the two transmitted streams interfere with each other and thereby influence CQI as well as PCI choice, the precoding matrix must be computed after joint equalization of both streams.

2.2 Receiver Designs for MIMO HSDPA: Part I

In this section we analyze performance of a variety of receiver designs for unitary precoded D-TxAA MIMO in HSDPA. The receiver structures proposed here are based on combining chip-level and symbol-level processing for enhanced performance. For each of these receivers, we derive the per-stream SINR expressions. We use the SINR to compute the sum-capacity, which can be interpreted as the upper bound for achievable rates. This will form the basis for comparing the performance of the proposed receivers. The precoding matrix in D-TxAA will influence the achievable sum-rate of the MIMO channel through its influence on the SINR of streams at the RX output. For D-TxAA with unitary precoding, there exists an optimal choice of the precoding matrix that maximizes the sum rate across the two streams. In principle, the receiver can evaluate the SINR corresponding to all precoding choices and request the application of the SINR-maximizing weights for the next TX frame. We will show that precoding choice and the extent of its impact depends on the MIMO receiver.

For the spatial multiplexing case in MIMO HSDPA, Figure 2.3 illustrates the equivalent baseband downlink signal model. The received signal vector (chip rate) at the UE can be modeled as:

$$\underbrace{\mathbf{y}[j]}_{2p\times1} = \underbrace{\mathbf{H}(z)}_{2p\times2}\underbrace{\mathbf{x}[j]}_{2\times1} + \underbrace{\boldsymbol{\eta}[j]}_{2\times1} \qquad (2.2)$$

In this model, j is the chip index, $\mathbf{H}(z)$ is the frequency selective MIMO channel the output of which is sampled p times per chip and $\boldsymbol{\eta}[j]$ represents

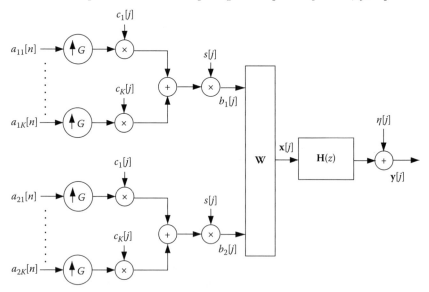

Figure 2.3   MIMO signal model with precoding.

the vector of noise samples that are zero-mean circular Gaussian random variables. The sequence $\mathbf{x}[j]$ introduced into the channel is itself a linear combination (D-TxAA, see [1]) of the two steams and is expressed as

$$\mathbf{x}[j] = \underbrace{\mathbf{W}}_{2 \times 2} \mathbf{b}[j] = \mathbf{W} \cdot \sum_{k=1}^{K} \underbrace{s[j]c_k[j \mod \mathcal{G}]\mathbf{a}_k[n]}_{\mathbf{b}_k[j]} \qquad (2.3)$$

where k is the code index, $n = \lfloor \frac{j}{\mathcal{G}} \rfloor$ is the symbol index, $\mathcal{G}$ is the spreading factor ($\mathcal{G} = 16$ for HSDPA), $\mathbf{W} = [\mathbf{w}_1 \ \mathbf{w}_2]$ is the 2×2 precoding matrix with $\mathbf{w}_1 = [\frac{1}{\sqrt{2}} \ w]^T$ and $\mathbf{w}_2 = [\frac{1}{\sqrt{2}} - w]^T$. The symbol vector $\mathbf{a}_k[n] = [a_{1k}[n] \ a_{2k}[n]]^T$ represents two independent symbol streams, $\mathbf{c}_k = [c_k[0] \cdots c_k[\mathcal{G} - 1]]^T$, where $\mathbf{c}_k^T \mathbf{c}_{k'} = \delta_{kk'}$ are unit-norm spreading codes common to the two streams, and $s[j]$ the common scrambling sequence element at chip time j, which is zero-mean *i.i.d.* with elements from $\frac{1}{\sqrt{2}}\{\pm 1 \pm \sqrt{-1}\}$.

2.2.1 Chip-Level Equalization

It is well known that orthogonal codes used in WCDMA downlink experience a loss of orthogonality when the transmission is over multipath channels with significant time-dispersion. Such multipath channels induce inter-code-interference in the classical correlator based receivers, rendering them suboptimal in such scenarios. It is shown in [9] that for single-input, single-output (SISO) links, linear minimum mean square error (LMMSE) equalizers can be used to restore the orthogonality of these codes prior to the despreading operation, thereby providing superior performance in multipath channels with large delay spreads. In frequency-selective MIMO channels, multipath mixes up signals in space and time, calling for alternative reception strategies. It is shown in [14] that linear chip-level MMSE equalizers not only are able to restore orthogonality of codes, but are also able to efficiently achieve spatial separation in MIMO frequency-selective channels. In the following section, we discuss receivers for MIMO HSDPA based on chip-level equalization.

2.2.1.1 LMMSE Chip-Equalizer Correlator

The classical MMSE chip-equalizer correlator receiver is an SINR-maximizing chip equalizer followed by code correlation and soft symbol estimate generation at the output of the correlator.

Consider the linear (MMSE) FIR estimation of the 2×1 chip sequence assuming the channel to have a finite impulse response (FIR). In the spatial multiplexing context, the LMMSE equalization tries to suppress not only all inter-chip interference (ICI), but also all inter-stream interference (ISI). In Figure 2.3, $\mathbf{b}[j]$ is the input chip vector defined as $\mathbf{b}[j] = [b_1[j] \ b_2[j]]^T$, where $b_i[j]$ is the jth chip of the ith input stream. Each chip stream is the sum of K spread and scrambled CDMA substreams (one user per CDMA code). Thus, $b_i[j] = \sum_{k=1}^{K} b_{ik}[j]$. The 2×2 matrix $\mathbf{H}[j]$ is the jth MIMO tap of the FIR channel and $\mathbf{W}$ is the precoding matrix. Denoting by L, the maximum delay spread of the frequency-selective channel (in chips) and assuming an arbitrary oversampling factor p at the receiver, the $2p \times 1$ received signal at the jth time instant is given as

$$\mathbf{y}[j] = \sum_{l=0}^{L-1} \mathbf{H}[l] \, \mathbf{W} \, \mathbf{b}[j-l] + \boldsymbol{\eta}[j] = \mathbf{H} \mathcal{W}_L \mathbf{b}_L[j] + \boldsymbol{\eta}[j] \qquad (2.4)$$

where $\mathbf{H} = [\mathbf{H}_1 \, \mathbf{H}_2]$, with $\mathbf{H}_i$ being the $2p \times L$ FIR channel from the ith transmit antenna to the two RX antennas. $\mathcal{W}_L = \mathbf{W} \otimes \mathbf{I}_L$ and $\mathbf{b}_L[j] = [\mathbf{b}_{1,L}^T[j] \, \mathbf{b}_{2,L}^T[j]]^T$ where $\mathbf{b}_{i,L}[j] = [b_i[j-L+1] \cdots b_i[j]]^T$ is the chip sequence vector of the ith stream. Stacking E successive samples of the received signal $\mathbf{y}[j]$, we can express the received signal as

$$\mathbf{Y}[j] = \mathcal{T}_E(\mathbf{H}) \mathcal{W}_{L+E-1} \mathbf{b}_{L+E-1}[j] + \boldsymbol{\Xi}[j] \qquad (2.5)$$

where $\mathcal{T}_E(\mathbf{H}) = [\mathcal{T}_E(\mathbf{H}_1) \, \mathcal{T}_E(\mathbf{H}_2)]$ and $\mathcal{T}_E(\mathbf{H}_i)$ is a block Toeplitz matrix with $[\mathbf{H}_i \, \mathbf{0}_{2p \times E-1}]$ as the first block row. Let us assume a $2 \times 2pE$ LMMSE equalizer $\mathbf{F} = [\mathbf{f}_1^T \, \mathbf{f}_2^T]^T$. The output of the equalizer is a linear estimate of the chip sequence given by

$$\hat{\mathbf{x}}[j] = \mathbf{F} \, \mathbf{Y}[j] = \mathbf{B} \underbrace{\mathbf{W} \, \mathbf{b}[j]}_{\mathbf{x}[j]} + \underbrace{\overline{\mathbf{B}} \mathcal{W}_{L+E-1} \overline{\mathbf{b}}_{L+E-1}[j] + \mathbf{F} \boldsymbol{\Xi}[j]}_{-\tilde{\mathbf{x}}[j]} \qquad (2.6)$$

Defining $\boldsymbol{\alpha}^{(ij)} = \mathbf{f}_i \mathcal{T}_E(\mathbf{H}_j)$, we have

$$\mathbf{B} = \begin{bmatrix} \alpha_d^{(11)} & \alpha_d^{(12)} \\ \alpha_d^{(21)} & \alpha_d^{(22)} \end{bmatrix} \quad \text{and} \quad \overline{\mathbf{B}} = \begin{bmatrix} \overline{\alpha}^{(11)} & \overline{\alpha}^{(12)} \\ \overline{\alpha}^{(21)} & \overline{\alpha}^{(22)} \end{bmatrix}$$

which are, respectively, the 2×2 matrices that represent the *joint bias* in the equalizer output and the residual ICI. The $\overline{\alpha}^{(ij)}$ are the same as $\alpha^{(ij)}$, with the $\alpha_d^{(ij)}$ term replaced by 0, and d is the equalization delay (in chips) associated with $\mathbf{F}$.

We can thus write the equalizer output as the sum of an arbitrarily scaled desired term and an error term:

$$\hat{\mathbf{x}}[j] = \mathbf{B}\mathbf{x}[j] - \tilde{\mathbf{x}}[j] \qquad (2.7)$$

In Equation (2.7), an estimate of the chip sequence $\mathbf{b}[j]$ can be obtained after a further stage of processing where the precoding is undone to separate streams. The latter, represented by $\mathbf{W}^H$, is a linear operation and

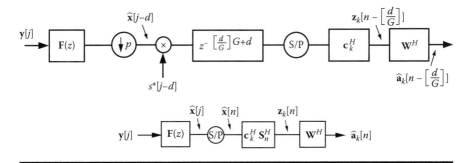

Figure 2.4 LMMSE equalizer and correlator. The second figure is a simplified representation used as chip-equalizer/correlator front-end stage for other receiver structures.

can be carried out before or after despreading (the latter case is shown in Figure 2.4). The joint-bias can also be interpreted as a spatial mixture at the chip-equalizer correlator output facilitating formulation of the spatial signal model to be treated henceforth. It must be pointed out that the spatial channel **B** is so definable, assuming the scrambler to be a random sequence. The resulting spatial channel is per-code, while still being the same for all codes. The error covariance matrix corresponding to the error term is denoted by $\mathcal{R}_{\widetilde{\widetilde{xx}}}$, from which the MSE can be obtained as follows:

$$
\mathcal{R}_{\widetilde{\widetilde{xx}}} = \begin{bmatrix} r_{11} & r_{12} \\ r_{21} & r_{22} \end{bmatrix} \tag{2.8}
$$

$$
r_{11} = \sigma_b^2 \left(\|\overline{\boldsymbol{\alpha}}^{(11)}\|^2 + \|\overline{\boldsymbol{\alpha}}^{(12)}\|^2 \right) + \mathbf{f}_1 \mathcal{R}_{nn} \mathbf{f}_1^H
$$

$$
r_{22} = \sigma_b^2 \left(\|\overline{\boldsymbol{\alpha}}^{(21)}\|^2 + \|\overline{\boldsymbol{\alpha}}^{(22)}\|^2 \right) + \mathbf{f}_2 \mathcal{R}_{nn} \mathbf{f}_2^H \tag{2.9}
$$

$$
r_{12} = r_{21}^* = \sigma_b^2 \left(\overline{\boldsymbol{\alpha}}^{(11)} \cdot \overline{\boldsymbol{\alpha}}^{(21)H} + \overline{\boldsymbol{\alpha}}^{(12)} \cdot \overline{\boldsymbol{\alpha}}^{(22)H} \right) + \mathbf{f}_1 \mathcal{R}_{nn} \mathbf{f}_2^H
$$

After despreading (for the kth code), the 2×1 signal at the symbol level is written as

$$
\mathbf{z}_k[n] = \mathbf{W}\mathbf{a}_k[n] - \widetilde{\mathbf{z}}_k[n] = \mathbf{B}\mathbf{W}\mathbf{a}_k[n] - \widetilde{\mathbf{z}}_k[n] \tag{2.10}
$$

Because the LMMSE joint bias is accounted for in **B**, the quantity $\widetilde{\mathbf{z}}_k[n]$ contains no desired symbol contribution. Note that in this receiver structure, we assume $\mathbf{W}^H \mathbf{z}_k[n]$ to be the decision statistic. Considering the scrambler as a random sequence and taking expectation over the scrambler as well as the input data symbol sequence, one can show that the covariance matrix of the estimation error $\mathcal{R}_{\widetilde{z}\widetilde{z}}$ is similar to the chip-equalizer output error covariance matrix $\mathcal{R}_{\widetilde{x}\widetilde{x}}$ with the scaling of the interference quantities by the number of users (codes). The elements of $\mathcal{R}_{\widetilde{z}\widetilde{z}}$ are given by

$$r_{11} = \sigma_a^2 \frac{K}{G} \left(\|\overline{\boldsymbol{\alpha}}^{(11)}\|^2 + \|\overline{\boldsymbol{\alpha}}^{(12)}\|^2 \right) + \mathbf{f}_1 \mathcal{R}_{\eta\eta} \mathbf{f}_1^H$$

$$r_{22} = \sigma_a^2 \frac{K}{G} \left(\|\overline{\boldsymbol{\alpha}}^{(21)}\|^2 + \|\overline{\boldsymbol{\alpha}}^{(22)}\|^2 \right) + \mathbf{f}_2 \mathcal{R}_{\eta\eta} \mathbf{f}_2^H$$

$$r_{12} = r_{21}^* = \sigma_a^2 \frac{K}{G} \left(\overline{\boldsymbol{\alpha}}^{(11)} \cdot \overline{\boldsymbol{\alpha}}^{(21)H} + \overline{\boldsymbol{\alpha}}^{(12)} \cdot \overline{\boldsymbol{\alpha}}^{(22)H} \right) + \mathbf{f}_1 \mathcal{R}_{\eta\eta} \mathbf{f}_2^H$$

The SINR for the ith stream at the output of the output of the LMMSE chip-equalizer/correlator is therefore

$$SINR_i = \frac{\sigma_a^2}{\left(\mathbf{W}^H \mathbf{B}^{-1} \mathcal{R}_{\widetilde{z}\widetilde{z}} \mathbf{B}^{-H} \mathbf{W} \right)_{ii}} - 1 \qquad (2.11)$$

Once MIMO joint bias is properly taken into account, the expression for the LMMSE chip-equalizer output SINR is exact. The situation is different at the symbol level, where the bias, in practice, varies over time. We address this issue later in this chapter.

The corresponding per-code capacity of the ith data stream can now be expressed as

$$\mathcal{C}_i = \log(1 + SINR_i)$$

$$\mathcal{C}_i = \log \left(\frac{\sigma_a^2}{MMSE_i} \right) \qquad (2.12)$$

Our objective is to choose the precoding matrix $\mathbf{W}$ to maximize the sum-capacity of two streams. This boils down to the following optimization problem:

$$\mathbf{W}_{opt} = \arg \max_{\mathbf{W}} \left[\log \left(\frac{\sigma_a^4}{MMSE_1 \cdot MMSE_2} \right) \right] \qquad (2.13)$$

The optimum precoding matrix can be seen to minimize the product of MMSEs of the streams. By exploiting the structure of the matrices in the unitary codebook specified in the HSDPA standard [1], the optimum precoding matrix $\mathbf{W}_{opt}$ maximizes $\Re(|wr_{12}|)$, where r_{12} is the top-right off-diagonal term of the error covariance matrix $\mathcal{R}_{\widetilde{z}\widetilde{z}}$. In other words, the $\mathbf{W}_{opt}$ attempts to maximize the SINR difference between the two streams.[1]

---

[1] To its best abilities given the limited resolution of $\mathbf{W}$.

2.2.1.2 Chip-Level Successive Interference Cancellation

Consider now a chip-level successive interference cancellation (SIC) receiver that detects data symbols from one stream, say, stream 1, and re-spreads, re-scrambles, and re-channelizes the detected data so that the contribution of the detected stream can be subtracted from the received signal (Figure 2.5). The second stream can now be detected using a new FIR LMMSE chip-level receiver obtained as

$$\mathbf{H}_{sic} = \overline{\mathbf{W}}\,\mathbf{H} \tag{2.14}$$

where $\overline{\mathbf{W}}$ is a diagonal matrix with $\mathbf{w}_2$ on its diagonal, and $\mathbf{H}_{sic}$ is the equivalent channel seen by the stream detected last due to cancellation effected at the receiver

$$\overline{\mathbf{Y}}_2[j] = \mathcal{T}(\mathbf{H}_{sic})\mathbf{b}_{2,L+E-1}[j] + \Xi[j] \tag{2.15}$$

and $\mathbf{b}_{2,L+E-1}[j] = [b_2[j-L-E+2]\cdots b_2[j]]^T$ is the chip sequence vector of the second stream. This case, assuming perfect cancellation of stream 1, is analogous to single-stream TxAA communications and the SINR achieved for stream 2 is much improved. The SINR expressions for this SIC receiver are straightforward. The SINR expression for the first stream remains the same as that of the chip-equalizer correlator receiver, and the expression for SINR for the second stream is similar to that for the MISO LMMSE chip-level equalizer/correlator case. One further consideration in this receiver is that if stream 1 symbol estimates are obtained at the output of a spatial MMSE, this would also imply spatial processing for stream 2 (because spatial processing, by nature, is simultaneous). Such treatment increases the complexity but may be well worth the effort in terms of SINR gains. Before moving on to combined chip-level and symbol-level processing, we would like to draw the reader's attention to the fact that the receiver structure discussed here can be classified as a symbol-level DFE (decision feedback equalizer). It is true that the interference cancellation is performed at chip level after re-spreading. However, the fact that feedback is based on decisions taken at the symbol level makes this equivalent to cancellation at the symbol level.

2.2.2 Combined Chip-Level and Symbol-Level Equalization

We start by motivating the need for combined chip-level and symbol-level processing and then move on to some receiver designs based on this approach.

Optimal linear receivers for WCDMA are symbol-level (deterministic), time-varying, multi-user receivers that are known to be prohibitively complex. One class of such receivers is based on symbol-level multi-user detection (MUD), where linear or nonlinear transformations can be applied to the output of the channel matched filter (RAKE). Linear methods in this category are decorrelating and MMSE MUD, both requiring inverses of large time-varying code cross-correlation matrices across symbols, thus leading to impractical computational complexities. Nonlinear MUD methods focus on estimating, reconstructing, and subtracting signals of interfering codes and are, in general, called interference canceling (IC) receivers. A less complex alternative is *dimensionality reducing* linear chip equalization followed by further linear or nonlinear interference canceling or joint detection stages to improve symbol estimates [2]. The basis of these receivers is that interference arises from the loss of orthogonality due to the multipath channel, and this problem is effectively solved by attempting to restore orthogonality through an SINR maximizing LMMSE equalizer. In MIMO WCDMA, in addition to inter-code interference due to loss of orthogonality, the MIMO channel also introduces inter-stream interference. The spatial separation effected by the LMMSE chip equalizer in this context is not perfect and therefore mandates additional processing that can be performed at the chip or symbol level. This type of processing can be intuitively treated as a dimensionality reduction stage in MUD. It may take, for example, the form of a general chip-level filter carrying out functions of a channel *sparsifier* or indeed a more specific spatio-temporal $\rightarrow$ spatial channel-shortener (e.g., $2N \times 2$ to 2×2 in MIMO HSDPA) [21]. This stage precedes either per-code joint detection of data streams at symbol level [14] or can be followed up by one of the several possible decision-feedback approaches [21] and [3]. In general, for MIMO, if the scrambler is treated as *i.i.d.* random, the resulting symbol-rate spatial channel can now be seen as a per-code spatial mixture and is constant. To this mixture, simplified (per-code) processing can now be applied. In this section we investigate such class of MIMO HSDPA receivers. To be precise, the chip-level processing stage will always consist of the MIMO LMMSE chip equalizer, which will be followed by the correlator. We then consider various symbol-level processing stages that can be employed at the receiver.

2.2.2.1 LMMSE Chip Equalizer: Symbol-Level LMMSE

Consider a receiver structure where the output of the chip equalizer is fed into a symbol-level (spatial) LMMSE filter after the descrambler/correlator block. This is shown in Figure 2.6. As discussed in the previous subsection, "Chip-Level Equalization," the output of the correlator is $\mathbf{z}_k[n]$, given by Equation (2.10). $\mathcal{F}_{sp}$ denotes the spatial MMSE at the output of which we have a linear estimate of the symbol vector as

$$\widehat{\mathbf{a}}_k[n] = \mathbf{a}_k[n] - \widetilde{\mathbf{a}}_k[n] \tag{2.16}$$

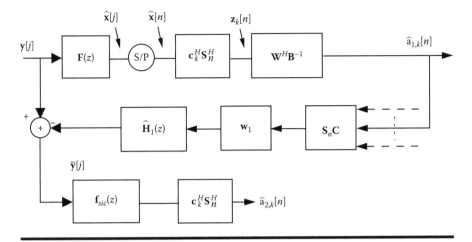

Figure 2.5   LMMSE chip equalizer/correlator and chip-level SIC for stream 2.

The error covariance matrix for the LMMSE estimate of $\mathbf{a}_k[n]$ is given by

$$\mathcal{R}_{\widetilde{aa}} = \mathcal{R}_{aa} - \mathcal{R}_{az'}\mathcal{R}_{z'z'}^{-1}\mathcal{R}_{z'a} \tag{2.17}$$

$$= \sigma_a^2\mathbf{I} - \sigma_a^4\mathbf{W}^H\left(\sigma_a^2\mathbf{I} + \mathbf{B}^{-1}\mathcal{R}_{\widetilde{zz}}\mathbf{B}^{-H}\right)^{-1}\mathbf{W} \tag{2.18}$$

Expressing the above relation in terms of the correlator output covariances, $\mathbf{B}\mathcal{R}_{\widetilde{zz}}\mathbf{B}^{-H}$, and using some algebra leads to the expression

$$\mathcal{R}_{\widetilde{aa}} = \sigma_a^2\mathbf{I} - \sigma_a^4\mathbf{W}^H\left(\sigma_a^2\mathbf{I} + \left(\mathcal{R}_{\widetilde{zz}}^{-1} - \mathcal{R}_{zz}^{-1}\right)^{-1}\right)^{-1}\mathbf{W} \tag{2.19}$$

where $\mathcal{R}_{\widetilde{zz}}$ in the above expression is related to the joint-bias $\mathbf{B}$ through

$$\mathbf{B} = \mathbf{I} - \mathcal{R}_{\widetilde{zz}}\mathcal{R}_{zz}^{-1} \tag{2.20}$$

as with the LMMSE chip-level equalizer/correlator receiver, this translates to a sum-capacity expression similar to the one derived in the previous

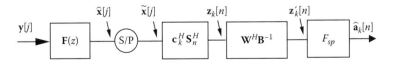

Figure 2.6   Chip LMMSE equalizer and correlator followed by symbol-level (spatial) MMSE.

subsection:

$$\mathcal{C}_1 + \mathcal{C}_2 = \log\left(\frac{\sigma_a^4}{\det(\mathrm{diag}(\mathcal{R}_{\widetilde{a}\widetilde{a}}))}\right) \tag{2.21}$$

The throughput maximizing precoding matrix can therefore be shown to be the one with element w that maximizes

$$\Re\left(\left|w\left[\left(\sigma_a^2\mathbf{I} + \left(\mathcal{R}_{\widetilde{z}\widetilde{z}}^{-1} - \mathcal{R}_{zz}^{-1}\right)^{-1}\right)^{-1}\right]_{12}\right|\right)$$

One may remark that spatial MMSE processing after the equalizer/correlator stage should lead to further suppression of residual interference and lends itself to low-complexity per-code implementation. The spatial channel sees a non-negligible contribution from the kth code (desired code); therefore this receiver does improve on the MMSE chip-equalizer correlator receiver, but its performance is limited by temporal (inter-chip) interference that is still sufficiently strong at the correlator output.

2.2.2.2 LMMSE Chip Equalizer: Predictive DFE

A noise-predictive decision feedback equalizer (DFE) [4] uses past noise estimates to predict the current noise sample. This is readily applied to our spatial multiplexing problem where once one stream is detected, spatial correlation of noise (spatial interference) can be exploited to improve estimation of the stream detected last (second in this case). With some abuse of terminology, this can be branded successive interference cancellation (SIC).

The SIC receiver is shown in Figure 2.7. Denote the output of the correlator as $\mathbf{u}_k[n]$, written as

$$\mathbf{u}_k[n] = \mathbf{W}^H\mathbf{B}^{-1}\mathcal{F}_{sp}\mathbf{z}'_{k,n} = \mathbf{a}_k[n] - \underbrace{\mathcal{F}_{sp}\mathbf{W}^H\mathbf{B}^{-1}\widetilde{\mathbf{z}}_k[n]}_{\widetilde{\widetilde{\mathbf{u}}}_k[n]} \tag{2.22}$$

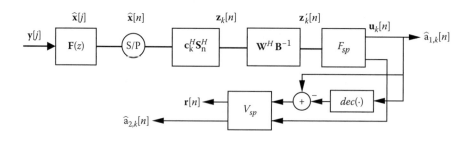

Figure 2.7 Chip LMMSE equalizer/correlator followed by spatial MMSE and symbol-level SIC for stream 2.

The covariance matrix $\mathcal{R}_{\widetilde{\widetilde{uu}}}$, the diagonal bias matrix $\mathbf{B}$, and $\mathcal{R}_{\widetilde{\widetilde{zz}}}$, the covariance matrix of $\widetilde{\mathbf{z}}$, can be related as

$$\mathcal{R}_{\widetilde{\widetilde{uu}}} = \mathcal{F}_{sp}\mathbf{W}^H\mathbf{B}^{-1}\mathcal{R}_{\widetilde{\widetilde{zz}}}\mathbf{B}^{-H}\mathbf{W}\mathcal{F}_{sp}^H \tag{2.23}$$

Assume a 2×2 lower triangular filter $\mathcal{V}_{sp}$ with unit diagonal and the remaining element υ_{21} such that $\widetilde{\mathbf{r}}[n] = \mathcal{V}_{sp}\widetilde{\mathbf{u}}_k[n]$. Then the new error covariance matrix is given as

$$\mathcal{R}_{\widetilde{r}\widetilde{r}} = \mathcal{V}_{sp}\mathcal{R}_{\widetilde{\widetilde{uu}}}\mathcal{V}_{sp}^H \tag{2.24}$$

which is minimized if $\mathcal{R}_{\widetilde{r}\widetilde{r}} = \mathbf{D}$, which is a diagonal matrix, and the problem boils down to the estimation of the error term in stream 2 from stream 1. Toward this end, consider LDU factorization of $\mathcal{R}_{\widetilde{\widetilde{uu}}} = \mathbf{LDL}^H$. Then, $\mathcal{V}_{sp} = \mathbf{L}^{-1}$ minimizes Equation (2.24). Denoting elements of $\mathcal{R}_{\widetilde{\widetilde{uu}}}$ as r_{ij}, the elements of $\mathbf{D}$ are given as $\sigma_{\widetilde{r}_1}^2 = r_{11}$ and

$$\sigma_{\widetilde{r}_2}^2 = r_{22} - r_{21}r_{11}^{-1}r_{12}$$
$$= \det(\mathcal{R}_{\widetilde{\widetilde{uu}}}) \tag{2.25}$$
$$= \det(\mathcal{F}_{sp})\det\left(\mathbf{B}^{-1}\mathcal{R}_{\widetilde{\widetilde{zz}}}\mathbf{B}^{-H}\right)\det\left(\mathcal{F}_{sp}^H\right),$$

Thus, MMSE for stream 1 is $\sigma_{\widetilde{r}_1}^2$ and that of stream 2 is $\sigma_{\widetilde{r}_2}^2$. As depicted in Figure 2.7, this can be interpreted as stream 1 achieving the same performance as for the chip-level LMMSE/correlator-spatial MMSE, while stream 2 benefits from stripping (and thus achieves the spatial natched filter bound).

An interesting observation is that the SINR expression for stream 2 in the symbol-level SIC case is independent of the precoding $\mathbf{W}$ applied. In this receiver, stream 1 should exhibit better performance than in the case of the chip equalizer/correlator receiver. An alternative receiver structure proposed in [21] is also possible, where stream 1 processing is just limited to the chip-equalizer correlator cascade and stream 2 is subjected to symbol-level SIC as above. However, the receiver discussed above is a better alternative to [21], because in this case, stream 1 should get an additional boost in SINR due to the spatial MMSE processing. This should not only amplify stream 1 rate, but also have the desirable effect of improving stream 1 detection. This improved reliability, although not relevant in this discussion where we assume ideal suppression of stream 1 is all-important in practical implementations, thus reducing chances of error propagation during the interference cancellation stage and hence directly impacting the detection performance of stream 2.

It should, however, be noted that any low-complexity, symbol-level processing is hardly comparable to the chip-level SIC receiver in any other way except that symbols on streams are detected in the order of decreasing SINR. While the former exploits the noise-plus-interference correlation between streams to improve the SINR of the symbol detected last, the latter benefits from stripping the spatiotemporal interference of the entire detected stream, where for the stream detected last, all streams can henceforth be considered nonexistent (assuming perfect cancellation). Not only do streams see different levels of interference, a new chip equalizer can be calculated at each stage that benefits from a larger noise subspace to cancel any remaining interference. For SIC, the stream detected last is known to attain the matched filter bound (MFB).

2.2.2.3 Spatial ML Receiver

Figure 2.8 shows another possible receiver structure, where the chip-equalizer correlator front end is followed, as before, by the spatial MMSE stage. The resulting spatial mixture

$$\mathbf{u}_k[n] = \mathcal{F}_{sp}\mathbf{z}'_k[n] = \mathbf{a}_k[n] - \widetilde{\widetilde{\mathbf{u}}}_k[n] \tag{2.26}$$

is later processed for joint detection (code-wise ML detection) of the two symbol streams. The ML metric is given as follows:

$$\mathcal{D} = \{\mathbf{u}_k[n] - \mathbf{a}_k[n]\}^H \mathcal{R}_{\underset{\widetilde{uu}}{}}^{-1} \{\mathbf{u}_k[n] - \mathbf{a}_k[n]\}$$

This metric can be solved for $\mathbf{a}_k[n]$. It is shown in [21] that joint detection outperforms SIC. However, the SIC structure in [21] addresses an SIC applied directly at the output of the chip equalizer-correlator output. Thus, stream 1 gets the same SINR as the chip-equalizer while in our case, stream 1 would also reap the benefits of spatial MMSE processing. For joint detection, the SINR for the ith stream corresponds to the MFB of

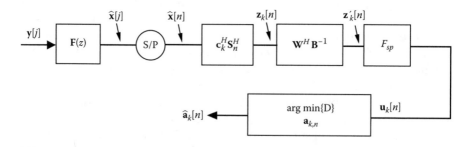

Figure 2.8 Chip LMMSE equalizer/correlator followed by spatial MMSE and joint detection.

spatial channel resulting from the cascade of $\mathcal{F}_{sp}$ and **B**. The MFB can be interpreted as the SNR of the ith stream when it is detected, assuming that the symbols of the other stream(s) are known. $\mathcal{R}_{\widetilde{\widetilde{u}}\widetilde{\widetilde{u}}}$ is the noise variance.

2.2.3 Numerical Results

We present here the simulation results and compare the performance of the different receiver structures that were discussed in this section based on their sum-capacity. For a fixed SNR and over several realizations of a frequency-selective $2p \times 2$ FIR MIMO channel **H**(z), we compute the optimal precoding matrices and use the corresponding SINRs of both streams at the output of the receivers to calculate an upper bound on the sum-capacity. The channel coefficients are complex valued, zero-mean Gaussian of length 20 chips. We assume FIR MIMO equalizers of length comparable to the channel. The sum-capacity cumulative distribution function (CDF) is thus used as a performance measure for all receivers. The structure of the precoding matrices used in HSDPA is such that two out of the four possible precoding matrices give the same SINR (and thus the sum-rate) for the LMMSE/correlator design. The difference between them is that one favors stream 1 by bestowing a higher SINR for stream 1, and the other matrix does just the reverse. This means that one can not only achieve the same sum-rate by choosing either of the two matrices, but one can also choose which stream of the two contributes a larger fraction of the sum. Without loss of generality, in all our simulations, we choose the matrix that maximizes the SINR of stream 1. Figure 2.9 shows the distribution of sum-capacity at the output of the MMSE chip-equalizer correlator receiver and that of the spatial MMSE receiver. With an additional processing stage of very small complexity, we are able to see some gain in the achievable rates of the receiver.

In Figure 2.10 we compare the performance of LMMSE chip-equalizer correlator receiver with the receiver that performs spatial MMSE as well as predictive DFE and the per-code ML receiver. As before, optimal precoding matrices are used at the base station. The receiver that performs spatial LMMSE and DFE benefits slightly from the additional spatial processing for both streams and a nonlinear equalization stage for stream 2. That the gain is not considerable is due to the fact that stream 1 does not benefit from nonlinear equalization. Because the performance measure is the sum-capacity of both streams, the performance of this receiver is limited by the performance of stream 1. By performing spatial ML detection, one is able to obtain much better performance. The chip-level SIC in Figure 2.11, as can be expected, outperforms all other receivers at the

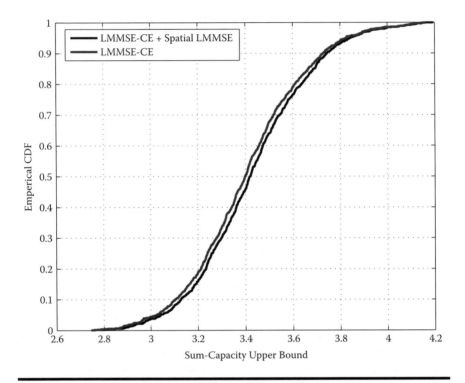

Figure 2.9 Performance of LMMSE chip equalizer/correlator receiver and LMMSE chip equalizer and spatial MMSE receiver.

cost of a significant processing delay and architectural complexity at the receiver.

2.3 Receiver Designs for MIMO HSDPA: Part II

Thus far, we have discussed various receiver designs that have assumed that the scrambler was random *i.i.d.* Modeling the scrambler as random *i.i.d* leads to a time-invariant spatial signal model, which in turn leads to intuitively pleasing RX solutions. While this assumption simplifies receiver designs for the second stage of the two-step processing employed in the receivers, it limits the performance of these receivers. Because the first step in the two-stage approach can be interpreted as a dimensionality reduction step, the limitation on the gain obtained by this design over classical chip equalization can be linked to the efficacy of the dimensionality reduction achieved at the output of the chip equalizer and also the type of processing at the symbol level. In the general MIMO case, the resulting symbol-rate

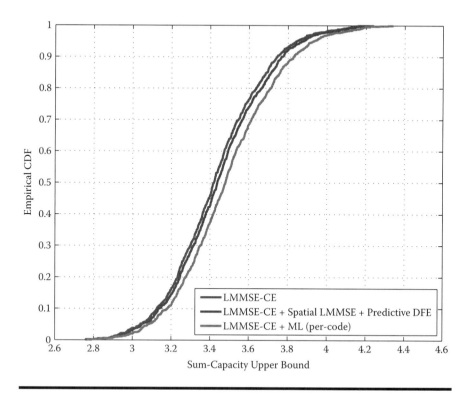

Figure 2.10   Comparison of sum-capacity upper bounds for different receiver structures.

spatial channel can now be seen as only a per-code spatial mixture. When the scrambler is treated as random, this mixture becomes time invariant, and therefore simplified (per-code) processing can be applied. For a processing gain $\mathcal{G}$, assuming that N_t is the number of TX streams, N_r the number of RX antennas, and p the oversampling factor *w.r.t.* the chip rate, this can be seen as a dimensionality reduction from $p \cdot \mathcal{G} \cdot N_r$ to N_t. Given this drastic reduction, it is not surprising to see performance falling well short of optimal time-varying, symbol-level processing (linear and nonlinear MUD solutions). In the previous section, we chose to trade off performance in the interest of reduced-complexity, symbol-level processing in order to point out that despite their shortcomings, their complexity/performance equation encourages use of these solutions. In this section, in an attempt to further increase the performance of our receiver designs, we put forth the idea of deterministic treatment of the scrambler and focus on the resulting spatial channel model [8]. Such a treatment mandates time-varying processing after the equalizer-correlator stage but offsets some of the performance losses of the dimensionality reduction stage and random scrambler assumption.

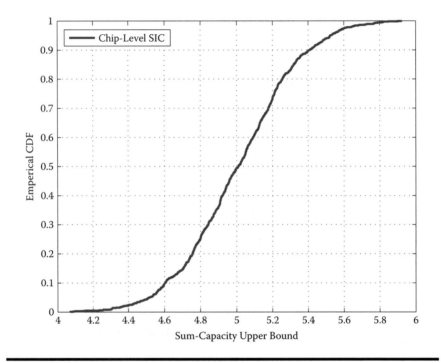

Figure 2.11 Upper bound for sum-capacity for the chip-level SIC receiver.

We show here that deterministic treatment of the scrambler allows us to retrieve the time-varying contribution of joint-bias, which would have otherwise been relegated to noise.

In the interest of simplicity, we do not consider here the precoding aspect of DL transmission. However, we do stress that introduction of precoding does not in any way alter the results obtained in this section. The DL signal model remains exactly the same as before, apart from the absence of linear precoding before transmission and we illustrate it here for convenience (Figure 2.12).

The received signal vector (chip rate) at the UE is now modeled as

$$\underbrace{\mathbf{y}[j]}_{2p \times 1} = \underbrace{\mathbf{H}(z)}_{2p \times 2} \underbrace{\mathbf{b}[j]}_{2 \times 1} + \underbrace{\eta[j]}_{2 \times 1} \tag{2.27}$$

where

$$\mathbf{b}[j] = \sum_{k=1}^{K} s[j] c_k[j \mod \mathcal{G}] \mathbf{a}_k[n] \tag{2.28}$$

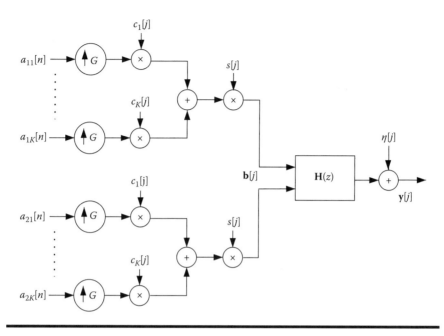

Figure 2.12 **MIMO signal model without precoding.**

2.3.1 Chip-Level Equalization

When the scrambler is treated as deterministic, the desired signal contribution at the correlator output not only is concentrated in one tap of the channel-equalizer cascade (as is the case when the scrambler is treated as random), but also contains a scrambler-dependent, time-varying component (thus not only a mean, but also a variance). This leads to a relationship linking the LMMSE chip equalizer output joint bias and the time-varying correlator output joint bias. Here we first consider the LMMSE chip-equalizer correlator receiver and then derive an analytical expression for the bias term and evaluate the SINR, including the explicit contribution of this quantity.

2.3.1.1 MMSE Chip-Equalizer Correlator Revisited

As before, we start by deriving the expression for the output energy of this receiver. Without loss of generality, we consider linear MMSE estimation of the 2×1 MIMO symbol sequence, $\mathbf{a}_k[n]$, of the kth code among K codes (each stream has K codes). In any case, this corresponds to the 2×2 MIMO case in HSDPA. Refer to Figure 2.13 for a vectorized TX signal model where $\mathbf{b}[n]$ is the $2\mathcal{G} \times 1$ chip vector defined as $\mathbf{b}[n] = [\mathbf{b}_0^T[n] \cdots \mathbf{b}_{\mathcal{G}-1}^T[n]]^T$, where $\mathbf{b}_m[n]$ is the mth multi-code (K codes) MIMO (2×1) chip corresponding to the nth MIMO symbol vector, $\mathbf{a}[n]$ of size $2K \times 1$ and is given

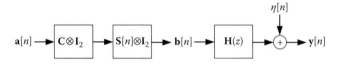

Figure 2.13 **MIMO TX signal model.**

by $\mathbf{a}[n] = [\mathbf{a}_1^T[n] \cdots \mathbf{a}_K^T[n]]^T$. Here, $\mathbf{a}_k[n]$ represents the symbol on kth code. Figure 2.13 depicts the vectorized transmit signal model where $\mathbf{C}$ represents the $G \times G$ spreading matrix, and the diagonal matrix $\mathbf{S}[n]$ of the same dimension represents multiplication of the scrambling sequence for the nth symbol instant. Then, at the symbol level, the input–output relationship at the symbol can be compactly represented using the q operator as

$$\mathbf{y}[n] = \mathbf{H}(q)\mathbf{a}[n] + \boldsymbol{\eta}[n] \tag{2.29}$$

where

$$\mathbf{H}(q) = \sum_{m=0}^{\lceil L/G \rceil - 1} \mathbf{H}(m)q^{-m} \tag{2.30}$$

This expression represents the convolution equation in a compact fashion because $q^{-m}\mathbf{a}[n] = \mathbf{a}[n-m]$. To disambiguate the above expression from the z-transform operation, we use q^{-1} to denote the unit delay operator. Assuming an oversampling factor of p, the symbol-level channel $\mathbf{H}(z) = \sum_m z^{-m}\mathbf{H}[m]$ consists of $pG \times G$ matrix taps. If the delay spread is L chips, then there are $\lceil L/G \rceil$ *pseudo-circulant* matrices that fully represent the channel. These matrices are defined as

$$\mathbf{H}[m] = \begin{bmatrix} \mathbf{h}[mG] & \mathbf{h}[mG+1] & \cdots & \mathbf{h}[(m+1)G-1] \\ \mathbf{h}[mG-1] & & & \vdots \\ \vdots & & \ddots & \\ \mathbf{h}[(m-1)G+1] & \cdots & \cdots & \mathbf{h}[mG] \end{bmatrix}$$

with $\mathbf{h}[.]$ being the $2p \times 2$ chip-level MIMO channel coefficients. The LMMSE equalizer $\mathbf{F}(z)$ in Figure 2.14 can be represented in a similar fashion and visualized to be composed of $\mathbf{f}[.]$, which would be the $2 \times 2p$ equalizer coefficients defined at the chip level. The channel-equalizer cascade is then

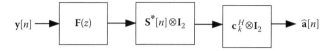

Figure 2.14   MIMO RX model.

given by

$$G(z) = F(z)H(z) \tag{2.31}$$

$$= \sum_{\kappa=0}^{N-1} F[\kappa]z^{-\kappa} \sum_{m=0}^{M-1} H[m]z^{-m} \tag{2.32}$$

$$= \sum_{\nu=0}^{N+M-2} G[\nu]z^{-\nu} \tag{2.33}$$

where, assuming that the chip equalizer length is E chips, we have $M = \lceil L/\mathcal{G} \rceil$ and $N = \lceil E/\mathcal{G} \rceil$. The channel-equalizer cascade at the symbol level can therefore be similarly defined as composed of 2×2 chip-level matrix coefficients $g[k] = \sum_{l=0}^{L-1} f[k-l]h[l]$.

Let the equalizer delay be d. Define the corresponding channel-equalizer cascade matrix at d as

$$G[0] = F(q)H(q)|_{[0]} = \begin{bmatrix} g[0] & g[1] & \cdots & g[\mathcal{G}-1] \\ g[-1] & & & \vdots \\ \vdots & & \ddots & \\ g[-\mathcal{G}+1] & \cdots & \cdots & g[0] \end{bmatrix} \tag{2.34}$$

Henceforth, we consider $G[0]$ as the $2\mathcal{G} \times 2\mathcal{G}$ zeroth MIMO matrix tap of the channel-equalizer cascade. $\overline{G}(z) = \sum_{m \neq 0} z^{-m} G[m]$ thus represents the MIMO inter-symbol interference (ISI). We can now write

$$\hat{a}_k[n] = \left(c_k^H \otimes I_2\right)(S^*[n] \otimes I_2) \{G(q)(S[n] \otimes I_2)(C \otimes I_2) a[n] + F(q)\eta[n]\} \tag{2.35}$$

Defining

$$B_{n,k}(z) = \left(c_k^H \otimes I_2\right)(S^*[n] \otimes I_2)G(z)(S[n] \otimes I_2)(C \otimes I_2)$$

as the symbol rate channel at time instant n [also a $\overline{\mathbf{B}}_{n,k}(z)$ corresponding to $\overline{\mathbf{G}}(z)$], we can write the correlator output as

$$\mathbf{z}_k[n] = \underbrace{\mathbf{B}_{n,k}[0]\mathbf{a}_k[n]}_{kth\ code} + \underbrace{\mathbf{B}'_{n,k}[0]\overline{\mathbf{a}}[n]}_{other\ codes} + \underbrace{\sum_m \mathbf{B}_{n,k}[m]\mathbf{a}[n+m]}_{all\ codes\ other\ symbols} + \underbrace{\mathbf{F}(z)\boldsymbol{\eta}[n]}_{noise} \quad (2.36)$$

In this expression, $\mathbf{B}_{n,k}[0]$ is the desired user channel at symbol time n (time-varying channel), which one can split into a time-invariant part $\mathbf{E}_n\mathbf{B}_{n,k}[0] = \mathbf{B}[0] = \mathbf{B} \cdot I_{\mathcal{G}}$ (assuming the scrambler to be white) and a time-varying part (if scrambler is treated as deterministic). When the scrambler is treated as white, we refer to the 2×2 channel as a spatial channel or even as joint MIMO bias and denote it as $\mathbf{g}_0 = \mathbf{B}$. As discussed in [8], treating the scrambler as white has the effect of capturing the mean signal energy (corresponding to the $\mathbf{g}[0]$ contribution) at the output of the per-code MIMO channel while consigning the variance (off-diagonal part in $\mathbf{G}[0]$) definitively and irrecoverably to the interference term.

It may be noticed that each element of $\mathbf{G}[m]$ is a 2×2 MIMO matrix coefficient. The former can therefore be split into four $\mathcal{G} \times \mathcal{G}$ SISO sub-matrices $\mathbf{G}_{r\kappa}[m]$, for $r, \kappa \in \{1, 2\}$. A corresponding $\mathcal{G} \times \mathcal{G}$ matrix coefficient $\overline{\mathbf{G}}_{r\kappa}[m] = \mathbf{G}_{r\kappa}[m] - \mathbf{g}_{r\kappa}[m] \cdot I_{\mathcal{G}}$ is also defined and so is $\mathbf{g}_{r\kappa}[m]$, the $r\kappa$th element of the spatial channel $\mathbf{g}[m]$.

Taking the expectation over the scrambler, we can express the output energy of the receiver as

$$\mathcal{R}_{zz} = \mathcal{R}_{des} + \mathcal{R}_{MUI} + \underbrace{\sum_m \mathcal{R}_{m,ISI} + \mathbf{F}\mathcal{R}_{\eta\eta}\mathbf{F}^H}_{\mathcal{R}_{\widetilde{zz}}} \quad (2.37)$$

where

$$\mathcal{R}_{des} = \begin{bmatrix} |g_{11}[0]|^2 + |g_{12}[0]|^2 & \sum_{\kappa=1}^{2} g_{1\kappa}[0]g_{2\kappa}^*[0] \\ \sum_{\kappa=1}^{2} g_{2\kappa}[0]g_{1\kappa}^*[0] & |g_{21}[0]|^2 + |g_{22}[0]|^2 \end{bmatrix} +$$

$$\frac{1}{\mathcal{G}^2} \cdot \begin{bmatrix} \sum_{\kappa=1}^{2} tr\{\overline{\mathbf{G}}_{1\kappa}[0]\overline{\mathbf{G}}_{1\kappa}^H[0]\} & \sum_{\kappa=1}^{2} tr\{\overline{\mathbf{G}}_{1\kappa}[0]\overline{\mathbf{G}}_{2\kappa}^H[0]\} \\ \sum_{\kappa=1}^{2} tr\{\overline{\mathbf{G}}_{2\kappa}[0]\overline{\mathbf{G}}_{1\kappa}^H[0]\} & \sum_{\kappa=1}^{2} tr\{\overline{\mathbf{G}}_{2\kappa}[0]\overline{\mathbf{G}}_{2\kappa}^H[0]\} \end{bmatrix}$$

$$
\mathbf{R}_{MUI} = \frac{K-1}{\mathcal{G}^2} \cdot
\begin{bmatrix}
\sum\limits_{\kappa=1}^{2} tr\{\overline{\mathbf{G}}_{1\kappa}[0]\overline{\mathbf{G}}_{1\kappa}^{H}[0]\} & \sum\limits_{\kappa=1}^{2} tr\{\overline{\mathbf{G}}_{1\kappa}[0]\overline{\mathbf{G}}_{2\kappa}^{H}[0]\} \\
\sum\limits_{\kappa=1}^{2} tr\{\overline{\mathbf{G}}_{2\kappa}[0]\overline{\mathbf{G}}_{1\kappa}^{H}[0]\} & \sum\limits_{\kappa=1}^{2} tr\{\overline{\mathbf{G}}_{2\kappa}[0]\overline{\mathbf{G}}_{2\kappa}^{H}[0]\}
\end{bmatrix}
$$

where the superscript $*$ represents complex conjugation. The ISI contribution from the mth symbol can be expressed as

$$
\mathcal{R}_{m,ISI} = \frac{K}{\mathcal{G}^2} \cdot
\begin{bmatrix}
\sum\limits_{\kappa=1}^{2} tr\{\overline{\mathbf{G}}_{1\kappa}[m]\overline{\mathbf{G}}_{1\kappa}^{H}[m]\} & \sum\limits_{\kappa=1}^{2} tr\{\overline{\mathbf{G}}_{1\kappa}[m]\overline{\mathbf{G}}_{2\kappa}^{H}[m]\} \\
\sum\limits_{\kappa=1}^{2} tr\{\overline{\mathbf{G}}_{2\kappa}[m]\overline{\mathbf{G}}_{1\kappa}^{H}[m]\} & \sum\limits_{\kappa=1}^{2} tr\{\overline{\mathbf{G}}_{2\kappa}[m]\overline{\mathbf{G}}_{2\kappa}^{H}[m]\}
\end{bmatrix}
$$

In these relations, $\mathcal{R}_{des}$ is composed of two contributions shown above as the sum of two 2×2 matrices. When the scrambler is treated as random, the term scaled by $1/\mathcal{G}^2$ is the quantity that ceases being part of the signal energy contribution and is associated instead with the interference, for reasons explained earlier.

At the output of the despreader for the kth code, one can therefore express the signal as

$$
\mathbf{z}_k[n] = \mathbf{B}_{n,k}[0]\mathbf{a}_k[n] - \tilde{\tilde{\mathbf{z}}}_k[n] \tag{2.38}
$$

where the time varying MIMO joint-bias $\mathbf{B}_{n,k}[0]$ is no longer constant and varies for each symbol. The per-user SINR of stream i is given by Equation (2.39):

$SINR_{k,i}$

$$
= \frac{\sigma_{a_k}^2\left(|g_{ii}[0]|^2 + \dfrac{1}{\mathcal{G}^2}tr\{\overline{\mathbf{G}}_{ii}[0]\overline{\mathbf{G}}_{ii}^{H}[0]\}\right)}{\sigma_{a_k}^2\left(\dfrac{(K-1)}{\mathcal{G}^2}\sum\limits_{\kappa=1}^{2} tr\{\overline{\mathbf{G}}_{i\kappa}[0]\overline{\mathbf{G}}_{i\kappa}^{H}[0]\} + \dfrac{K}{\mathcal{G}^2}\sum\limits_{m}\sum\limits_{\kappa=1}^{2} tr\{\overline{\mathbf{G}}_{i\kappa}[m]\overline{\mathbf{G}}_{i\kappa}^{H}[m]\}\right) + \sigma_{\eta}^2\|\mathbf{f}_i\|^2}
$$

$$\tag{2.39}$$

2.3.1.2 Chip-Level SIC Revisited

We now consider the effect of treatment of the scrambler as deterministic on the chip-level SIC. The SIC structure in [21] qualifies as a symbol-level SIC, feeding back symbol decisions to the output of equalizer correlator. This assumes a time-invariant symbol-level channel **B** resulting from treating the scrambler as random. The suboptimalities introduced in all earlier stages (i.e., dimensionality reduction through chip equalization and reducing the time-varying, symbol-level channel to its mean value) take their toll; this SIC does not provide significant gains over the chip-equalizer correlator solution.

On the other end of the SIC spectrum is the chip-level SIC, discussed here, which is an entirely different solution that considers all stages of the channel-equalizer correlator stage as deterministic and recreates all components (ISI and MUI) of the stream detected first before subtracting it from the input signal. Subsequently, the second stream can be dealt with under much improved conditions where interference from the first stream is (ideally) entirely suppressed.

2.3.2 Combined Chip-Level and Symbol-Level Equalization

We will now briefly discuss the effect of deterministic treatment of scrambler on further linear or non-linear symbol level processing stages when the receiver design is based on combined chip and symbol level equalization. For the spatial MMSE receiver, in order to claim the quantity $\frac{1}{G^2} tr\{\overline{\mathbf{G}}_{rr}[0]\overline{\mathbf{G}}_{rr}^{H}[0]\}$ in Equation (2.39) as part of signal energy, it suffices to put in place time-varying processing at the correlator output, where the nth symbol vector on the kth code, $z_k[n]$ is given by Equation (2.36). As a result of time-varying symbol level joint-bias, the 2×2 MMSE equalizer will now have to be computed for each symbol. This will indeed provide higher gains than the spatial MMSE receiver above, which treats the time-varying signal contribution as noise. [In the case of the spatial-ML receiver, treating the scrambler as a random sequence, the spatial channel **B** in the ML metric is time-invariant.] A continuous processing matched filter bound can therefore be defined per stream. The ith stream MFB is therefore proportional to the energy in the corresponding SIMO channel. On the contrary, if a deterministic scrambler is assumed, time variation in the channel must be accounted for in ML metrics. Strictly speaking, the MFB is only defined per symbol as the SINR of the nth symbol considering all other symbols are known (correctly detected). We can, nevertheless, argue that deterministic treatment of the scrambler leads to reduced interference variance $\mathcal{R}_{\tilde{z}\tilde{z}}$ and increased recoverable signal power that will lead to performance improvement for the ML solution.

2.3.3 Numerical Results

As before, we compare the performance of different receiver structures based on their sum-capacity. We simulate here a single-user situation where 15 codes are assigned to the same user. Furthermore, we assume code reuse across antennas. The length of FIR MIMO equalizers is comparable to the channel delay spread in chips. Figure 2.15 plots the capacity bounds for two cases. In the first instance, we treat the scrambler as random. The symbol energy for code k is therefore given by the symbol variance for the code scaled by an arbitrary time-invariant scale factor. In the second case, we treat the scrambler as a known sequence. In this case, the signal power now is time-varying at symbol rate. This time-varying signal power can be seen as the sum of a "mean" power contribution equal to the signal power when the scrambler is assumed to be random, and time-varying contribution due to deterministic treatment of the scrambler. Note that the SINR

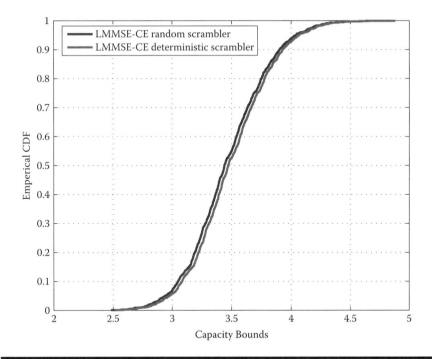

Figure 2.15   Performance of LMMSE chip-equalizer correlator with random and deterministic scrambler.

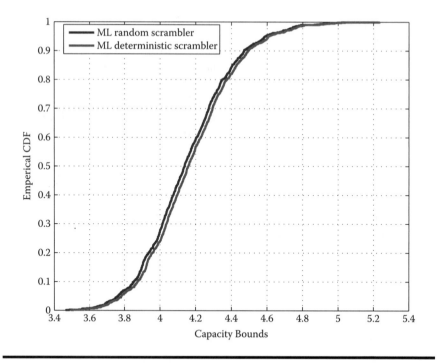

Figure 2.16 Sum-capacity at the output of spatial-ML receiver with deterministic and random scrambler.

distribution for the deterministic treatment of the scrambler in Figure 2.16 represents the *average* gains and not the true gain. The actual gain will be higher than that seen in Figure 2.16.

2.4 Multi-User Extensions to HSDPA MIMO

In this section we shift our focus to extending MIMO in HSDPA to support multiple users in the downlink (MU-MIMO). In its present form, the standard only supports 2 × 2 SU-MIMO in the downlink (DL) in the form of D-TxAA. It is possible for the BS to employ spatial division multiple access (SDMA) and service multiple UEs in DL instead. In this case, the limitation of two transmit antennas implies that a maximum of two spatially separated users can be simultaneously served by the BS with the same code. In general, MU extensions for closed-loop transmit diversity schemes (both TxAA and D-TxAA) introduce multi-user interference in downlink because there exists the possibility of different users feeding back different beamforming vectors in TxAA or different precoding matrices in D-TxAA.

There is a large amount of literature available for multi-user MIMO communication in the general case. It was studied previously in [18] and more recently in [7], where multi-user transmission techniques were classified into linear and nonlinear transmission algorithms. Nonlinear algorithms involving multi-user signal designs that avoid interference generation to other users based on dirty paper coding techniques remain currently impractical due to the requirement of perfect channel state information at the transmitter (CSIT). They also suffer from all the drawbacks associated with outdated CSIT due to scheduling delays at the base station and/or rapidly changing downlink channels. Linear processing of transmitted signals, such as multi-user beamforming, remain by far the most practical solution for multi-user transmission. Theoretical research in multi-user communications tends to consider frequency-flat channels: In reality, most mobile communication channels are frequency selective. There exists some literature on multi-user extension of HSDPA. In [11] the authors propose code reuse in D-TxAA based on a multi-user beamforming (MUB) scheme that schedules users with orthogonal weight vectors to separate them in space. They, however, limit their analysis to flat channels. In [20], the authors consider MU-TxAA for frequency-selective channels and propose the so-called "interference-aware" receiver, which in addition to requiring multiple antennas at the receiver, also assumes knowledge of beamforming weight vectors of all the users at the receiver. On the other hand, in this section we look at the problem of maximizing system capacity in the frequency-selective MISO/MIMO downlink channels, assuming the receivers select weights that maximize receive SINR (and thus increase their individual data rates). In the HSDPA context, the BS is equipped with two transmit antennas, that is, $N_{tx} = 2$. In our treatment, we do not assume any explicit knowledge of beamforming weight vectors of other users; for single stream transmission, we consider single-antenna UE and study different beamforming strategies that can be adopted by the BS, and for dual-stream transmission, we consider UE with two antennas and compare the performance of SDMA against spatial multiplexing to a single user by extending D-TxAA to an MU configuration where, at most, N_{tx} users can be synchronously served by the BS. Each transmit stream is assigned to a different user. This rules out simultaneously serving any two users that feed back the same beamforming weight vector. Users who request linearly independent weight vectors can, however, be served simultaneously.

2.4.1 Multi-User TxAA

We consider a 2-transmit, 1-receive antenna configuration for TxAA. For the remainder of this section, whenever we refer to a MU-TxAA system, we consider U separate UEs each having a single receive antenna. The number

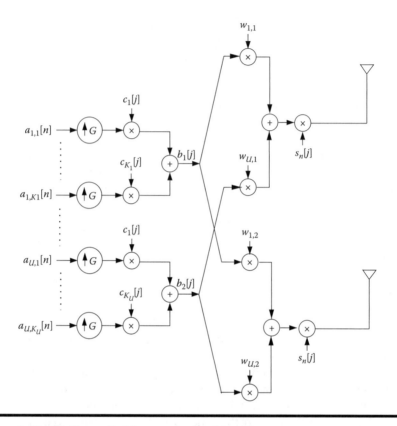

Figure 2.17   Multi-user TxAA transmit signal model.

of codes assigned to each user is denoted by $K_1, K_2, \ldots, K_U$ and $K = \sum_{u=1}^{U} K_u$. Then, for TxAA, from Figure 2.17, the transmit and beamformed chip sequence is given by

$$\mathbf{x}[j] = \sum_{u=1}^{U} \mathbf{w}_u \cdot s_n[j \mod \mathcal{G}] \sum_{k \in K_u} c_k[j \mod \mathcal{G}] a_{u,k}[\lfloor \frac{j}{\mathcal{G}} \rfloor n] \qquad (2.40)$$

where j is the chip index, n is the symbol index, u is the user index, k is the code index, $\mathcal{G}$ is the spreading gain, s_n denotes the scrambler for the nth symbol, c_k denotes the kth spreading code, $\mathbf{w}_u = [w_{u,1} \; w_{u,2}]^T$ is the weight vector corresponding to uth user, and finally, $a_{u,k}[n]$ is the uth user's symbol on code index k given that $k \in K_u$. The transmitted signal propagates through a multipath channel, which we denote here by $\mathcal{H}_u^0, \mathcal{H}_u^1, \ldots, \mathcal{H}_u^{L-1}$. For an oversampling factor of p at the receiver, each $\mathcal{H}_u^l$ matrix is a $p \times 2$ matrix corresponding to the lth tap of the uth user's multipath channel. For simplicity we assume that all UEs see a channel with a maximum delay spread of L chips and employ an equalizer of length E (in chips). The chip-rate received signal at each UE is given by

$$\mathbf{y}_u = \mathbf{H}_u \mathbf{x} + \eta \tag{2.41}$$

where $\mathbf{H}_u$ is the channel convolution matrix for the uth user given by

$$\mathbf{H}_u = \begin{bmatrix} \mathcal{H}_u^0 & \mathcal{H}_u^1 & \cdots & \mathcal{H}_u^{L-1} & \mathbf{0} & \mathbf{0} \\ \mathbf{0} & \mathcal{H}_u^0 & \cdots & \cdots & \mathcal{H}_u^{L-1} & \vdots \\ \mathbf{0} & \mathbf{0} & \ddots & \ddots & \ddots & \mathbf{0} \\ \mathbf{0} & \mathbf{0} & \ddots & \mathcal{H}_u^0 & \ddots & \mathcal{H}_u^{L-1} \end{bmatrix} \tag{2.42}$$

$\mathbf{x}$ is the transmit chip-vector formed by stacking $L + E - 1$ vectors and can be expressed as

$$\mathbf{x} = [\mathbf{x}^T[j], \mathbf{x}^T[j-1], \dots, \mathbf{x}^T[j - L - E + 2]] \tag{2.43}$$

and η is zero mean, circularly symmetric, Gaussian distributed, additive white noise of variance σ_η^2. In addition, we also define the $p \times 1$ vector $\mathbf{r}_{u,v}^l = \mathcal{H}_u^l \mathbf{w}_v$, $v \in 1, 2, \dots, U$ and use this to define the lth beamformed channel tap of user u, due to the beamforming weight of another synchronous DL user v. We denote this by $\mathbf{R}_{u,v}$ and express this as

$$\mathbf{R}_{u,v} = \begin{bmatrix} \mathbf{r}_{u,v}^0 & \mathbf{r}_{u,v}^1 & \cdots & \mathbf{r}_{u,v}^{L-1} & \mathbf{0} & \mathbf{0} \\ \mathbf{0} & \mathbf{r}_{u,v}^0 & \cdots & \cdots & \mathbf{r}_{u,v}^{L-1} & \vdots \\ \mathbf{0} & \mathbf{0} & \ddots & \ddots & \ddots & \mathbf{0} \\ \mathbf{0} & \mathbf{0} & \ddots & \mathbf{r}_{u,v}^0 & \ddots & \mathbf{r}_{u,v}^{L-1} \end{bmatrix} \tag{2.44}$$

2.4.1.1 Beamforming Strategies at the Transmitter

Consider the case where the base station serves U simultaneous users in the downlink. We assume standard MMSE chip-equalizer correlator receivers. Let $\mathbf{f}_u$ represent the MMSE filter of length E applied at user u, then the equivalent channel-equalizer cascade at the output of the chip equalizer for user u is given by

$$\alpha^{(u)} = \mathbf{f}_u \mathbf{R}_{u,u} + \mathbf{f}_u \sum_{v \neq u}^{U} \mathbf{R}_{u,v} \tag{2.45}$$

which can be represented by

$$\alpha^{(u)} = \alpha_{u,u} + \sum_{v \neq u}^{U} \alpha_{u,v} \tag{2.46}$$

where $\alpha_{u,u}$, is the channel-equalizer cascade for codes assigned to user u and $\alpha_{u,v}$ is the channel-equalizer cascade for codes assigned to user v at user u. $\alpha_{u,u}$ can, in turn, be split into the desired equalizer response and the residual inter-chip interference and be represented as

$$\alpha_{u,u} = \alpha_{u,u}^d + \overline{\alpha}_{u,u} \tag{2.47}$$

$$\alpha_{u,u}^d = \left[\overbrace{0 \ldots 0}^{d-1} \ \alpha_{u,u}^d \ \overbrace{0 \ldots 0}^{L+E-2-d} \right] \tag{2.48}$$

where d is the equalizer delay. The LMMSE equalizer is considered to be followed by a stacking operation allowing despreading and symbol decision.

2.4.1.1.1 Simple Multi-User Beamforming

To understand the effect of multiple users with distinct beamforming weights in DL, it is insightful to derive the per-code SINR at the receiver for the case where multiple users are served in the downlink with different beamforming weights. When the BS employs different user-defined beamforming weights in the downlink for MU transmission, at each receiver, codes assigned to different users propagate through U distinct *beamformed channels* even though the physical channel through which they propagate is the same. Without explicit knowledge of all beamforming weights used in the downlink, which is the so-called interference-aware [20] receiver, the receiver will not be able to effectively mitigate the effect of MUI. Because each user is aware only of beamforming weights that will be applied for codes assigned to itself and not of other users, the equalizer at each user is only matched to the beamformed channel seen by the codes assigned to this user. In computing the ideal beamforming weights for itself, a UE has to make some hypothesis on the beamforming weight vectors of other users in DL and choose the weight vector that maximizes the SINR corresponding to that hypothesis. For the general case where there exist U different users, defining K_u as the index set containing code indices of the uth user, the SINR per-code $SINR_{k \in K_u}$ that is seen by the code assigned to the user is given by

$$\frac{\sigma_k^2 |\alpha_{u,u}^d|^2}{\frac{1}{G} \sum_{k \in K_u} \sigma_k^2 \|\overline{\alpha}_{u,u}\|^2 + \sum_{v \neq u} \frac{1}{G} \sum_{k \in K_v} \sigma_k^2 \|\alpha_{u,v}\|^2 + \sigma_\eta^2 \mathbf{f}_u \mathbf{f}_u^H} \tag{2.49}$$

where σ_k^2 denotes the chip variance of the *k*th code. In a simple extension of beamforming with multiple users with different beamforming weight vectors, each UE makes the assumption that all users in DL have the same beamforming weight vectors and compute the ideal beamforming weight vector under this assumption. The BS, however, makes no attempt to group users with the same beamforming weights. As a result, it is expected that the downlink capacity drops significantly.

2.4.1.1.2 Weight Optimization by Average Interference Criterion

Alternatively, a UE can anticipate that, in reality, any of the four weights may be chosen by the other users in the DL. Assuming that other users choose one of four beamforming weights with equal likelihood, it is reasonable to choose the beamforming weight which has the maximum SINR when averaged over all four hypotheses for the other users' weights. Each UE therefore computes the ideal beamforming weight by plugging into Equation (2.49) all possible combinations of weight vectors and feeds back the weight vector with the best average SINR over all the hypotheses for all the other users in DL. The idea is that while the true SINR at the receiver may still not be the same as the expected SINR, the resulting SINR is higher than obtained by assuming that the same beamforming weight is requested by all users scheduled in the DL. Thus, this beamforming vector must perform better on average, and increase the average data rate per user when compared to the simple multi-user beamforming case.

2.4.1.1.3 Cooperative Beamforming

If the BS were to have the knowledge of the SINR seen by a particular user for all possible combinations of weight vectors applied at the base station, then the BS could choose the optimal combination of weights that maximizes the downlink capacity. We call this *cooperative beamforming* because, in this case, all the users compute all possible SINRs corresponding to the weight vectors in the codebook. From Equation 2.49 we see that for a given weight vector, the SINR is highest when all other users also have the same beamforming weight vector. Each user therefore feeds back as many SINRs as the codebook size. Thus, it is a form of cooperation between the users and the BS to maximize system capacity. In practice, this involves a considerable amount of receiver processing and also a lot of feedback to the BS. Nonetheless, the gains in such a case are worth investigating.

2.4.1.1.4 Scheduled Beamforming

The practical and, indeed, the best solution to this problem with least complexity is for the BS to schedule in the DL only those users that request the same beamforming weights. Each user assumes that same weights are applied to all codes in DL and computes the weight vector that maximizes the per-code SINR. For this case, the user can then restore the orthogonality of all codes with the MMSE chip-equalizer correlator receiver. The per-code SINR for the uth user is then given by

$$\frac{\sigma_k^2 |\alpha_{u,u}^d|^2}{\frac{K}{G}\sigma_k^2 \|\overline{\alpha}_{u,u}\|^2 + \sigma_\eta^2 \mathbf{f}_u \mathbf{f}_u^H} \tag{2.50}$$

The combination of scheduling at BS and the choice of weight vector that maximizes the individual SINR at the receiver results in maximization of DL capacity.

2.4.2 Multi-User D-TxAA

For the MU-D-TxAA system, we consider two separate UEs with N_{rx} receive antennas each. In an MU-D-TxAA system, the BS transmits two transport blocks for as many users scheduled in the DL. All codes of a single stream are assigned to one user and reused across the two streams. From Figure 2.18 we see that the transmit signal vector in downlink can be modeled as

$$\mathbf{x}[j] = \underbrace{\mathbf{W}}_{2\times 2} \mathbf{b}[j] = \mathbf{W} \cdot \sum_{k=1}^{K} s[j]c_k[j \mod \mathcal{G}]\mathbf{a}_k[n] \tag{2.51}$$

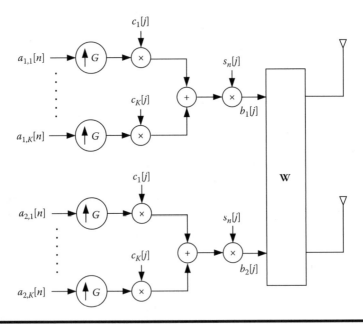

Figure 2.18 Multi-user D-TxAA transmit signal model.

$\mathbf{W} = [\mathbf{w}_1 \, \mathbf{w}_2]$ is the 2×2 unitary precoding matrix. The columns of $\mathbf{W}$ are composed of the beamforming weight vectors corresponding to the two downlink users. The symbol vector $\mathbf{a}_k[n] = [a_{1k}[n] \; a_{2k}[n]]^T$ represents two independent symbol streams belonging to two different users. The spreading codes are common to the two streams and so is the scrambling sequence $s[j]$.

2.4.2.1 Spatial Multiplexing versus SDMA

In the spatial multiplexing context, there is only a single user in the downlink, and the precoding matrix corresponds to the weight vectors applied to the two separate streams transmitted to the same user. For such a case, we can write the equalizer output as the sum of an arbitrarily scaled desired term and an error term:

$$\widehat{\mathbf{x}}[j] = \mathbf{x}[j] - \widetilde{\mathbf{x}}[j] \tag{2.52}$$

The error $\widetilde{\mathbf{x}}[j]$ is a zero-mean complex normal random variable. The error covariance matrix is denoted by $\mathcal{R}_{\widetilde{x}\widetilde{x}}$.

In Equation (2.52), an estimate of the chip sequence can be obtained after a further stage of processing where the precoding is undone to separate streams. The latter represented by $\mathbf{W}^H$ is a linear operation and can be carried out before or after despreading.

Under the assumption of a FIR signal model, the estimation error covariance matrices $\mathcal{R}_{\widetilde{x}\widetilde{x}}$ (chip level) and $\mathcal{R}_{\widetilde{z}\widetilde{z}}$ (symbol level) are derived in [16]. It can be shown that the SINR for the qth stream at the output of the output of the LMMSE chip equalizer/correlator is given in [16] as

$$SINR_q = \frac{\sigma_a^2}{\left(\mathbf{W}^H \mathcal{R}_{\widetilde{z}\widetilde{z}} \mathbf{W}\right)_{qq}} - 1 \tag{2.53}$$

where σ_a^2 corresponds to the symbol variance.

In the SDMA context, the BS transmits a single stream for each of the two downlink users. The BS applies the precoding matrix $\mathbf{W}$ whose columns correspond to the weight vectors fed back by the two users. It is obvious that two users who feedback the same weightvector cannot be scheduled simultaneously for transmission in the downlink. At the receiver, each UE receives both the streams but processes only the stream assigned to itself. In HSDPA, 2×2 unitary precoding is used; this implies that the two columns of the precoding matrix are orthogonal. Moreover, knowledge of a single column automatically fixes the other column of $\mathbf{W}$. Thus, the BS does not have to explicitly inform one UE of the weight vector applied for the other UE. The SINR for the stream assigned to the user in question is therefore the same as in Equation (2.53).

2.4.3 Numerical Results

In this section we present Monte-Carlo simulation results and performance comparison of different beamforming strategies proposed in this chapter. We consider a multipath channel with a maximum delay spread L of 10 chips with uniform power in all channel taps. At any given time, the BS simultaneously serves two users. The beamforming weights are calculated to maximize the per-code SINR at the output of the equalizer correlator combination. Simulations were carried out for a fixed SNR at each receive antenna while keeping the total transmit power normalized to 1. The cumulative distribution function (CDF) of the sum-capacity upper bound in DL is then used as a performance metric to compare different strategies. Depending on the number of independent transport blocks at the transmitter, the other simulation parameters are given as below.

2.4.3.1 TxAA

Each UE is assumed to have a single receive antenna. Normally, each UE feeds back only its preferred weight vector index, only in the case of cooperative-operative beamforming, it feeds back SINR values to the BS. For the sake of simplicity, we assume that each UE is allocated 7 of the 15 codes in the DL, all with the same power.

2.4.3.2 D-TxAA

Each independent transport block is assumed to be allocated to a different user. Thus, all codes of a stream are allocated to one user. For SDMA with single antenna receivers, we assume users with orthogonal weights are scheduled together. For SDMA with two-antenna receivers, users with different beamforming weight vectors are assumed to be scheduled together. In the spatial multiplexing case, a 2×2 MIMO system is assumed, with all codes and both streams transmitted to a single user. Figure 2.19 compares the sum-capacity in the DL for the case of TxAA. The DL capacity is worst for the case of beamforming without scheduling. This is because of the inability of the receivers to effectively restore orthogonality for all codes and hence effectively mitigate MUI because they do not know the actual beamforming weight of the other user. When the beamforming weight is optimized by the average interference criterion, the weights are not chosen based solely on the channel seen by each user, but also on the capability of these weights to reduce the average multi-user interference due to different beamforming weights of the other user. The downlink capacity is thus better than that in the case of simple multi-user beamforming. At the cost of an increase in complexity and feedback, cooperative-operative beamforming performs better than that of the earlier schemes. Even so, it is does not do better than the scheduled beamforming because the UEs

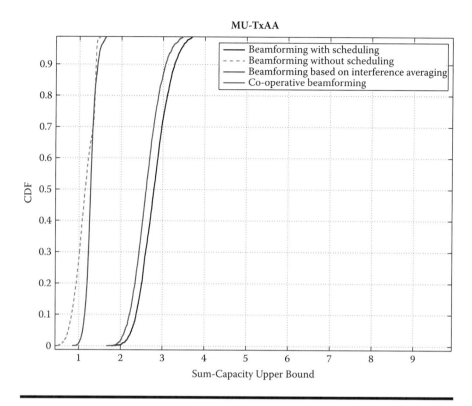

Figure 2.19   Performance of different beamforming schemes for MU-TxAA.

need not necessarily be assigned the weight vector that maximizes their individual SINR. Scheduled beamforming thus outperforms all the other schemes because in this case each user is able to effectively mitigate MUI due to the same beamformed channel seen by all codes in the downlink. It should be noted that for the case where the total number of users in DL far exceeds the number of users actually scheduled in the DL, the performance of cooperative-operative beamforming is expected to improve. In Figure 2.20 we compare the performance of D-TxAA in spatial multiplexing mode with that of the multi-user (SDMA) mode. Simulation results show that the DL sum-capacity is greater for the case of SDMA with single stream transmission to both users.

2.5 Concluding Remarks

In this chapter we first discussed advanced receiver designs for MIMO HSDPA based on the concept of combined chip-level and symbol-level processing. In particular, the chip-level processing stage was the SINR

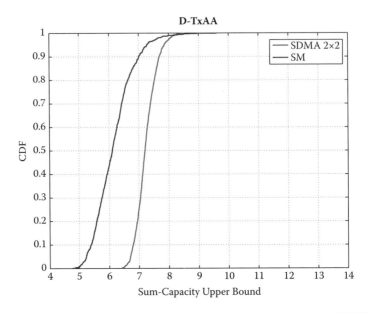

Figure 2.20 DL sum-capacity for MU-D-TxAA.

maximizing LMMSE chip equalizer, which in addition to restoring the orthogonality of the codes also achieves spatial separation to a certain degree. Further processing stages at the symbol level were introduced to enhance the performance of the receivers. When MIMO HSDPA receivers are based on MMSE designs, we showed that there exists an optimal choice of precoding matrix to be employed at the transmitter that maximizes the sum-capacity of these receivers and derived analytical expressions for the choice of optimal precoding matrix. The receiver designs discussed in the first part of the chapter operated on the assumption that the scrambler used at the transmitter can be modeled as a random sequence. Because such a treatment contributes to the degradation of receiver performance, in the second section, we looked at receiver designs that treat the scrambler as deterministic. We saw that such receivers can then resort to time-varying symbol-level processing after the equalizer-correlator stage in order to regain the time-varying signal contribution that would otherwise be treated as noise. This leads to additional gains in SINR, which ultimately affects the achievable capacity of the receivers. Simulation results show that, indeed, these receivers outperform conventional receivers that treat the scrambler as random. Finally we discussed multi-user extensions to closed-loop transmit diversity techniques that have been standardized in [1] and proposed multi-user beamforming strategies that can be employed at the BS in order to maximize the downlink capacity. Simulation results show that for MIMO

HSDPA, downlink capacity is maximized by using the MIMO channel to service multiple single-stream users (SDMA) instead of single-user spatial multiplexing, which is currently supported in the standards.

References

[1] 3GPP. *TS 25.214 Physical Layer Procedures (FDD) (Release 7)*, May 2007. Version 7.5.0.

[2] A. Baştuğ and D.T.M. Slock. Downlink WCDMA receivers based on combined chip and symbol level equalization. *European Trans. on Telecoms.*, 16: 51–63, 2005.

[3] J. Choi, S.R. Kim, Y. Wang, and C.C. Lim. Receivers for chip-level decision feedback equalizer for CDMA downlink channels. *IEEE Trans. on Wireless Comm.*, 3(1): 300–313, 2004.

[4] J.M. Cioffi, G.P. Dudevoir, M.V. Eyuboglu, and G.D. Forney. MMSE Decision-feedback equalizers and coding. 1. Equalization results. *IEEE Trans. on Comm.*, 43(10): 2582–2594, 1995.

[5] G.J. Foschini and M.J. Gans. On limits of wireless communications in a fading environment When using multiple antennas. *Wirel. Pers. Commun.*, 6(3): 311–335, 2009.

[6] G.J. Foschini. Layered Space-Time Architecture for Wireless Communication in a Fading Environment When Using Multielement Antennas. Technical Report, Bell Labs, Autumn 1996.

[7] D. Gesbert, M. Kountouris, R. Heath, C.B.W. Chae, and T. Slzer. Shifting the MIMO paradigm. *IEEE Signal Processing Mag.*, 24(5): 36–46, September 2007.

[8] I. Ghauri, S.P. Shenoy, and D.T.M. Slock. On LMMSE bias in CDMA SIMO/MIMO receivers. In *Proc. IEEE Int. Conf. Acoustics, Speech and Signal Processing*, Las Vegas, NV, March 2008.

[9] I. Ghauri and D.T.M. Slock. Linear receivers for the DS-CDMA downlink exploiting orthogonality of spreading sequences. In *Proc. 32nd Asilomar Conf. Signals, Systems & Computers*, 1: 650–654, Pacific Grove, CA, November 1998.

[10] Global Mobile Suppliers Association. 3G/WCDMA-HSPA Fact Sheet. Technical Report, GSA, March 2009.

[11] V. Haikola, M. Lampinen, and M. Kuusela. Practical multiuser beamforming in WCDMA. In *Proc. IEEE Vehicular Technology Conference*, Montreal, Quebec, Canada, Fall 2006.

[12] K. Ko, D. Lee, M. Lee, and H. Lee. Novel SIR to channel-quality indicator (CQI) mapping method for HSDPA system. In *Proc. IEEE Vehicular Technology Conf.*, Montreal, Quebec, Canada, Fall 2006.

[13] A. Lozano and C. Papadias. Layered space-time receivers for frequency-selective wireless channels. *IEEE Trans. Commun.*, 50(1): 65–73, January 2002.

[14] L. Mailaender. Linear MIMO equalization for CDMA downlink signals with code reuse. *IEEE Trans. Wireless Commun.*, 4(5): 2423–2434, September 2005.

[15] A.J. Paulraj and C.B. Papadias. Space-time processing for wireless communications. *IEEE Signal Processing Mag.*, pp. 49–83, November 1997.

[16] S.P. Shenoy, I. Ghauri, and D.T.M. Slock. Optimal precoding and MMSE receiver designs for MIMO WCDMA. In *Proc. IEEE 67th Vehicular Technol. Conf.*, Singapore, May 2008.

[17] D.K.C. So and R.S. Cheng. Detection techniques for V-BLAST in frequency selective fading channels. *Wireless Commun. Networking Cont.*, 1(17–21): 487–491, March 2002.

[18] Q.H. Spencer, C.B. Peel, A.L. Swindlehurst, and M. Haardt. An introduction to multi-user MIMO downlink. *IEEE Commun. Mag.*, 42(10): 60–67, October 2004.

[19] I.E. Telatar. Capacity of multi-antenna gaussian channels. *Eur. Trans. Telecommun.*, 10: 585–595, 1999.

[20] M. Wrulich, C. Mehlführer, and M. Rupp. Interference aware MMSE equalization for MIMO TxAA. In *Proc. 2008 3rd Int. Symp. Commun., Control and Signal Processing (ISCCSP2008)*, St. Julians, Malta, March 2008.

[21] J. (Charlie) Zhang, B. Raghothaman, Y. Wang, and G. Mandyam. Receivers and CQI measures for MIMO-CDMA systems in frequency-selective channels. *EURASIP J. Appl. Signal Processing*, (11): 1668–1679, November 2005.

Chapter 3

Advanced Receivers for MIMO HSDPA

Martin Wrulich, Christian Mehlführer, and Markus Rupp

Contents

3.1 Overview

The performance of Wideband Code Division Multiple Access (WCDMA) networks is limited due to interference more than by any other single effect. Frequency-selective channels cause a loss of orthogonality between the utilized spreading codes and impose restrictive throughput constraints.

Modern receiver design has to face these difficulties and tries to circumvent them without increasing the complexity to unacceptable levels.

This chapter covers the development of a multi-user, intra-cell, interference-aware minimum mean squared error (MMSE) equalizer for the transmit antenna array (TxAA) mode of multiple-input, multiple-output (MIMO) High-Speed Downlink Packet Access (HSDPA). The resulting receiver is able to exploit the special structure imposed by the precoded TxAA transmission to efficiently suppress the interference by only moderately increasing the complexity (compared to the classical single-user (SU) equalizer). The solution can be interpreted as a multi-user extension of the classical MMSE equalizer for HSDPA systems.

In this chapter we define a suitable system model to derive the intra-cell interference-aware MMSE equalizer and investigate its capabilities to suppress the multi-user intra-cell interference for TxAA HSDPA-operated networks. The resulting throughput performance is then investigated by means of physical layer as well as system-level simulations.

3.2 Introduction

HSDPA has been standardized as an extension of the Universal Mobile Telecommunications System (UMTS) as a part of the 3rd Generation Partnership Project (3GPP) Release 5 [1]. It is spectrally the most efficient WCDMA system commercially available at the moment. To satisfy the need for higher data rates and new services with the current base station sites, even higher cell capacities and spectral efficiencies must be achieved. Correspondingly, 3GPP has considered numerous proposals that incorporate MIMO techniques for the enhancement of frequency division duplex (FDD) HSDPA [2]. 3GPP has chosen the dual-stream transmit antenna array (D-TxAA) as the MIMO scheme for FDD because of its backward compatibility with TxAA [3]. In D-TxAA, a second data stream is transmitted via spatial multiplexing when user equipment with two receive antennas experiences high channel quality. If channel quality is low, the transmission mode is switched to TxAA, which is also the supported mode when the user equipment has only one receive antenna available. In contrast to D-TxAA, TxAA allows for scheduling multiple users simultaneously, exploiting multi-user diversity.

The intra-cell interference caused by the loss of orthogonality between the spreading codes due to the frequency-selective channel imposes restrictive throughput constraints. To combat this effect in HSDPA MMSE equalizers are disposed [4–7]. However, these solutions do not take multi-user operation or the interference structure in the cell into account.

A second approach to combat interference is the so-called interference mitigation that can be performed at the transmitter and/or the receiver side. In particular, in the downlink, each receiver needs to detect a *single* desired signal, while experiencing two types of interference. These are caused by the serving Node-B (intra-cell interference) and by a few dominant neighboring Node-Bs (inter-cell interference). Handling the interference on the receiver side is a difficult job, especially in the multi-user case in which classical approaches result in complex receiver structures [8–10]. To ease the job of the receiver, the precoding has been optimized for specific receiver structures; see, for example, [6,7,11]. Because HSDPA implicitly utilizes multi-code operation in the downlink, care also must be taken to design the receiver appropriately [12]. For D-TxAA, the multiple stream operation can be further utilized by successive interference cancellation (SIC) receivers [13]. More practical investigations on this subject were also conducted by 3GPP [14], but so far none of these recommendations have been implemented. Given the limited battery capabilities of today's handsets, complexity is an important issue. Despite the available research, the combined effects in the case of the precoded multi-user multi-code operation have not been addressed thus far.

In the uplink, on the other hand, the base station receiver has to detect *all* desired users in the cell, while also suppressing neighboring cell interference from many different sources [15]. A good overview about the interference situations in the uplink and downlink case, together with some well-known solutions for them, can be found in [16].

When interference mitigation is performed at the transmitter, accurate channel state information from all users is needed [17]. This requires lots of signaling and feedback information exchange, which typically is not available in cellular contexts.

A novel approach to cope with the problems of TxAA HSDPA is an interference-aware MMSE equalizer that utilizes the spatial structure of the intra-cell interference, that is, the precoding state of the cell. In particular, the receiver introduced in this chapter is able to effectively suppress the multi-user interference with moderate complexity. Further details on the explanations here can also be found in [18–20].

3.3 Transmit Antenna Array (TxAA) Mode

TxAA was introduced with UMTS in 1999 and now builds the foundation of the spatial multiplexing enhancement in MIMO HSDPA [3]. In contrast to the double-stream operation of MIMO HSDPA, TxAA allows for multiple users being served in the downlink simultaneously, which is called *multi-code scheduling*. Multi-code scheduling is well suited to work optimally in terms of the sum-rate throughput and short-term fairness trade-off [21]. Future-generation mobile networks implicitly build on similar concepts;

for example, Long-Term Evolution (LTE) [22] puts a strong emphasis on multi-user scheduling in its time-frequency downlink frames.

In TxAA HSDPA, every user is assigned a specific number of orthogonal spreading sequences of length 16. At a maximum, 15 spreading sequences can be assigned to all users, the 16th orthogonal spreading sequence is reserved for transmitting the pilot channels and other control channels. Depending on the link-adaptation feedback in the form of channel quality indicator (CQI) values, the scheduler can decide which users should be served in parallel. Furthermore, note that the TxAA scheme allows for an arbitrary number of receive antennas being utilized at the user equipment (UE).

3.3.1 System Model

Figure 3.1 shows the TxAA transmission scheme for one receive antenna when U users are simultaneously served. We define the spread and scrambled chip stream of user u at time instant i as

$$\mathbf{s}_i^{(u)} \triangleq \left[s_i^{(u)}, \dots, s_{i-L_b-L_f+2}^{(u)} \right]^T \tag{3.1}$$

where L_b and L_f are the length of the channel impulse response and the equalizer length, respectively. Thus, the vector $\mathbf{s}_i^{(u)}$ contains the L_b+L_f-1 recent chips. We assume that the power σ_s^2 of the chip stream $\mathbf{s}_i^{(u)}$ of each user u is normalized to one.

By multiplying $\mathbf{s}_i^{(u)}$ by a factor $\alpha^{(u)}$, the base-station can allocate a certain amount of transmit power to each served user. After the power

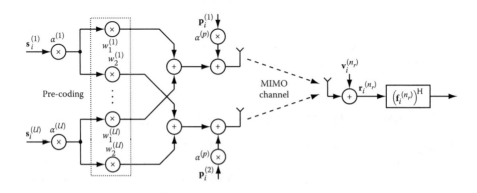

Figure 3.1 **Multi-user transmission in TxAA for a total number U of simultaneously served users. The precoding is conducted individually for every user. At the receiver, only one receive antenna is depicted, although the scheme allows for an arbitrary number of receive antennas.**

allocation, the chip streams are weighted by the user-dependent complex precoding coefficients $w_1^{(u)}$ and $w_2^{(u)}$ at the first and second transmit antennas, respectively. Our modeling holds for arbitrary precoding weights, but we want to note that due to standardization, these coefficients are strongly quantized [2], which we will take advantage of later. The weighted chip streams of all users are then added to the sequences $\alpha^{(p)}\mathbf{p}_i^{(1)}$ and $\alpha^{(p)}\mathbf{p}_i^{(2)}$, representing the sum of all channels that are transmitted without precoding, that is, the Common Pilot CHannel (CPICH), the High-Speed Shared Control CHannel (HS-SCCH), and other signaling channels—to which we will refer as non-data channels.

The frequency-selective channel between the n_tth transmit and the n_rth receive antennas is represented in Figure 3.2 by the vector $\mathbf{h}^{(n_r, n_t)}$, $n_t = 1, 2 = N_T$, $n_r = 1, \ldots, N_R$, composed of the taps of the channel. If we neglect for the moment the non-data channels, $\alpha^{(p)}\mathbf{p}_i^{(1)}$ and $\alpha^{(p)}\mathbf{p}_i^{(2)}$, the multi-user transmission in Figure 3.1 can be represented by U virtual antennas, one for each active user, as illustrated in Figure 3.2. The resulting equivalent (virtual) channels between user u and receive antenna n_r are then given by $\tilde{\mathbf{h}}^{(u, n_r)} = w_1^{(u)}\mathbf{h}^{(n_r, 1)} + w_2^{(u)}\mathbf{h}^{(n_r, 2)}$, $u = 1, 2, \ldots, U$. From this description it can be seen immediately that the intra-cell interference observed by user u can be treated as being transmitted over U different channels. The classical MMSE equalizer, however, would be determined only by the precoding weights $w_1^{(u)}$ and $w_2^{(u)}$ and would not consider the special structure of the interference. Thus, the degraded transmission scheme of TxAA imposes an interference situation that cannot be handled well by the classical MMSE equalizer, which is matched only to the channel of the desired user. This is in contrast to the single-input, single-output (SISO) HSDPA case, where due to the lack of precoding, the interference of simultaneously served users is transmitted over the same channel as the one of the desired user. Thus, in SISO HSDPA, equalization of the desired user's signal also equalizes the signal of the simultaneously served users.

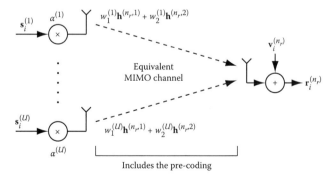

Figure 3.2 **Equivalent representation of the multi-user TxAA transmission, representing U virtual antennas and U virtual channels.**

For the derivation of our intra-cell interference-aware MMSE equalizer, let us define the $L_f \times (L_b + L_f - 1)$-dimensional band matrix modeling the channel between the n_tth transmit and the n_rth receive antenna:

$$\mathbf{H}^{(n_r, n_t)} = \begin{bmatrix} b_0^{(n_r, n_t)} & \cdots & b_{L_b-1}^{(n_r, n_t)} & 0 \\ & \ddots & & \ddots \\ 0 & b_0^{(n_r, n_t)} & \cdots & b_{L_b-1}^{(n_r, n_t)} \end{bmatrix}, \quad n_t = 1, 2, \quad n_r = 1, \ldots, N_R$$

(3.2)

The full frequency-selective MIMO channel can be modeled by a block matrix $\mathbf{H}$ consisting of $N_R \times 2$ band matrices:

$$\mathbf{H} = \begin{bmatrix} \mathbf{H}^{(1,1)} & \mathbf{H}^{(1,2)} \\ \vdots & \vdots \\ \mathbf{H}^{(N_R,1)} & \mathbf{H}^{(N_R,2)} \end{bmatrix}$$

(3.3)

where we explicitly put in the assumption that for TxAA, only two transmit antennas ($N_T = 2$) are utilized.

By stacking the received signal vectors of all N_R receive antennas,

$$\mathbf{r}_i = \left[\left(\mathbf{r}_i^{(1)} \right)^{\mathrm{T}}, \cdots, \left(\mathbf{r}_i^{(N_R)} \right)^{\mathrm{T}} \right]^{\mathrm{T}}$$

(3.4)

and by stacking the transmitted signal vectors of all U users and the vectors $\mathbf{p}_i^{(1)}$ and $\mathbf{p}_i^{(2)}$,

$$\mathbf{s}_i = \left[\left(\mathbf{s}_i^{(1)} \right)^{\mathrm{T}}, \cdots, \left(\mathbf{s}_i^{(U)} \right)^{\mathrm{T}}, \left(\mathbf{p}_i^{(1)} \right)^{\mathrm{T}}, \left(\mathbf{p}_i^{(2)} \right)^{\mathrm{T}} \right]^{\mathrm{T}}$$

(3.5)

we obtain the compact system description:

$$\mathbf{r}_i = \underbrace{\mathbf{H}(\mathbf{W} \otimes \mathbf{I}_{L_b+L_f-1})}_{\mathbf{H}_w} \mathbf{s}_i + \mathbf{v}_i = \mathbf{H}_w \mathbf{s}_i + \mathbf{v}_i$$

(3.6)

Here, $\otimes$ denotes the Kronecker product, and $\mathbf{v}_i$ is an additive noise vector that can incorporate both the thermal noise and the interference from other base-stations (inter-cell interference). The $2 \times (U+2)$ dimensional matrix $\mathbf{W}$ contains the precoding coefficients $w_1^{(u)}$ and $w_2^{(u)}$ of all users, as well as their power coefficients $\alpha^{(u)}$ ($u = 1, 2, \ldots, U$) and the power coefficients $\alpha^{(p)}$, and is defined as

$$\mathbf{W}^{(\mathrm{MU})} \triangleq \begin{bmatrix} \alpha^{(1)} w_1^{(1)} & \cdots & \alpha^{(U)} w_1^{(U)} & \alpha^{(p)} & 0 \\ \alpha^{(1)} w_2^{(1)} & \cdots & \alpha^{(U)} w_2^{(U)} & 0 & \alpha^{(p)} \end{bmatrix} \tag{3.7}$$

This matrix reflects the premise that the non-data channels are not pre-coded; thus the two columns on the right side are specified solely by the single parameter $\alpha^{(p)}$, which controls the total power spent on these channels. In general, we also assume that the power available at the base station is fully spent; thus the coefficients α are prone to a sum-power constraint:

$$\sum_{u=1}^{U} \left(\alpha^{(u)}\right)^2 + 2 \left(\alpha^{(p)}\right)^2 = P \tag{3.8}$$

3.4 Interference-Aware MMSE Equalization

Having our system model being specified as in the previous section, we are now able to derive the resulting MMSE equalizer. Without loss of generality, we assume in the following that the sequence of User 1 is to be reconstructed. The MMSE equalizer coefficients can be calculated by minimizing the quadratic cost function [23]:

$$J(\mathbf{f}) = \mathbb{E} \left\{ \left| \mathbf{f}^{\mathrm{H}} \mathbf{r}_i - s_{i-\tau}^{(1)} \right|^2 \right\} \tag{3.9}$$

with τ specifying the delay of the equalized signal, fulfilling $\tau \geq L_h$ due to causality. In this work, we assume the channel to be known at the receiver site.

This cost function minimizes the distance between the equalized chip stream and the transmitted chip stream. In Equation (3.9), the vector $\mathbf{f}$ defines N_R equalization filters:

$$\mathbf{f} = \left[\left(\mathbf{f}^{(1)}\right)^{\mathrm{T}}, \cdots, \left(\mathbf{f}^{(N_R)}\right)^{\mathrm{T}} \right]^{\mathrm{T}} \tag{3.10}$$

Each filter $\mathbf{f}^{(n_r)} = [f_0^{(n_r)}, \ldots, f_{L_f-1}^{(n_r)}]^{\mathrm{T}}$ has a length L_f. Note that because of the definition of $\mathbf{f}$ and $\mathbf{r}_i$, the inner product $\mathbf{f}^{\mathrm{H}} \mathbf{r}_i$ can be implemented by summing the outputs of the N_R equalization filters $(\mathbf{f}^{(n_r)})^{\mathrm{H}}$. This sum then yields the MMSE estimate of the transmitted chip sequence.

The minimization of the cost function can be performed by deriving Equation (3.9) with respect to $\mathbf{f}^*$ [24], evaluating the expectation operation, and setting the derivative equal to zero:

$$\frac{\partial J}{\partial \mathbf{f}^*} = \left(\mathbf{H}_w \mathbf{R}_{ss} \mathbf{H}_w^{\mathrm{H}} + \mathbf{R}_{vv}\right) \mathbf{f} - \sigma_s^2 \mathbf{H}_w \mathbf{e}_\tau = 0 \qquad (3.11)$$

The matrices $\mathbf{R}_{ss}$ and $\mathbf{R}_{vv}$ are the signal and noise covariance matrices, respectively, and the vector $\mathbf{e}_\tau$ is a zero vector of length $(U+2)(L_b+L_f-1)$ with a single "1" at cursor position τ. The equalizer coefficients for the data stream of the first user are therefore given by

$$\mathbf{f} = \sigma_s^2 \left(\mathbf{H}_w \mathbf{R}_{ss} \mathbf{H}_w^{\mathrm{H}} + \mathbf{R}_{vv}\right)^{-1} \mathbf{H}_w \mathbf{e}_\tau \qquad (3.12)$$

If the transmitted data signals of the users are uncorrelated with equal power σ_s^2, the covariance matrix $\mathbf{R}_{ss}$ becomes $\sigma_s^2 \mathbf{I}$, and if we assume the noise vector $\mathbf{v}_i$ white with variance σ_v^2, the noise covariance matrix becomes $\sigma_v^2 \mathbf{I}$. Without losing generality, we can assume that the signal covariance σ_s^2 is equal to 1 because the individual transmit powers of the users are determined by $\alpha^{(u)}$. The variance σ_v^2 is composed of the thermal noise and the power received from the other base stations. Note that if the receiver takes the structure of the inter-cell interference into account, effort must be expended to obtain an accurate estimation of the covariance matrix $\mathbf{R}_{vv}$.

Because this equalizer considers the interference of all users in the cell due to the full knowledge of the matrix $\mathbf{W}^{(\mathrm{MU})}$, we call it an intra-cell interference-aware MMSE equalizer. The standard equalizer is a special case of our solution and neglects the interference from other users, which we consequently call single-user (SU) equalizer in the following. It can be calculated from Equation (3.12) using the single-user precoding weight matrix of rank 1

$$\mathbf{W}^{(\mathrm{SU})} = \begin{bmatrix} \alpha^{(1)} w_1^{(1)} \\ \alpha^{(1)} w_2^{(1)} \end{bmatrix} \mathbf{e}_1^{\mathrm{T}} \qquad (3.13)$$

instead of the multi-user precoding weight matrix $\mathbf{W}^{(\mathrm{MU})}$. Here, $\mathbf{e}_1$ is a zero column-vector of length $U+2$ and a "1" at the first position. If only a single user is receiving data in the cell, both equalizers are very similar, with the only difference being that the intra-cell interference-aware equalizer also considers the interference generated by the non-data channels.

3.4.1 Interference Suppression

Before assessing the throughput performance of our proposed MMSE equalizer, we first want to look at the interference suppression capabilities. To do so, we adapted the model in [25], which describes the post equalization and despreading signal-to-interference-plus-noise ratio (SINR) for arbitrary linear receivers in a multi-stream closed-loop MIMO Code Division Multiple Access (CDMA) system. The remaining intra-cell interference after equalization—for a specific channel realization and precoding state—generated by the desired user and all other active users is explicitly given by

$$P_{\text{intra}} = \sum_{\substack{m=0 \\ m\neq\tau}}^{L_b+L_f-2} \left| \mathbf{f}^{\text{H}} \boldsymbol{\gamma}_m^{(1)} \right|^2 + \sum_{u=2}^{U} \sum_{\substack{m=0 \\ m\neq\tau}}^{L_b+L_f-2} \left| \mathbf{f}^{\text{H}} \boldsymbol{\gamma}_m^{(u)} \right|^2 \tag{3.14}$$

with $\boldsymbol{\gamma}_m^{(u)}$ denoting the mth column of the user-dedicated channel matrix

$$\mathbf{H}^{(u)} = \mathbf{H} \left(\left[\alpha^{(u)} w_1^{(u)}, \alpha^{(u)} w_2^{(u)} \right]^{\text{T}} \otimes \mathbf{I}_{L_b+L_f-1} \right) \tag{3.15}$$

Because Equation (3.14) depends on the current realization of the precoding state, we cannot use it directly to evaluate the performance of our proposed equalizer for general conditions. Thus, we approximate the remaining intra-cell interference by its expectation over the precoding state of the interfering users. Given a certain number of active users in the cell, the remaining intra-cell interference after equalization then becomes

$$\mathbb{E}_w\{P_{\text{intra}}\} = \underbrace{\sum_{\substack{m=0 \\ m\neq\tau}}^{L_b+L_f-2} \left| \mathbf{f}^{\text{H}} \boldsymbol{\gamma}_m^{(1)} \right|^2}_{f_{\text{self}}} + \underbrace{\sum_{u=2}^{U} \mathbb{E}_w\left\{ \sum_{\substack{m=0 \\ m\neq\tau}}^{L_b+L_f-2} \left| \mathbf{f}^{\text{H}} \boldsymbol{\gamma}_m^{(u)} \right|^2 \right\}}_{f_{\text{other}}} \tag{3.16}$$

with f_{self} and f_{other} denoting the determinative factors for the self and other-user intra-cell interference remaining after equalization, and $\mathbb{E}_w\{\cdot\}$ being the expectation with respect to the precoding coefficients w_1, w_2 of the interfering users.

As already mentioned, the 3GPP specifies a quantized codebook of possible precoding vectors [2]. The UE is responsible for evaluating and signaling the precoding vector that leads to the best pre-equalization SINR. It is important to note that the CQI feedback must be evaluated jointly with the precoding to obtain good throughput results [26]. Given the precoding codebook of [2], in [27] we are also able to show that in D-TxAA—even if multi-code scheduling would be implemented—an equalizer has no possibility to exploit information about the precoding state of the cell to suppress intra-cell interference.

Table 3.1 lists the simulation parameters we applied to evaluate the capability of our proposed equalizer to suppress the intra-cell interference caused by the other active users in the cell. Figures 3.3 and 3.4 show the performance in terms of the interference suppression capabilities, both for the self interference f_{self} and the other-user interference f_{other}, assuming

Table 3.1 Simulation Parameters for Figures 3.3 and 3.4

Parameter	Value
Simulated slots	1000
Slot time	2/3 ms
Receive antennas N_R	2
Precoding codebook	3GPP TxAA [2]
Equalizer span L_f	40 chips
Equalizer delay τ	20 chips
Precoding delay	11 slots
Mobile speed	3 km/h
Channel profile	ITU Pedestrian B (PedB) [28]
Active users U	4
Fading model	Improved Zheng model [29,30]

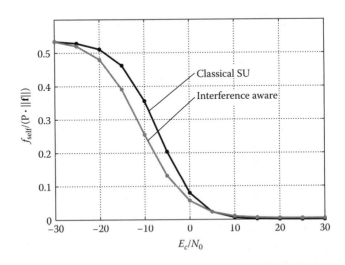

Figure 3.3 Self intra-cell interference f_{self} performance comparison of the proposed intra-cell interference aware equalizer and the classical single-user equalizer, assuming perfect knowledge of the cell's precoding state.

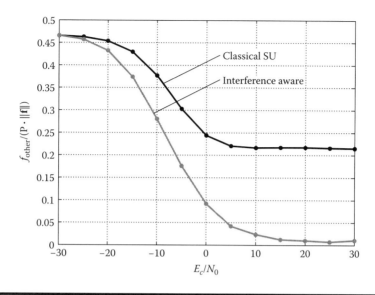

Figure 3.4 **Other-user intra-cell f_{other} performance comparison of the proposed intra-cell interference-aware equalizer and the classical single-user equalizer, assuming perfect knowledge of the cell's precoding state.**

perfect knowledge of the cell's precoding state. We normalize the two coefficients by the total received interference power (that is, because the channel is by assumption already normalized to 1) by dividing by the norm of the equalizer $\|\mathbf{f}\|$ and the total transmitted intracell power P, defined in Equation (3.8). It can be observed that our proposed equalizer is able to outperform the classical single-user equalizer, especially in terms of the other-user intra-cell interference when the chip-level signal-to-noise ratio (SNR) E_c/N_0 becomes high. The classical single-user equalizer is not able to cancel the interference generated by the transmission to other users, leading to a saturation of f_{other}. In contrast, our proposed equalizer can effectively decrease this interference term. The self intra-cell interference suppression is similar for higher E_c/N_0 values because the single-user equalizer is also matched to the channel of the desired user, which is sufficient to be able to restore the spreading code orthogonality, that is, lower f_{self}.

3.4.2 Complexity

Given the solution to the interference-aware MMSE equalization in Equation (3.12), it is interesting to assess the additional complexity needed to compute our proposed filter. Assuming uncorrelated transmit sequences with constant power, that is, $\mathbf{R}_{ss} = \sigma_s^2 \mathbf{I}$, the additional complexity can be evaluated by looking at the product $\mathbf{H}_w \mathbf{R}_{ss} \mathbf{H}_w^H \propto \mathbf{H}_w \mathbf{H}_w^H$ that is needed in the inverse part of $\mathbf{f}$. By writing

$$\mathbf{H}_w \mathbf{H}_w^H = \mathbf{H}\left(\mathbf{W}^{(MU)}\left(\mathbf{W}^{(MU)}\right)^H \otimes \mathbf{I}_{L_b+L_f-1}\right)\mathbf{H}^H \qquad (3.17)$$

it can easily be seen that the additional cost of our proposed equalizer is determined only by the larger matrix multiplication of $\mathbf{W}^{(MU)}(\mathbf{W}^{(MU)})^H$ instead of $\mathbf{W}^{(SU)}(\mathbf{W}^{(SU)})^H$, which is low compared to the cost of, for example, the inverse.

In particular, if we assume the complexity of matrix multiplications to be of order $\mathcal{O}\{K^3\}$ (with K being the larger matrix dimension), the added complexity of considering the multi-user precoding matrix is of order $\mathcal{O}\{(U+2)^3\}$. The matrix inverse, however, is of order $\mathcal{O}\{(N_R L_f)^3\}$, and because U is typically much smaller than L_f in practical systems, the multiplication of the muli-user precoding matrix is negligible.

3.5 Precoding State

In TxAA HSDPA, the MIMO channel is estimated by utilizing the CPICH, similar to UMTS. To be able to calculate the receive filter for the data channel, that is, the High-Speed Downlink Shared CHannel (HS-DSCH), however, the mobile needs to know (1) the *power offset* of the individual High-Speed Physical Downlink Shared CHannel (HS-PDSCH) compared to the CPICH, and (2) the *precoding coefficients* that the base station applied for the transmission. The power offset is signaled by higher layers [31], and the precoding coefficients of all simultaneous transmissions are signaled on the according HS-SCCH [32,33], where every active user has its own channel. This unfortunately makes things difficult for our proposed equalizer because the HS-SCCHs are scrambled with user-specific scrambling sequences[1], thus making it impossible to monitor the precoding state of the other users.

To overcome this problem, three different solutions are possible:

1. Change the signaling scheme in the HS-SCCH such that all active users know about the whole precoding state in the cell.
2. Include some training data in the HS-PDSCHs of the users to estimate the precoding state.
3. Blindly estimate the precoding state.

---

[1] The scrambling sequence in HSDPA is a function of the user identification number, known only to the base station and the particular user.

From this list, the first solution would require a change in the transmission standard [34], and thus is unfeasible for a practical implementation. Also the second solution, the training data-based approach, is not directly portable and would require network equipment vendors to implement algorithms of additional complexity in the base station, as well as the reservation of spreading code resources. The blind estimation solution seems thus far the most practicable approach that does not require any changes in the current standard and does not reduce the available resources in the CDMA cell.

Blind estimation is generally quite a challenging task, in particular in a multi-user context. Typical approaches treat the unknown inputs—in this case the unknown transmit data **s**—as *nuisance parameters* that the estimator must cope with in order to supply blind estimates of the parameters of interest. The Maximum Likelihood (ML) principle provides a systematic way to deduce the Minimum Variance Unbiased (MVU) estimator, maximizing the joint likelihood function [35]. There exist a number of possibilities to avoid the joint estimation of all parameters, that is, the power coefficients $\alpha^{(u)}$ and the data **s**. The *unconditional* or stochastic ML criterion models the vector of nuisance parameters as a random vector and maximizes the marginal of the likelihood function conditioned to **s**. Unfortunately, the unconditional ML estimator is generally unknown because the expectation with respect to **s** cannot be solved in closed form. However, in the low SNR regime, the unconditional likelihood function becomes quadratic in the observation with independence of the statistical distribution of the nuisance parameters. Nevertheless, this estimator class is generally difficult to solve and works only reasonably well in the low SNR regime [35].

Another modern approach for this kind of problems is based on random set theory [9], which can be utilized to find optimum estimators for random vectors when both the length and the values of the vector components are unknown. In the context of the precoding state estimation, the random vector to be estimated would be the vector of power coefficients $\alpha^{(u)}$, $u = 1, \ldots, U$, and $\alpha^{(p)}$, where the values of the coefficients as well as the length of the vector are unknown, because U is unknown. Applying random set theory leads to optimum Bayesian ML estimators. However, these require a joint estimation of the data sequence and the precoding state, which is typically computationally very demanding and thus disadvantageous for battery-powered mobile devices.

In [20] we propose a second-order Gaussian ML estimator that is able to estimate the precoding state of the cell with reasonable complexity. Due to the difficulty of the estimation problem, the estimator performance is far from optimum, but is sufficient to allow the multiuser interference-aware MMSE equalizer to closely achieve the performance when the precoding state is perfectly known. As the treatment of the precoding state estimation would go beyond the scope of this chapter, we refer the reader to [20] for further details.

3.6 Performance Evaluation

We split our performance evaluation into two different parts: (1) physical-layer simulations for a fixed transmission setup of TxAA HSDPA and (2) system-level simulations with adaptive feedback and scheduling. Each simulation approach has a different focus, with the physical-layer simulations covering channel encoding and decoding, W-CDMA processing, as well as the receiver processing in detail. On the other hand, system-level simulations represent a whole HSDPA network with adaptive feedback, scheduling, and Radio Resource Control (RRC) algorithms.

3.6.1 Physical-Layer Simulation Results

We conducted physical-layer simulations utilizing a standard compliant HSDPA simulator. The simulation assumptions in Table 3.2 correspond

Table 3.2 Simulation Parameters for Physical-Layer Simulations

Parameter	Value
Active users U	4
Desired user CQI	13
Interfering HS-PDSCH E_c/I_{or}	$[-6, -8, -10]$ dB
Interfering user CQIs	$\{16, 11, 8\}$
Interfering user precoding	$\left\{ \begin{bmatrix} 1 \\ \frac{1}{\sqrt{2}}(+1-j) \end{bmatrix}, \begin{bmatrix} 1 \\ \frac{1}{\sqrt{2}}(-1+j) \end{bmatrix}, \begin{bmatrix} 1 \\ \frac{1}{\sqrt{2}}(-1-j) \end{bmatrix} \right\}$
Precoding codebook	3GPP TxAA [2]
CPICH E_c/I_{or}	-10 dB
Other non-data channel E_c/I_{or}	-12 dB
UE capability class	6
Channel profile	ITU Pedestrian A (PedA), , Pedestrian B (PedB)
UE speed	3 km/h

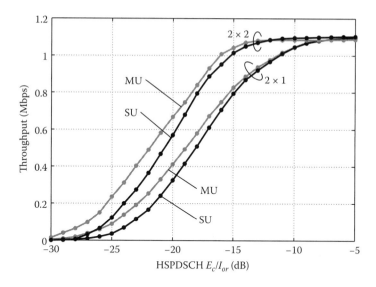

Figure 3.5   Throughput of desired user in a spatially uncorrelated ITU PedA channel at CQI 13, corresponding to a maximum throughput of 1.14 Mbps.

to a cell in which four users are receiving data simultaneously. User 1 is moving through the cell and obtains the precoding coefficients as adaptively requested, according to the definition in the standard [2]. The three interfering users are assumed to be stationary; thus, their precoding coefficients and transmit power do not change. In these simulations we assume that all users are always scheduled with the same CQI value, that is, no link adaptation other than the precoding is performed.

The achieved data throughput of User 1 in PedA and PedB environment is plotted in Figure 3.5 and Figure 3.6, respectively. In both scenarios, the interference-aware equalizer with perfect knowledge of the precoding state significantly outperforms the single-user equalizer. In the figures, E_c/I_{or} denotes the ratio of the average transmit energy per chip (E_c) to the total transmit power spectral density (I_{or}) [32].

The gain in the PedB channel in Figure 3.6 is much larger than the gain in the PedA channel, which has a much shorter maximum delay spread. This is caused by the larger loss of orthogonality in the PedB environment and the subsequently larger post equalization interference.

For a larger number of receive antennas, the simulation results show larger performance gains. The interference-aware equalizer can effectively utilize the spatial information to suppress the interfering signals. The largest performance increase of the interference-aware equalizer was found for the 2×2 PedB environment with 4 dB.

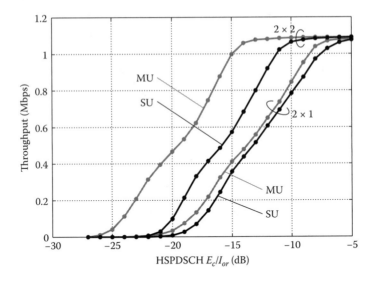

Figure 3.6 **Throughput of desired user in a spatially uncorrelated ITU PedB channel at CQI 13, corresponding to a maximum throughput of 1.14 Mbps.**

3.6.2 System-Level Results

To assess performance on the network level, we also conducted a set of system-level simulations with the simulator described in [25,27]. The simulation assumptions in Table 3.3 correspond to a 19-site scenario with a homogenous network load in which the multi-code scheduler serves four active users simultaneously. All 25 simulated users are moving through the cell with random directions and are adaptively reporting their CQI and precoding feedback according to their capability class [3]. The feedback delay was set to 11 slots.

The distributions of the SINR for the PedA and PedB channels, averaged over all active users in the cell, are plotted in Figure 3.7. It can be observed that the interference-aware (multi-user (MU)) equalizer is able to deliver significantly higher SINRs for PedB channels. In the PedA environment, the gain is negligible.

Figure 3.8 shows the average sector throughput. The interference-aware (MU) equalizer outperforms the classical (SU) equalizer significantly, again with more remarkable gains in the PedB environment—up to 11.7%.

Table 3.3   Simulation Parameters for System-Level Simulations

Parameter	Value
Simultaneously active users U	4
Transmitter frequency	1.9 GHz
Base station distance	1000 m
Total power available at Node-B	20 W
CPICH power	0.8 W
Power of other non-data channels	1.2 W
Spreading codes available for HSDPA	15
Macro-scale pathloss model	Urban micro [36]
Scheduler	Round Robin
Stream power loading	Uniform
Users in the cell	25
Cell deployment	Layout type 1 [37]
Precoding codebook	3GPP TxAA [2]
UE capability class	10
Equalizer span	40 chips
Feedback delay	11 slots
Channel profile	ITU Pedestrian A (PedA), Pedestrian B (PedB)
UE speed	3 km/h, random direction
Simulation time	25,000 slots, each 2/3 ms

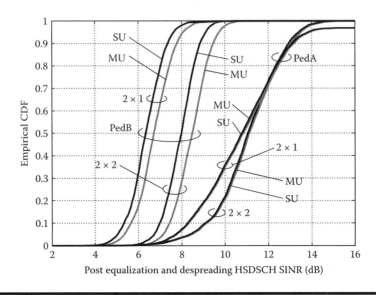

Figure 3.7 **Empirical CDFs of the post equalization and despreading SINR of the HS-DSCH.**

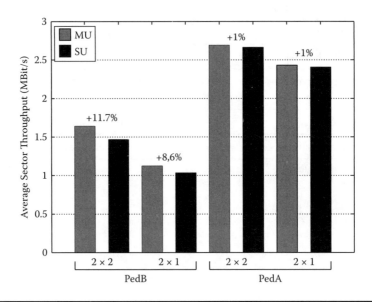

Figure 3.8 Average sector throughput results in the network specified by Table 3.3.

3.7 Summary

This chapter introduced a system model for TxAA HSDPA that takes the structure of the intra-cell interference in case of multi-code scheduling into account. The consideration of all simultaneously served users in the derivation of the MMSE equalizer leads to an interference-aware equalizer that has only slightly increased complexity compared to the classical SU MMSE equalizer. Simulations showed greatly reduced post equalization interference for the interference-aware equalizer. Finally, this chapter presented the performance gain of the introduced equalizer by means of physical-layer and system-level simulations. The results showed that the performance gains of the interference-aware MMSE equalizer increase with the frequency selectivity or, equivalently, the delay spread of the channel and the number of receive antennas. Both simulation types identified the new receiver structure as superior to the classical approach.

Acknowledgments

This work was funded by mobilkom austria AG and the Christian Doppler Laboratory for Wireless Technologies for sustainable Mobility. The views expressed are those of the authors and do not necessarily reflect the views within mobilkom austria AG.

References

[1] H. Holma and A. Toskala, *WCDMA for UMTS—Radio Access for Third Generation Mobile Communications*, 3rd ed., John Wiley & Sons, 2005.
[2] Technical Specification Group Radio Access Network, Multiple-Input Multiple-Output UTRA, 3rd Generation Partnership Project (3GPP), Tech. Rep. TR 25.876 Version 7.0.0, March 2007.
[3] Technical Specifiation Group Radio Access Network, Physical Layer Procedures (FDD), 3rd Generation Partnership Project (3GPP), Tech. Rep. TS 25.214 Version 7.4.0, March 2007.
[4] L. Mailaender, Linear MIMO equalization for CDMA downlink signals with code reuse, *IEEE Trans. Wireless Commun.*, 4(5): 2423–2434, September 2005.
[5] M. Melvasalo, P. Janis, and V. Koivunen, MMSE equalizer and chip level inter-antenna interference canceler for HSDPA MIMO systems, in *Proc. IEEE 63rd Vehicular Technology Conf. Spring (VTC)*, 4: 2008–2012, 2006.
[6] S. Shenoy, M. Ghauri, and D. Slock, Receiver designs for MIMO HSDPA, in *Proc. IEEE Int. Conf. Commun. (ICC)*, 2008, pp. 941–945.
[7] H. Zhang, M. Ivrlac, J.A. Nossek, and D. Yuan, Equalization of multiuser MIMO high speed downlink packet access, in *Proc. IEEE 19th Int. Symp. Personal, Indoor and Mobile Radio Commun. (PIMRC)*, 2008, pp. 1–5.

[8] F. Petre, M. Engels, A. Bourdoux, B. Gyselinckx, M. Moonen, and H.D. Man, Extended MMSE receiver for multiuser interference rejection in multipath DS-CDMA channels, in *Proc. IEEE VTS 50th Vehicular Technol. Conf. Fall (VTC)*, 3: 1840–1844, 1999.

[9] E. Biglieri and M. Lops, Miltiuser detection in a dynamic environment. I. User identification and data detection, *IEEE Trans. Inf. Theory*, 53(9): 3158–3170, September 2007.

[10] E. Virtej, M. Lampinen, and V. Kaasila, Performance of an intra- and inter-cell interference mitigation algorithm in HSDPA system, in *Proc. IEEE 67th Vehicular Technology Conference Spring (VTC)*, 2008, pp. 2041–2045.

[11] S. Shenoy, I. Ghauri, and D. Slock, Optimal precoding and MMSE receiver designs for MIMO WCDMA, in *Proc. IEEE 67th Vehicular Technology Conference Spring (VTC)*, 2008, pp. 893–897.

[12] B.-H. Kim, X. Zhang, and M. Flury, Linear MMSE space-time equalizer for MIMO multicode CDMA systems, *IEEE Trans. Commun.*, 54(10): 1710–1714, 2006.

[13] D. Bosanska, C. Mehlführer, and M. Rupp, Performance evaluation of intra-cell interference cancelation in D-TxAA HSDPA, in *Proc. ITG International Workshop on Smart Antennas (WSA)*, Darmstadt, Germany, February 2008, pp. 338–342.

[14] Technical Specification Group Radio Access Network, Feasibility Study on Interference Cancellation for UTRA FDD User Equipment (UE), 3rd Generation Partnership Project (3GPP), Tech. Rep. TR 25.963 Version 7.0.0, April 2007.

[15] A. Nordio and G. Taricco, Linerar receiver interfaces for multiuser MIMO communications, in *Proc. Eur. Signal Processing Conf. (EUSIPCO)*, 2006.

[16] J. Andrews, Interference cancellation for cellular systems: A contemporary overview, *IEEE Wireless Commun. Mag.*, 12(2): 19–29, April 2005.

[17] H. Zhang and H. Dai, Cochannel interference mitigation and cooperative processing in downlink multicell multiuser mimo networks, *EURASIP J. Wireless Commun. Netw.*, 2: 222–235, December 2004.

[18] M. Wrulich, C. Mehlführer, and M. Rupp, Interference aware MMSE equalization for MIMO TxAA, in *Proc. IEEE 3rd Int. Symp. Commun. Control and Signal Processing (ISCCSP)*, 2008, pp. 1585–1589.

[19] C. Mehlführer, M. Wrulich, and M. Rupp, Intra-cell interference aware equalization for TxAA HSDPA, in *Proc. IEEE 3rd Int. Symp. Wireless Pervasive Computing*, 2008, pp. 406–409.

[20] M. Wrulich, C. Mehlführer, and M. Rupp, Managing the interference structure of MIMO HSDPA: A multi-user interference aware MMSE receiver with moderate complexity, *IEEE Transactions on Wireless Communications*, Vol. 9, No. 3, Mar. 2010.

[21] D.I. Kim and S. Fraser, Two-best user scheduling for high-speed downlink multicode CDMA with code constraint, in *Proc. IEEE Global Telecommun. Conf. (GLOBECOM)*, 4: 2569–2663, 2004.

[22] Technical Specification Group Radio Access Network, Evolved Universal Terrestrial Radio Access (E-UTRA); LTE Physical Layer—General Description, 3rd Generation Partnership Project (3GPP), Tech. Rep. TS 36.201 Version 8.3.0, March 2009.

[23] J. Proakis, *Digital Communications*, 4th ed., McGraw-Hill Science Engineering, August 2000.

[24] S. Haykin, *Adaptive Filter Theory*, 4th ed., Prentice-Hall, 2002, ISBN 978-0-13-090126-2.

[25] M. Wrulich, S. Eder, I. Viering, and M. Rupp, Efficient link-to-system level model for MIMO HSDPA, in *Proc. IEEE 4th Broadband Wireless Access Workshop*, 2008.

[26] C. Mehlführer, S. Caban, M. Wrulich, and M. Rupp, Joint throughput optimized CQI and precoding weight calculation for MIMO HSDPA, in *Proc. 42nd Asilomar Conf. Signals, Systems and Computers*, October 2008.

[27] M. Wrulich and M. Rupp, Computationally efficient MIMO HSDPA system-level evaluation, *EURASIP Journal on Wireless Communications and Networking*, vol. 2003, Article ID382501, 14 pages, 2009. doi: 10.1155/2009/382501.

[28] Members of ITU, Recommendation ITU-R M.1225: Guidelines for Evaluation of Radio Transmission Technologies for IMT-2000, International Telecommunication Union (ITU), Tech. Rep., 1997.

[29] Y. Zheng and C. Xiao, Simulation models with correct statistical properties for rayleigh fading channels, *IEEE Trans. Commun.*, 51(6): 920–928, June 2003.

[30] T. Zemen and C. Mecklenbräuker, Time-variant channel estimation using discrete prolate spheroidal sequences, *IEEE Trans. Signal Process.*, 53(9): 3597–3607, September 2005.

[31] Technical Specification Group Radio Access Network, Radio Link Control (RLC) Protocol Specification, 3rd Generation Partnership Project (3GPP), Tech. Rep. TS 25.322 Version 8.4.0, March 2009.

[32] Technical Specification Group Radio Access Network, Multiplexing and Channel Coding (FDD), 3rd Generation Partnership Project (3GPP), Tech. Rep. TS 25.212 Version 8.5.0, March 2009.

[33] Technical Specification Group Radio Access Network, Medium Access Control (MAC) Protocol Specification, 3rd Generation Partnership Project (3GPP), Tech. Rep. TS 25.321 Version 7.0.0, March 2006.

[34] Technical Specification Group Radio Access Network, UTRA High Speed Downlink Packet Access (HSDPA); Overall Description; Stage 2, 3rd Generation Partnership Project (3GPP), Tech. Rep. TS 25.308 Version 5.0.0, September 2001.

[35] S.M. Kay, *Fundamentals of Statistical Signal Processing. Estimation Theory*, vol. 1, Prentice-Hall, 1993.

[36] D.J. Cichon and T. Kürner, *COST 231—Digital Mobile Radio towards Future Generation Systems*. COST, 1998, ch. 4.

[37] Technical Specification Group Radio Access Network, Spatial Channel Model for Multiple Input Multiple Output (MIMO) Simulations, 3rd Generation Partnership Project (3GPP), Tech. Rep. TS 25.996 Version 7.0.0, June 2007.

Chapter 4

Interference Cancellation in HSDPA Terminals

Ahmet Baştuğ, Irfan Ghauri, and Dirk T.M. Slock

Contents

4.1 Introduction

Over-the-air communication is interference limited. Deployment of wireless networks therefore needs resource planning. Neighboring base stations in a cellular network could, for instance, transmit in different frequency bands. For such deployment, hexagonal cell geometry is assumed in cellular systems (see Figure 4.1), and resource planning is done by allocation of disjoint chunks of spectrum to neighboring base stations. The number of neighboring cells in which a certain frequency can be used only once is known in the cellular literature as the *frequency reuse factor*. A frequency reuse factor of 7 results when isotropic antennas are deployed at cell-site (base station) and of 3 when using antenna sectorization of three sectors per hexagonal cell. Such frequency planning obviously aims interference avoidance, but it makes network deployment cumbersome and is wasteful of spectral resources. A direct sequence (DS)-CDMA (Code Division Multiple Access) system, in principle, boasts a frequency reuse factor of 1. In fact, this factor was purportedly one of the main advantages of the first CDMA-based cellular network, IS-95 [41]. Users can coexist in the same frequency spectrum. A resource is a spreading code and in the specific case of downlink communications, codes belong to a binary orthogonal set, the Walsh-Hadamard set. A base station has the entire set of codes at its disposal.

All base stations employ the same code-set as resources, and each base station overlays a frame containing a number of spread symbols by a specific long pseudo-random code with quarternary alphabet called the scrambling code. The code helps the mobile station distinguish between base station signals.

Standard CDMA mobile receivers are matched filters, matched to the spreading code of the signal (user) of which one needs to detect transmitted symbols. When the signal passes through a multipath channel, the receiver is matched to the cascade of the spreading code and the channel. The matched filter is also called a RAKE receiver in the context of multipath signal due to one standard correlator-based implementation that resembles an agricultural rake. In this receiver energy in delayed multipath signals is collected by correlating the delayed copies of the signal with the code of the user of interest.

User-specific spreading and base station–specific scrambling are the key elements of the general structure of downlink communications in early-day CDMA systems such as IS-95 and even circuit-switched UMTS (Universal Mobile Telecommunication System) [4], both of which support low-rate applications. In the latter, maximum service data rates of 384 kbps per base station using all physical channels resources are achievable. Such rates are sufficient for applications such as video streaming.

Spreading and scrambling at the transmitter and corresponding descrambling and despreading at the receiver do not fundamentally change the communication paradigm but are simply elements of the multiple access method (CDMA). It is hoped that noise-like user signals rendered thus due to bandwidth expansion (spreading) will be rejected by the matched filter. All other transceiver stages, such as error-control coding and interleaving to mitigate block-fading, are similar to those in single-user communications.

A mobile station can be affected by two types of interference known in the literature as intra-cell and inter-cell interference. In the particular case of downlink CDMA communications, which we shall henceforth address, the former comes about due to multipath propagation in the channel. Indeed, the channel distorts signals in transit that were orthogonal upon transmission. The RAKE receiver is then limited in performance due to this interference, sometimes also referred to as self-interference because even in the case of one trasmitted code, copies of the signal interfere mutually. This is no different from the inter-symbol-interference (ISI) problem in band-limited channels. It is shown in [38] that the signal-to-interference-plus-noise ratio (SINR) of the RAKE receiver contains a per-code interference term that is the sum of energies of all multipath components at the RAKE output scaled by the inverse of the spreading factor. As the number of codes increases for a given spreading factor (SF), as is the case in High-Speed Downlink Packet Access (HSDPA) where up to 15 codes with SF 16 can co-exist, the SINR at the RAKE output may degrade sufficiently to render communication unreliable. Mitigation of intra-cell interference can be done through equalization prior to despreading receivers [13].

In the downlink (HSDPA) context, inter-cell interference is the signal a mobile station sees from one or more neighboring base stations. Due to propagation factors, the cell boundary is not really a regular contour (e.g., a hexagon) and is only of figurative interest. When frequency reuse is unity, a good way of delimiting a cell is the strength of the signal. However, with full frequency reuse as in the case in CDMA, the signal-to-interference ratio (SIR) experienced at the mobile station is a more appropriate cell boundary notion. Users experience increasing levels of interference as they move from the base station/cell site toward the cell boundary. As interference could be a sum of interferences from several neighboring cell sites, it is customary [5] to define SIR as the signal to total-intercell-interferences ratio, and each interfering station to total inter-cell interference ratio is referred to as the dominant interference portion (DIP) ratio. Some reference numbers for these quantities are SIR of 0 dB with two interfering base stations with DIP of −2.75 and −7.64 dB respectively [5].

This chapter is organized as follows. The section entitled "UMTS Downlink and HSDPA" revisits the UMTS downlink signal and discusses HSDPA in relation with early-day non-packet UMTS. The section entitled "Downlink Channel and HSDPA Signal Models" describes the propagation environment and the channel model. The section entitled "Suppression of Intra-cell Interference in HSDPA" addresses the RAKE receiver, and some linear chip-level equalizers capable of dealing with intra-cell interference. The section entitled "Advanced Receivers for Interference Cancellation" discusses inter-cell interference and presents a variety of solutions for suppression of this interference.

4.2 UMTS Downlink and HSDPA

We limit ourselves to the discussion of the UMTS WCDMA standard, of which HSDPA is one component. This system is based on frequency division duplexing (FDD) so that uplink and downlink transmissions occur in non-overlapping frequency bands. Thus, a mobile terminal sees only signals from base stations. Another version of UMTS uses time division duplexing (TDD) where mobile terminals could effectively see interference from terminals talking to other base stations unless strict time-scheduling rules are introduced. Despite being an interesting problem itself, UMTS TDD is beyond the scope of this discussion.

4.2.1 Multiple Access in UMTS FDD Downlink

In the UMTS downlink, base stations transmit in frequency bands of 5 MHz around the 2.1-GHz frequency. Each UMTS operator in general has two or three bands for downlink transmission. The base station, which is known as Node B in the UMTS radio access network (RAN) context, is the source of transmissions for its *logical cell*. If a sectorized cell planning is deployed, then Node B is responsible for more than one logical cell, as shown in Figure 4.1.

The signals transmitted from different logical cells are differentiated from each other by the assignment of different pseudo-random *scrambling* codes that are repeated every UMTS *frame* of 38,400 chips, and hence are known as *long* overlay codes.

Multiple access of the users in the same logical cell is realized by a CDMA scheme that uses *short* orthogonal *channelization* codes from various levels of the OVSF code tree shown in Figure 4.2. Each level of the code tree

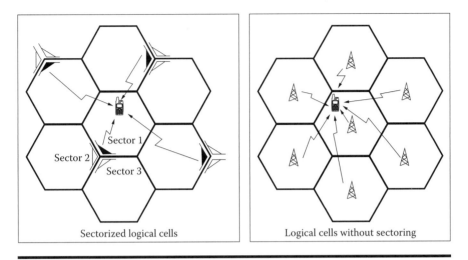

Figure 4.1 Logical cells differentiated by scrambling codes. (Also one frequency per cell in FDMA systems.)

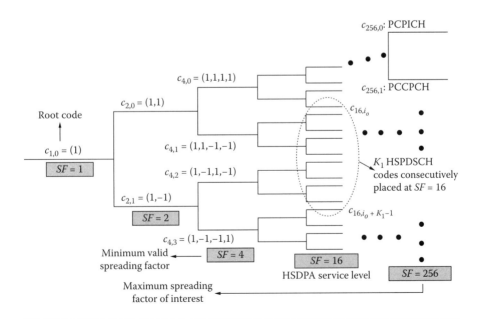

Figure 4.2 Partial schematic of the OVSF code tree.

contains codes corresponding to the columns of the Walsh-Hadamard transform (WHT) of relevant size. A channelization code assigned to a user is periodically used for the transmission of each symbol. Any particular code in the *i*th position at the SF level *t* is related to its two closest child codes at SF level $t + 1$ with the transformation

$$\left[\boldsymbol{c}_{2^{t+1},2^i} \; \boldsymbol{c}_{2^{t+1},2^i+1} \right] = \begin{bmatrix} 1 & 1 \\ 1 & -1 \end{bmatrix} \otimes \boldsymbol{c}_{2^t,i} \qquad (4.1)$$

where $\otimes$ stands for the Kronecker product. The valid code lengths are from the set $\{2^t, \; t \in 2, 3, \dots, 9\}$. The largest code length (i.e., the spreading factor 512) is very rarely used. When a particular code is assigned to a user, then all its parent or child codes are blocked for usage in order to preserve orthogonality among the used codes. These properties make UMTS FDD downlink a *code-limited* system. In case only a single spreading level is used from the OVSF tree, then the number of available codes for that particular scenario is upper bounded by the associated spreading factor.

4.2.2 UMTS Services

The flexibility of using different length codes makes UMTS a *multi-rate* system, enabling services with different data rates and thus different quality-of-service (QoS).

Just like any other cellular multi-access system, a UMTS network must support a large number of users with different QoS and have ubiquitous coverage. Large coverage area of individual cells (i.e., decreasing the number of deployed base stations) decreases network user capacity. On the other hand, more cells would mean more inter-cell interference. From this relationship it is easy to see that in order to meet both coverage and user capacity requirements, advanced transmission (diversity) or reception techniques are of use. The UMTS standard defines four QoS classes with differing delay and packet-ordering requirements [3]:

1. *Conversational:* low delay, strict ordering (e.g., voice)
2. *Streaming:* modest delay, strict ordering (e.g., video)
3. *Interactive:* modest delay, modest ordering (e.g., web browsing)
4. *Background:* no delay guarantee, no ordering (e.g., bulk data transfer)

4.2.3 HSDPA Features

Background and *interactive* UMTS service classes have a burst nature enabling time-divided multi-user scheduling, thus reaping the benefits of *multi-user diversity* [42]. This consideration triggered time-sharing system resources among users, most importantly the orthogonal codes in the downlink leading to the standardization of HSDPA in the UMTS Standard Release 5 [5].

- *Allocation of multiple access codes for HSDPA service:* Motivated by the bursty nature of the data, as shown in Figure 4.2, $K_1 \in \{1, 2, \dots, 15\}$, of the 16 channelization code resources at SF = 16 are allocated as High-Speed Physical Downlink Shared Channels (HS-PDSCHs) these codes are dynamically time multiplexed among users in order to achieve a higher spectral efficiency and a larger link adaptation dynamic range. The variable $i_o \in \{1, 2, \dots, 15\}$ denotes the position of the first HSPDSCH code.
- *Fast scheduling of allocated codes:* Multi-user diversity is obtained if there are independently varying temporal channel conditions for different users, leading to significantly increase in *sum capacity*, that is, the total delivered payload by the base station (BS). By one extreme approach, as demonstrated in Figure 4.3 in a simple two-user system context, one can preferably assign all the codes to a single user with the instantaneously best channel conditions, thus maximizing the throughput. At the other extreme, users might be served in a *fair* round-robin fashion. In this respect, operators are free to choose any set of schedulers compromising throughput and fairness by basing their decisions on the predicted channel quality, the cell load, and the traffic priority class. In order to reduce the delay in signaling and to better track the channel variations, scheduling is performed at Node B, which is closer to the air interface compared to the radio network controller (RNC), which was responsible for such tasks in the earlier UMTS releases. Moreover, the scheduling period decreases to 2-ms. sub-frame duration (i.e., 1 TTI, transmission time interval), from the

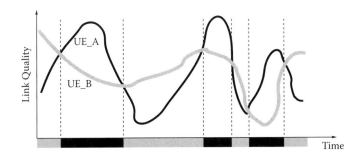

Figure 4.3   Principle of multi-user diversity.

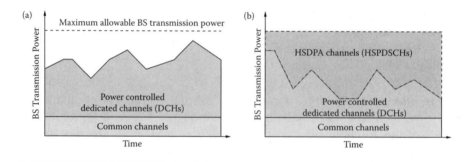

Figure 4.4   Power control with and without HSDPA. (a) UMTS FDD downlink *R99* power control mechanism, (b) UMTS FDD downlink *R5* power control mechanism.

10-ms frame duration of pre-HSDPA UMTS[1]. Soft handover is also replaced by fast *best-signaling-cell* selection, which can be considered a kind of *spatial* scheduling complementing the *temporal* scheduling.

■ *Link adaptation:* As schematically demonstrated in Figure 4.4, perhaps one of the most important differences between HSDPA and its packet-switched ancestor (Release 99 or R99 UMTS) is that there is *no fast power control* on HSPDSCHs, and all the instantaneously remaining *allowed* BS power is assigned to HSPDSCHs, which creates a high amount of power variation of HSPDSCH codes [2] over time. In this case, as can be interpreted from the figure, the system is also capable of utilizing the available BS power more efficiently than the power-controlled case. Furthermore, as shown in Figure 4.9, different user distances from the BS and different user mobility levels create a high amount of *inter-user link quality* differences. These two properties make Node B scheduling more versatile in deciding the number of allocated HSPDSCH codes, coding rate, puncturing rate, and the modulation scheme. 16-QAM (quadrature amplitude modulation) a possibility in addition to QPSK (quadrature phase-shift keying) high received power conditions, to maximize the throughput of the instantaneously scheduled user. For this purpose, Node B might use either the explicit channelquality indicator (CQI) measurement reports from the UE based on the SINR of PCPICH or the known transmit power of the power-controlled downlink dedicated physical channel (DPCH) associated with the HS-PDSCHs.

---

[1] Low-end and medium-end HSDPA UEs (user equipments) that do not have enough buffering capability are obliged to wait at least two or three TTI periods, respectively, between two consecutive TTI data scheduling [5].

■ *Hybrid Automatic Repeat reQuest (HARQ):* When transmission entities are identified to be erroneous by a standard protocol such as *selective-repeat* or *stop-and-wait,* fast retransmit request is done from Node B and combinations of soft information from the original transmission and previous retransmissions are combined to increase the probability of correct reception [19,39]. These operations compensate for errors in the channel quality estimates used for link adaptation. Two well-known methods are *chase combining,* where weighting of identical retransmissions is done, and the *incremental redundancy,* where additional parity bits are sent each retransmission.

To support the listed functionalities, two new channel types are introduced. In the downlink, one or more High-Speed Shared Control Channels (HSSCCHs) broadcast the scheduled UE identity, the transport format and the HARQ process identifier. The UE monitors up to four different HSSCCHs and tries to find out if it is going to be scheduled, or not. In the uplink, the High-Speed Dedicated Physical Control Channel (HSDPCCH) carries the status reports for HARQ and the CQIs. Figure 4.5 briefly demonstrates the order of events in the HSDPA transmission protocol. More detailed timing information and the slot structures are given in Figure 4.6, together with other UMTS channels relevant for this.

The HSSCCH is frame aligned with the PCPICH, which is generally used as a reference by several other UMTS channels and synchronization procedures as well.

HSPDSCHs are offset by two time slots with respect to HSSCCH, which gives the UE enough time to decode the time-critical control and supervision

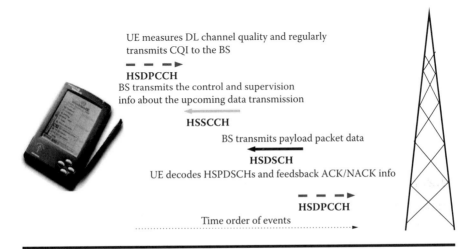

Figure 4.5   HSDPA transmission protocol.

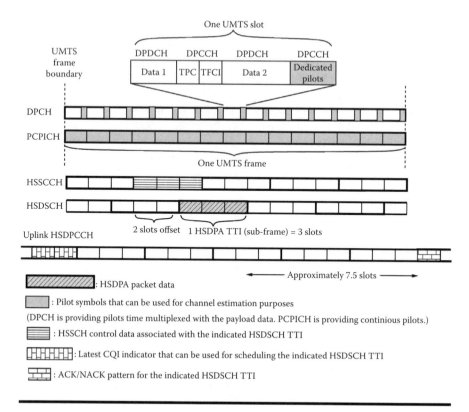

Figure 4.6 Slot structures and timings of UMTS channels of interest.

information carried by the first slot of HSSCCH before receiving the HSD-SCH payload data. Learning the scheduling of UE two slots beforehand is at the same time very useful for the *adaptive* equalizers that we discuss in subsequent sections. To do power savings, it is generally preferable to freeze the adaptation mechanism of an equalizer when the UE does not receive any HSDPA data. On the other hand, it is beneficial to start the adaptation process some time earlier than the start of the data reception to enable the equalizer to converge earlier.

For detailed description of HSDPA, see [1,6,12,17].

4.2.4 Downlink Transmission Model

The baseband downlink transmission model of the UMTS-FDD mode system with HSDPA support is given in Figure 4.7.

At the transmitter, the first group of K_1 i.i.d. QPSK or 16-QAM modulated symbol sequences $\{a_1[n], a_2[n], \ldots, a_{K_1}[n]\}$, which belong to the HSDPA

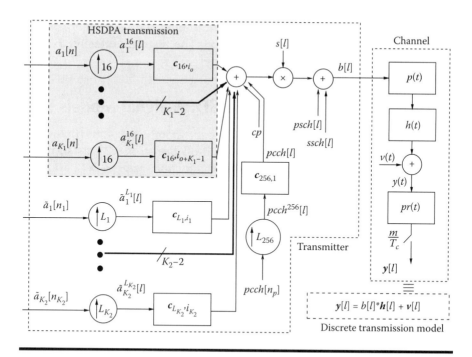

Figure 4.7   Baseband UMTS downlink transmission model.

transmission, are first up-sampled by a factor of 16 and then multiplied with their respective unit-amplitude channelization codes $\{c_{16,i_o}, c_{16,i_o+1}, \ldots, c_{16,i_o+K_1-1}\}$ shown in Figure 4.2. All HSPDSCH symbols have the same power and the same modulation scheme.

The second group of multi-rate transmissions $\{\tilde{a}_1[n_1], \tilde{a}_2[n_2], \ldots, \tilde{a}_{K_2} \times [n_{K_2}]\}^2$ representing the DPCHs, HSSCCHs, and other control channels are similarly up-sampled and convolved with their respective channelization codes:

$$\{c_{L_1,i_1}, c_{L_2,i_2}, \ldots, c_{L_{K_2},i_{K_2}}\}$$

The third group of chip sequences associated with PCCPCH, PCPICH, PSCH, and SSCH channels are denoted as $pcch[l]$, cp, $psch[l]$, and $ssch[l]$, respectively. The common pilot cp symbols are all $\frac{1+j}{\sqrt{2}}$.

The sum of all the generated chip sequences is multiplied with the unit-energy, BS-specific, aperiodic scrambling sequence $s[l]$. PSCH and SSCH are the exceptions, multiplexed after the scrambler, because as a first task

---

[2] different symbol indices such as $n, n_1, n_2, \ldots, n_{K_2}, n_p$ are used in the text and in Figure 4.7 to stress the *multi-rate* property of the transmission scheme.

in the receiver, they are utilized for determining (i.e., searching), which scrambling sequence is assigned to the BS. The resultant effective BS chip sequence $b[l]$ is transmitted on the channel.

4.3 Downlink Channel and HSDPA Signal Models

UMTS downlink channel has three cascade components: a *root-raised-cosine* (RRC) pulse shape $p(t)$ with a roll-off factor of 0.22, as shown in Figure 4.8; the *time-varying multipath propagation* channel $b(t)$; and a receiver front-end filter $p_r(t)$ that is generally chosen to be again an RRC pulse shape with a roll-off factor of 0.22 due to the fact that the *raised cosine* (rc) result of the 'c' cascade is a *Nyquist pulse* whose T_c-spaced disrete time counterpart is a single unit pulse at time instant 0. In this case, the only source of inter-chip interference (ICI) is $b(t)$. Alternatively, a low-pass anti-aliasing filter with a cutoff frequency between $\frac{1.22}{T_c}$ and $\frac{2}{T_c}$ may be considered as $p_r(t)$ in the case of twice the chip rate sampling. The latter case is a reasonable choice for fractionally spaced equalizers [20,21].

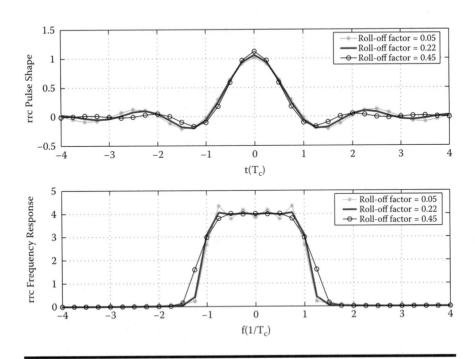

Figure 4.8 Root-raised cosine pulse shapes with different roll-off factors. Higher factors induce less ICI because the tails decay faster but they consume more bandwidth.

The *effective* continuous time channel is hence given as

$$h_{eff}(t) = p(t) * h(t) * p_r(t) \qquad (4.2)$$

When there is no transmitter beamforming, the propagation channel and the effective overall channel are unique for all the transmitted code channels from the same BS.

4.3.1 Channel Impairments and Mitigation

Modeling of the propagation channel $h(t)$ is a complex topic in itself [31,32]. We restrict this discussion to intuitive explanations of some aspects that are essential for this chapter.

The most observable effect of the propagation channel on the received signal quality is the time-varying signal amplitude attenuation, which is more often known as *fading* and is a combined consequence of different propagation effects.

The environment-dependent *large-scale* statistics of the UE received power at a distance d (in kilometers) is modeled as[3]

$$P_r \text{ (dBm)} = P_t \text{ (dBm)} - G \text{ (dB)} - 10n log_{10} d \text{ (dB)} + 10 log_{10} x (dB) \quad (4.3)$$

where P_t is the transmitted power,[4] G is the amount of path loss at a reference distance of 1 km, n is the path loss exponent, and x is the log-normal shadowing term with zero mean and standard deviation σ_x. Shadowing is a consequence of signal absorption by the obstacles in the terrain between the BS and UE, such as hills, trees, buildings, and cars, and it causes a *variance* around the distance-dependent *mean* path loss. It is a *slow-fading* parameter that varies only when the UE changes its position by a distance proportional to the dimensions of an obstructing object.

The most important propagation channel characteristic is the *multipath* effect. Many replicas of the transmitted signal are reflected from several objects and reach the UE with different delays and different complex attenuation factors. Specular replicas are clustered together to generate the effective multi-paths shown in Figure 4.9. The *sparse channel model*, which takes into account only the most dominant P paths, can be formulated as

$$h(t) = \sum_{i=1}^{P} h_i(t) \delta(t - \tau_i(t)) \qquad (4.4)$$

The difference between the largest and smallest delay elements, $\Delta_\tau = \tau_P - \tau_1$, is the *delay spread* of the channel. If $\Delta_\tau \geq T_c$, then the channel is

---

[3] dBm is a relative measure w.r.t. 1 mW power level.
[4] In UMTS terminology, $I_{or} = P_t$ and $\hat{I}_{or} = P_r$.

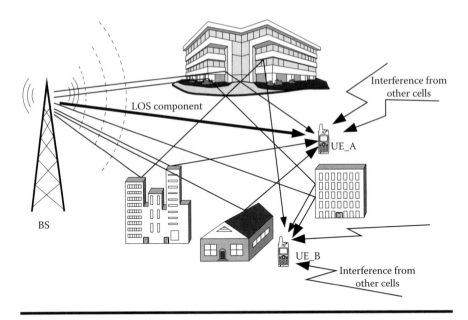

Figure 4.9   Multipath effect.

frequency selective. This comes from the inverse of the delay spread known as *coherence bandwidth* $B_o \approx \frac{1}{\Delta_\tau}$. The physical meaning of B_o is that when two different sinusoidal components with frequencies f_1 and f_2 are transmitted, they are impacted differently by the channel if $\Delta f = |f_1 - f_2| \geq B_o$. In other words, if the signal BW is larger than B_o, which is the case for UMTS *wideband* CDMA downlink, then signal spectrum is nonuniformly affected by the channel. On one hand, if no channel equalization is applied, this is a dispersive situation causing interchip interference (ICI) and driving communication unreliable. On the other hand, it is an opportunity to exploit the inherent *frequency diversity* coming from different sub-bands of the spectrum, which are considered to be *independently fading.* In the time domain, this property results in the *resolution of the paths* that are separated from each other by at least a distance of T_c. Conventional RAKE receiver exploits frequency diversity by collecting energy via multiple correlations at time instants corresponding to the path delays. Although exact resolution is lower bounded by the chip period, some more diversity is expected from decreasing this constraint to slightly lower values such as $\frac{3T_c}{4}$ [8]. If an opposite situation occurs, that is, if the signal BW is smaller than B_o, then the channel is *flat fading*, meaning that there is no ICI. At first glance this seems to be a good situation not requiring a complicated equalization procedure. However, when there are deep and long-lasting channel fades, as is the case in slow fading channels, other means such as transmit diversity or receive diversity are necessary to re-cover from the *outage* state. These techniques, however, complicate BS and UE design.

When all the specular components that generate one dominant path are modeled as i.i.d. complex random variables, then by central limit theorem, channel parameters turn out to be circularly symmetric Gaussian random variables with zero mean and variance $2\sigma_i^2$. Consequently, their complex envelope amplitudes are Rayleigh distributed.

$$p(|b_i(t)|) = \frac{|b_i(t)|}{\sigma_i^2} e^{-\frac{|b_i(t)|^2}{2\sigma_i^2}} \quad \text{for } |b_i(t)| \geq 0, \quad 0 \quad \text{otherwise} \qquad (4.5)$$

When there is one very dominant line-of-sight (LOS) path, as is the case for UE_A in Figure 4.9, its distribution is *Rician*, which is more desirable because in that case there are less frequent and less deep fades. In this thesis we are not considering LOS situations.

Sparse multipath channel parameters are modeled as wide-sense sta-tionary and uncorrelated with each other (WSS-US model) [31]. Therefore, each one of them experiences an independent *small-scale* fading due to the movement of the UE, and the movements of the objects that have impacts on that particular path. Previously mentioned shadowing is a *large-scale* complement manifesting itself as birth or death of a path.

The *time-variation* of sparse channel parameters is a metric associated with the amount of signal spectral broadening caused by a Doppler shift, which, in return, is proportional to the *effective* UE velocity in the direc-tion of the coming path ray. The dual relation of a broadening in the fre-quency domain transfer function is a narrowing of the non-zero channel autocorrelation window in the time domain from infinity to a finite quan-tity known as channel *coherence time* T_o. The physical meaning of T_o is that when a sinusoid is transmitted twice at times t_1 and t_2, the two are influenced differently by the channel if $\Delta t = |t_1 - t_2| \geq T_o$. The chan-nel is typically considered *fast fading* when $T_o < T_c$ because in this case different parts of a chip are influenced by different-valued channel pa-rameters. With this reasoning, CDMA channels always fall into *slow fading* category. A better criterion to judge whether a channel is fast or slow fading is to compare T_o with the design requirements of the considered application or receiver. If we consider a UE chip equalizer, for example, which recomputes its weights periodically from scratch using the channel estimates, then coherence time should be more than the chosen update period.[5]

---

[5] A typical requirement for the computation period of nonadaptive HSDPA equalizer weights is 512 chips.

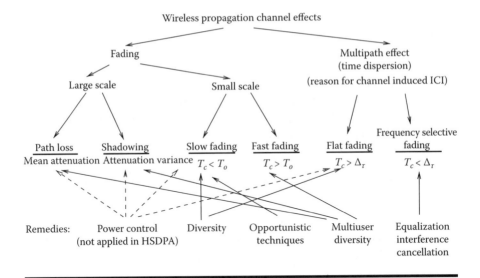

Figure 4.10 Summary of channel impacts and *most relevant* procedures against them.

One might think that slow fading is always a desired situation; however, as explained before, it causes the deep fades to last a long time. In some catastrophic cases, none of the diversity measures will help. Recently some work has been done to mitigate such situations by artificially generating fast-fading channel conditions via a transmission scheme called *opportunistic beamforming* [42].

Figure 4.10 gives a brief summary of the common channel impairments and the principal techniques for mitigating them.

4.3.2 HSDPA Signal Model

The discrete time counterpart of $h_{eff}(t)$ after the sampling operation becomes an FIR *multi-channel* $\mathbf{b}[l]$ or equivalently a *poly-phase* channel $\mathbf{b}_p[l]$ in the presence of multiple antennas and/or integer factor oversampling w.r.t. the chip rate as is shown in Figure 4.11. In such cases, the received *vector stationary*[6] signal can be modeled as the output of a $m \times 1$ single-input, multi-output (SIMO) system[7] with a past memory of $N - 1$ input

---

[6] Meaning each phase is stationary.

[7] Although stationarity holds only for time-invariant channels, we assume it also for the wireless channels considered in this text, which vary slowly.

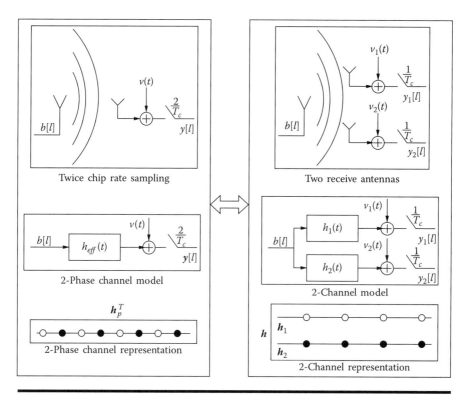

Figure 4.11 The equivalence of the poly-phase and the multi-channel models with a two-phase example. It is possible to pass from one form to the other by P/S and S/P operations.

elements and with the relations

$$y[l] = \sum_{i=0}^{N-1} b[i]b[l-i] + v[l] \tag{4.6}$$

$$y[l] = \begin{bmatrix} y_1[l] \\ \vdots \\ y_m[l] \end{bmatrix}, \quad b[l] = \begin{bmatrix} b_1[l] \\ \vdots \\ b_m[l] \end{bmatrix}, \quad v[l] = \begin{bmatrix} v_1[l] \\ \vdots \\ v_m[l] \end{bmatrix} \tag{4.7}$$

where $v[l]$ denotes the additive noise which represents the sum of the thermal noise and the inter-cell interference filtered by $p_r(t)$, and m denotes the product of the number of antennas and the oversampling factor. The multi-channel b spanning N chips with $m \times 1$ chip rate elements, its poly-phase equivalent[8] $\boldsymbol{b}_p$, the up-down flipped form $\overline{\boldsymbol{b}}_p$, and the poly-phase matched filter $\boldsymbol{b}_p^\dagger$ in row vector format, can be written as

$$\boldsymbol{b} = [\boldsymbol{b}[0], \boldsymbol{b}[1], \ldots, \boldsymbol{b}[N-1]] \tag{4.8}$$

$$\boldsymbol{b}_p = \begin{bmatrix} \boldsymbol{b}[0] \\ \boldsymbol{b}[1] \\ \vdots \\ \boldsymbol{b}[N-1] \end{bmatrix}, \quad \overline{\boldsymbol{b}}_p = \begin{bmatrix} \boldsymbol{b}[N-1] \\ \boldsymbol{b}[N-2] \\ \vdots \\ \boldsymbol{b}[0] \end{bmatrix}, \quad \boldsymbol{b}_p^\dagger = \overline{\boldsymbol{b}}_p^H \tag{4.9}$$

Assuming $Q-1$ interfering cells, we can write

$$\boldsymbol{y}[l] = \sum_{q=1}^{Q} \sum_{i=0}^{N-1} \boldsymbol{b}^{(q)}[i] b^{(q)}[l-i] + \boldsymbol{v}[l] \tag{4.10}$$

where index $q = 0$ denotes the desired BS.

4.4 Suppression of Intra-cell Interference in HSDPA

Given the channel model discussed above for a single-cell system, we are now ready to discuss some typical receivers for general downlink CDMA systems and, more specifically, HSDPA that are designed to suppress intra-cell interference.

4.4.1 RAKE Receiver and LMMSE Chip Equalizer

As shown in Figure 4.12, all the *linear* UE receivers can be mathematically represented in the form of a common chip-level filter followed by code-specific correlators.[9]

To motivate the discussion of this section, we consider detection of a *single* HSPDSCH user symbol $a_1[0]$ transmitted over the $L \times 1$ channelization code $\boldsymbol{c}_1 = \boldsymbol{c}_{16,0}$.

We consider a two-phase linear filter that has a length of N chips, which is the minimum length to deconvolve, that is, to zero force, a two-phase channel with a length of N chips.

We denote a block of the received signal as $\boldsymbol{Y}$ and denote a block of the *total* transmitted chip sequence as $\boldsymbol{B}$ whose elements are relevant

---

[8] Represented in column format for compatibility with later formulations.
[9] The order of the correlator and filtering can change, as in the conventional RAKE receiver.

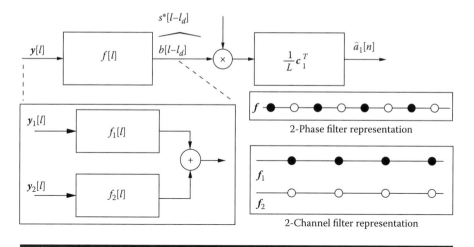

Figure 4.12 **Receivers with linear *chip-level filter/correlator* cascade. The order of the phases is reversed w.r.t. the channel phases order.**

to the estimation of the latter's subset $\boldsymbol{B}_0 = [b[L-1], \ldots, b[0]]^T$, which overlaps with the period of the $a_1[0]$ symbol. $\boldsymbol{Y}$ and $\boldsymbol{B}$ are related by the $2(L+N-1) \times (L+2N-2)$ channel convolution matrix $\boldsymbol{T}(\boldsymbol{b})$ with the term $\boldsymbol{V}$, which is modeled as an additive white Gaussian noise (AWGN) representing the sum of the thermal noise and the inter-cell interference.

$$\boldsymbol{Y} = \boldsymbol{T}(\boldsymbol{b})\boldsymbol{B} + \boldsymbol{V} \qquad (4.11)$$

$$= \begin{bmatrix} \boldsymbol{y}[L+N-2] \\ \vdots \\ \boldsymbol{y}[0] \end{bmatrix} \qquad (4.12)$$

$$\boldsymbol{T}(\boldsymbol{b}) = \begin{bmatrix} \boldsymbol{b}[0] & \ldots & \boldsymbol{b}[N-1] & \boldsymbol{0} & 0 \\ \boldsymbol{0} & \ddots & \ddots & \ddots & \boldsymbol{0} \\ 0 & 0 & \boldsymbol{b}[0] & \ldots & \boldsymbol{b}[N-1] \end{bmatrix}, \quad \boldsymbol{V} = \begin{bmatrix} \boldsymbol{v}[L+N-2] \\ \vdots \\ \boldsymbol{v}[0] \end{bmatrix} \qquad (4.13)$$

$$\boldsymbol{B} = \sum_{k=1}^{K} \begin{bmatrix} b_k[L+N-2] \ldots b_k[L-1] \ldots b_k[0] \ldots b_k[-N+1] \end{bmatrix}^T \qquad (4.14)$$

K is the number of codes. The linear filter $\boldsymbol{f}[l]$ is a 1×2 multi-input, single-output (MISO) system that turns the overall channel into a single-input, single-output (SISO) system $\tilde{g}[l]$.

$$f = [f[0], f[1], \ldots, f[N-1]], \quad f[l] = [f[0], f[1]], \quad \tilde{g}[l] = \sum_{i=0}^{l} f[i]b_p[l-i]$$

(4.15)

The estimated BS chip sequence $\boldsymbol{B}_0$ can be formulated by the following equations:

$$\widehat{\boldsymbol{B}}_0 = \boldsymbol{T}(f)\boldsymbol{Y} = \boldsymbol{T}(f)\boldsymbol{T}(b)\boldsymbol{B} + \boldsymbol{T}(f)\boldsymbol{V} = \boldsymbol{T}(g)\boldsymbol{B} + \tilde{\boldsymbol{V}}$$

(4.16)

$$\boldsymbol{T}(f) = \begin{bmatrix} f[0] \ldots f[N-1] & \boldsymbol{0} & & 0 \\ \boldsymbol{0} & \ddots & \ddots & \boldsymbol{0} \\ 0 & \boldsymbol{0} & f[0] & \ldots f[N-1] \end{bmatrix}, \quad \tilde{\boldsymbol{V}} = \begin{bmatrix} \tilde{v}[L-1] \\ \vdots \\ \tilde{v}[0] \end{bmatrix}$$

(4.17)

$$\boldsymbol{T}(g) =$$

$$\begin{bmatrix} g[-N+1]\,g[-N+2] & \cdots & g[0] & \cdots g[N-1] & \boldsymbol{0} & 0 \\ \boldsymbol{0} & \ddots & \ddots & \ddots & \ddots & \ddots & \boldsymbol{0} \\ 0 & \boldsymbol{0} & g[-N+1]\,g[-N+2] \cdots & g[0] & \cdots g[N-1] \end{bmatrix}$$

(4.18)

where $\boldsymbol{T}(f)$ denotes the $L \times 2(L+N-1)$ filter convolution matrix, $\boldsymbol{T}(g)$ denotes the $L \times (L+2N-2)$ overall channel convolution matrix, and $g[l] = \tilde{g}[l+N-1]$ reflects a variable change in order to better represent the precursor and postcursor parts of the overall channel, the central tap of which corresponds to $g[0] = f\bar{b}_p$. The channel matched filter (CMF) $f = \bar{b}_p^H$, which is an equivalent of the filtering part of the conventional RAKE receiver, maximizes the SNR, collecting all the channel energy to the central tap as $g[0] = \|\bar{b}_p\|^2 = \|b_p\|^2$. The unbiased form of CMF can be written as

$$f = \left(b_p^H b_p\right)^{-1} \bar{b}_p^H$$

Expanding Equation (4.16), $\widehat{\boldsymbol{B}}_0$ contains every user's chip sequences at different windows, each scaled by the associated tap of the overall channel

$$\widehat{\boldsymbol{B}}_0 = \sum_{i=-N+1}^{N-1} g[i] \underbrace{\sum_{k=1}^{K} \boldsymbol{B}_{k,i}}_{\boldsymbol{B}_i} + \widetilde{\boldsymbol{V}} \qquad (4.19)$$

where $\boldsymbol{B}_{k,i} = [b_k[i + L - 1], \ldots, b_k[i]]^T$ and $b_k[i] = a_k \lfloor \frac{i}{L_k} \rfloor c_k[mod(i, L_k)]s[i]$.

The second-stage correlation despreading part of the receiver can be written as

$$\widehat{a_1[0]} = \frac{1}{L} \boldsymbol{c}_1^T \boldsymbol{S}_0^* \widehat{\boldsymbol{B}}_0 \qquad (4.20)$$

where $\boldsymbol{S}_i$ denotes a diagonal matrix with the scrambling elements $[s[L - 1 + i], \ldots, s[i]]$ and the normalization by L is done to make up for the spreading–despreading gain.

Theorem 4.1 *The average SINR of the symbol estimate $\widehat{a_1[0]}$ by the usage of a general chip-level linear filter $\boldsymbol{f}$ is equal to*

$$\Gamma_1 = \frac{L|g(0)|^2 \sigma_{b_1}^2}{(\|\boldsymbol{g}\|^2 - |g(0)|^2)\sigma_b^2 + \|\boldsymbol{f}\|^2 \sigma_v^2} \qquad (4.21)$$

where L is the spreading factor of the first user, $\boldsymbol{g}$ is the impulse response of the channel filter cascade, $\sigma_{b_k}^2$ is the variance of the chips for user k, and $\sigma_b^2 = \sum_{k=1}^{K} \sigma_{b_k}^2$.

Proof. First we give a useful relation:

$$\frac{1}{L} \mathbb{E}\{\boldsymbol{c}_1^T \boldsymbol{S}_0^* \boldsymbol{S}_i \boldsymbol{c}_k^T\} = \begin{cases} 1 & i = 0, k = 1 \\ 0 & i = 0, k \neq 1 \\ \frac{1}{L} & i \neq 0 \end{cases} \qquad (4.22)$$

The symbol estimate can be partitioned into four groups:

$$\widehat{a_1[0]} = \underbrace{g(0)a_1[0]}_{\text{signal}} + \underbrace{\frac{1}{L}\boldsymbol{c}_1^T \boldsymbol{S}_0^* g[0] \sum_{k=2}^{K} \boldsymbol{B}_{k,0}}_{0} + \underbrace{\frac{1}{L}\boldsymbol{c}_1^T \boldsymbol{S}_0^* \left(\sum_{\substack{i=-N+1 \\ i \neq 0}}^{N-1} g[i] \sum_{k=1}^{K} \boldsymbol{B}_{k,i} \right)}_{\text{intracell interference}}$$

$$+ \underbrace{\frac{1}{L}\boldsymbol{c}_1^T \boldsymbol{S}_0^* \widetilde{\boldsymbol{V}}}_{\text{noise}} \qquad (4.23)$$

The first component, which represents the useful signal part, is the symbol of interest scaled by the central channel tap $g[0]$. The second component is zero because at the central tap instant $i = 0$, the scrambling and descrambling blocks are aligned, matching each other, $S_0^* S_0 = I_L$, and preserving the orthogonality among users:

$$c_1^T S_0^* B_{k,0} = c_1^T S_0^* S_0 c_k a_k[0] = c_1^T c_k a_k[0] = 0 \quad \forall k \neq 1 \tag{4.24}$$

The third *intra-cell* interference component represents the sum of ICI and MUI (multi-user interference) from the subcomponents with indices $k = 1$ and $k \neq 1$, respectively. The fourth component represents the noise contribution.

Taking the expected value of the symbol estimate power, we obtain

$$\mathbb{E}|\hat{a}_1[0]|^2 = |g(0)|^2 \sigma_{a_1}^2 + \frac{1}{L} \sum_{\substack{i=-N+1 \\ i \neq 0}}^{N-1} |g[i]|^2 \sum_{k=1}^{K} \sigma_{a_k}^2 + \frac{1}{L} \|f\|^2 \sigma_v^2 \tag{4.25}$$

where $\sigma_{a_k}^2$ represents the symbol variances, the noise power is amplified by the filter energy as is observed in the third component, and the cross terms in the second component disappear due to the expectation relation $\mathbb{E}\{a_k[0]\hat{a}_l^*[0]\} = 0$, $k \neq l$.

Using the equalities $\|f\|^2 \sigma_v^2 = f R_{vv} f^H$ and $\|g\|^2 = f T(b) T(b)^H f^H$, $\sigma_{b_k}^2 = \sigma_{a_k}^2$ due to the fact that the channelization codes are not normalized, we obtain

$$\mathbb{E}|\hat{a}_1[0]|^2 = |g(0)|^2 \sigma_{a_1}^2 + \frac{1}{L}(\|g\|^2 - |g(0)|^2)\sigma_b^2 + \frac{1}{L}\|f\|^2 \sigma_v^2$$

$$= |g(0)|^2 \sigma_{a_1}^2 + \frac{1}{L} f \underbrace{\left(\sigma_b^2 T(b) T(b)^H + R_{vv} \right)}_{R_{yy}} f^H - \frac{1}{L}|g(0)|^2 \sigma_b^2$$

Accordingly, we reach the SINR expression:

$$\Gamma_1 = \frac{|g(0)|^2 \sigma_{a_1}^2}{\frac{1}{L}(\|g\|^2 - |g(0)|^2)\sigma_b^2 + \frac{1}{L}\|f\|^2 \sigma_v^2} \tag{4.26}$$

$$= \frac{|g(0)|^2 \sigma_{b_1}^2}{\frac{1}{L}(\|g\|^2 - |g(0)|^2)\sigma_b^2 + \frac{1}{L}\|f\|^2 \sigma_v^2} \tag{4.27}$$

$$= \frac{L|g(0)|^2 \sigma_{b_1}^2}{f R_{yy} f^H - |g(0)|^2 \sigma_b^2} \tag{4.28}$$

□

Although the SINR expression is for symbol estimation, in reality the linear filter f estimates the BS chip sequence $b[l]$. Therefore, the modified SINR expression for the estimation of the BS chip sequence can be written as

$$\Gamma_c = \frac{|g(0)|^2 \sigma_b^2}{f R_{yy} f^H - |g(0)|^2 \sigma_b^2} \tag{4.29}$$

where at the numerator (i.e., the useful energy part), there is no spreading gain and $\sigma_{b_1}^2$ is replaced by σ_b^2.

The SINR metrics in Equations (4.26) and (4.29) are based on the estimation of R_{yy} statistics by taking expectation over the scrambler, which is modeled as a random sequence, and using the orthogonality property of the codes. The receiver that maximizes these SINR metrics is the Max-SINR receiver, which is more often known as the chip-level LMMSE (linear minimum mean square error) receiver [20,21].

Theorem 4.2 *The unbiased linear filter which maximizes SINR without exploiting the code and the power knowledge of the active users but by modeling the scrambling sequence as a random sequence and by taking expectations over it to approximate the received signal covariance matrix R_{yy} is equal to [21]*

$$f_o = \left(\overline{b}_p^H R_{yy}^{-1} \overline{b}_p\right)^{-1} \overline{b}_p^H R_{yy}^{-1} \tag{4.30}$$

Proof. We first define the unbiasedness constraint as $g[0] = f_o \overline{b}_p = 1$. Then the optimization problem can be formulated as

$$f_o = \arg_f \max_{f_o \overline{b}_p = 1} \Gamma_1 = \arg_f \min_{f_o \overline{b}_p = 1} f R_{yy} f^H \tag{4.31}$$

The solution can be obtained by the standard Lagrange multiplier technique as follows:

$$\Omega(f^H, f) = f R_{yy} f^H + 2\Re[\lambda(f \overline{b}_p - 1)]$$

$$\nabla_f \Omega(f^H, f) = f R_{yy} + \lambda \overline{b}_p^H$$

$$\Rightarrow f_o = -\lambda \overline{b}_p^H R_{yy}^{-1}$$

$$f_o \overline{b}_p = 1 \Rightarrow -\lambda \overline{b}_p^H R_{yy}^{-1} \overline{b}_p = 1$$

$$\Rightarrow \lambda = \frac{-1}{\overline{b}_p^H R_{yy}^{-1} \overline{b}_p}$$

$$\Rightarrow f_o = \frac{\overline{b}_p^H R_{yy}^{-1}}{\overline{b}_p^H R_{yy}^{-1} \overline{b}_p} = \frac{b_p^\dagger R_{yy}^{-1}}{b_p^\dagger R_{yy}^{-1} \overline{b}_p}$$

$\square$

By taking an approximation of Equation (4.30), we can obtain the biased (but simpler) chip-level LMMSE filter:

$$\tilde{f}_o = \sigma_b^2 b_p^\dagger R_{yy}^{-1} = R_{by} R_{yy}^{-1} \tag{4.32}$$

which fits the Wiener filtering format.

Similar to the update from Equations (4.6) to (4.10), if one has channel estimates of some other cells, then a better performing chip equalizer expression can be obtained as

$$\tilde{f}_o = \sigma_{b^{(0)}}^2 b_p^{(0)\dagger} \left(\sum_{q=0}^{Q-1} \sigma_{b^{(q)}}^2 T(b^{(q)}) T(b^{(q)})^H + R_{vv} \right)^{-1} \tag{4.33}$$

by modifying the R_{yy} term.[10]

4.4.2 HSDPA Performance Analysis of RAKE Receiver and Chip Equalizers

In this section we obtain the maximum achievable SINR and throughput performance metrics for various HSDPA service deployment scenarios while using the CMF and LMMSE equalizer-type UE receivers. The distributions of the radio channel parameters and the received powers from the own and surrounding base stations are modeled under correlated shadowing w.r.t. the mobile position, the cell radius, and the type of environment. From such modeling, more realistic performance figures might be obtained as compared to fixing them to a selected set of values.

---

[10] This equalizer has the inter-cell interference suppression capability, which will be further elaborated in later parts of this chapter.

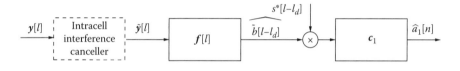

Figure 4.13   Hypothetical receiver model.

4.4.2.1 Hypothetical Receiver Models

We consider that possibly an interfering cancelling (IC) structure is used in the first stage to cancel out intra-cell interference contributions of the PCPICH and HSPDSCHs (Figure 4.13).

We assume that the residual BS signal $\tilde{b}[l]$ contained in the remaining sequence $\tilde{y}[l]$ after the IC block is still block stationary and the second-order inter-cell interference plus noise statistics σ_v^2 are the same as before.

The modified SINR expression at the unbiased linear filter output (i.e., when $g[0] = 1$), for the symbol estimates of a single HSPDSCH channel of a UE situated at a particular position of the cell can be written as

$$\Gamma_1 = \frac{16\rho P_0}{(\frac{1}{\alpha} - 1)\chi P_0 + \sum_{i=1}^{6} P_i \|\mathbf{g}_i\|^2 + \|\mathbf{f}\|^2 \sigma_n^2} \tag{4.34}$$

where 16 is the HSPDSCH spreading gain, P_0 is the received power of the desired BS (BS 1), there are K_1 HSPDSCH codes, ρ is the BS signal power portion of one HSPDSCH channel, χ is the remaining BS signal power portion after the IC block, P_i is the received powers from the ith cell among the six first-tier interfering unsectored cells[11] shown in Figure 4.1, $\mathbf{g}_i$ is the convolution of the linear filter $\mathbf{f}$ and the channel $\mathbf{b}_i$ originating from the ith surrounding cell as $\mathbf{g}_i = \mathbf{f} * \mathbf{b}_i$, and σ_n^2 is the AWGN variance. The AWGN term and inter-cell interference are treated separately for performance analysis purposes, whereas they are treated similarly for filter adaptation due to the fact that it is impractical to incorporate in the signal model channel estimates and signal variances for a large number (six here) of neighboring cells. As shown in the example of channel and CMF impulse responses in Figure 4.14, the term $\alpha = \frac{1}{\|\mathbf{g}\|^2}$ represents the ratio of the useful effective channel energy to the total effective channel energy and is known as the *orthogonality factor*, which was previously treated in the literature only for the RAKE receiver variants [25, 29].

---

[11] The analysis done in this chapter is also valid for the sectored case.

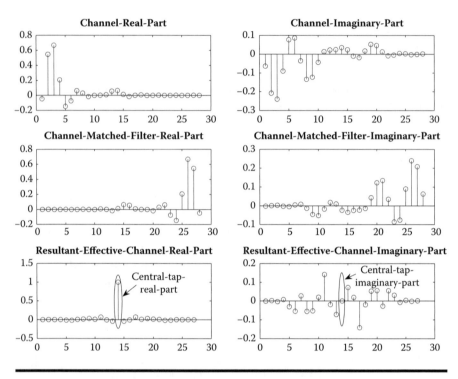

Figure 4.14 Orthogonality factor representation over an unbiased CMF. The central tap of the effective channel collects all the energy, which is 1 due to unbiasedness. The cumulative energy of all the taps is $\|g\|^2$.

4.4.2.2 Parameter Modeling

We model all the parameters that appear in the SINR expression and that implicitly or explicitly depend on one or more of the locations of the UE in the cell, the radius (r) of the cell, and the type of environment.

4.4.2.2.1 Modeling Received Powers

Received powers are calculated by the path loss and shadowing computations discussed previously. For shadowing we first randomly generate a vector of seven independent shadowing values $\tilde{x}$ of the own and first-tier six cells and turn it into a cross-correlated vector x by left multiplying with the lower triangular Cholesky factorization output matrix L_x of the symmetric shadowing correlation matrix R_{xx} whose elements $\rho_{x_i x_j}$ given in Table 4.1 are obtained from the distance ratio dr_{ij} and the angle values θ_{ij}

Table 4.1 Shadowing Correlation Matrix Elements

	$0 < \theta_{ij} < 30°$	$30° \leq \theta_{ij} < 60°$	$60° \leq \theta_{ij} < 90°$	$90° \leq \theta_{ij}$
$dr_{ij} \in [0, 2]$	$\rho_{x_i x_j} = 0.8$	$\rho_{x_i x_j} = 0.5$	$\rho_{x_i x_j} = 0.4$	$\rho_{x_i x_j} = 0.2$
$dr_{ij} \in [2, 4]$	$\rho_{x_i x_j} = 0.6$	$\rho_{x_i x_j} = 0.4$	$\rho_{x_i x_j} = 0.4$	$\rho_{x_i x_j} = 0.2$
$dr_{ij} \geq 4$	$\rho_{x_i x_j} = 0.4$	$\rho_{x_i x_j} = 0.2$	$\rho_{x_i x_j} = 0.2$	$\rho_{x_i x_j} = 0.2$

between the corresponding couples among the seven BSs and the UE as shown in Figure 4.15 [46].

$$\boldsymbol{R_{xx}} = \begin{bmatrix} \rho_{x_0,x_0} & \rho_{x_0,x_1} & \cdots & \rho_{x_0,x_6} \\ \rho_{x_1,x_0} & \rho_{x_1,x_1} & \cdots & \rho_{x_1,x_6} \\ \vdots & \vdots & \vdots & \vdots \\ \rho_{x_6,x_0} & \rho_{x_6,x_1} & \cdots & \rho_{x_6,x_6} \end{bmatrix} = \boldsymbol{L_x L_x}^T, \quad \boldsymbol{x} = \boldsymbol{L_x \tilde{x}} \qquad (4.35)$$

4.4.2.2.2 Modeling Channel Parameters

The linear filter $\boldsymbol{f}$, the orthogonality factor α, and the $\boldsymbol{g}_i$ terms depend on the channel parameters, for which we refer to Greenstein's channel model derived from the rms delay spread σ_{τ_i} and the power delay profile $P(\tau_i)$ [16]. Delay spread is equal to $\sigma_{\tau_i} = T_1 d^\epsilon y_i$, where T_1 is the reference delay spread at 1-km distance from the BS; ϵ is the model parameter, which is around 0.5 for almost all types of environments except very irregular mountainous terrains; and y_i is a coefficient, which is log-normally distributed with geometric mean 0 and geometric standard deviation σ_{y_i}. From field tests, $\log(y_i)$ is also observed to be correlated with $\log(x_i)$ by a factor $\rho_{x_i y_i} = -0.75$ [16]. So we obtain the value of y_i from the correlation with the obtained shadowing value in the previous section. From the obtained delay spread we generate the power delay profile as $P(\tau_i) \propto e^{-\tau_i/\sigma_{\tau_i}}$ where $\propto$ is

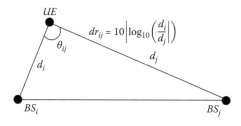

Figure 4.15 Distance and angle relations between two BSs and a UE.

the proportionality sign and τ_i values are the sampling instants. Because this is an infinite-length sequence, we truncate it at the position where the final significant tap has 15 dB less power than the first tap. Then we pass the discrete power delay profile through *Rayleigh* fading to generate the *propagation* channel. The *transmission* channels are obtained by convolving the obtained propagation channels with the pulse shape and normalizing the result to unit energy.

4.4.2.2.3 Modeling χ

Among the common downlink channels, the pilot tone PCPICH has the highest interference with 10% BS power portion and it can be cancelled with high accuracy [35]. However, it might be even more meaningful to consider cancelling the interference of HSPDSCH codes because, by a highly probable deployment scenario, they will carry the majority (if not all) of the data traffic. We contemplate this because it would be easier to manage for an operator to dedicate one of its two or three carriers of 5 MHz to the HSDPA service instead of distributing it over two or three available carriers. Furthermore, there is no justified advantage of carrying high-rate data on a DCH with a very low spreading factor instead of on multiple HSPDSCHs. So, in the reception chain for a single HSPDSCH, we define five *perfect* first-stage interference cancellation (IC) scenarios:

1. No interference canceller exists: $\chi = 1$
2. Pilot tone cancelled: $\chi = 0.9$
3. All the other HS-PDSCHs cancelled: $\chi = 1 - (K_1 - 1)\rho$
4. Pilot+HS-PDSCHs cancelled: $\chi = 0.9 - (K_1 - 1)\rho$
5. All intra-cell interference cancelled: $\chi = \rho$

4.4.2.3 Simulations Results

Five different environments are considered, the relevant parameters of which are shown in Table 4.2 and adopted from COST231 propagation

Table 4.2 Cellular Deployment Scenarios

Parameters	G_1	n	r	T_1	σ_x	σ_y
Indoor	138	2.6	0.2	0.4	12	2
Urbanmicro	131	3	0.5	0.4	10	3
Urbanmacro	139.5	3.5	1	0.7	8	4
Suburbanmacro	136.5	3.5	2	0.3	8	5
Rural	136.5	3.85	8	0.1	6	6

models [34] and from [16]. We fix the transmitted BS power and AWGN power to $P_t = 43$ dBm and $\sigma_n^2 = -102$ dBm, respectively. Three different low-end to high-end HSDPA service scenarios are considered with $\{K_1, \rho\}$ set as $\{1, 0.1\}$, $\{5, 0.06\}$, and $\{10, 0.06\}$. We uniformly position 10^4 UEs in one cell (home cell) and approximately that many more in an expanded region penetrating into other cells. This models the effect of shadowing, resulting in handing off some UEs to other base stations depite their relatively greater distance as compared to the BS closest to the user. We also exclude a few closest points to the BS because otherwise, the BS will serve only the closest UEs. For each node receiving the highest power from the BS of interest, we determine the relevant second-order statistics over ten Rayleigh fading channel realizations. At each realization we obtain the SINR and spectral efficiency bound $\mathcal{C} = \log_2(1 + \Gamma_1)$ results for the CMF and LMMSE equalizer-correlator type receivers under the five above-mentioned interference cancellation scenarios. It was shown that the interference at the output of multi-user detectors can be approximated by a Gaussian distribution [30,47]. Hence, $\mathcal{C}$ is an approximate Shannon capacity and is a more meaningful measure than SINR because it defines the overall performance bound that can be achieved by the use of efficient transmission diversity, modulation, and channel coding schemes. A number of spatiotemporal results on the order of 10^5 suffices to obtain the distribution of $\mathcal{C}$. Cumulative distributions of $\mathcal{C}$ for ten HSPDSCH codes deployment in the five reference environments are shown in Figures 4.16 through 4.20. The calculated median values of $\mathcal{C}$ for all the settings defined are tabulated in Table 4.3.

In the figures and the table, C represents CMF; E represents LMMSE equalizer-correlator receiver; suffixes to C and E ($\{1, \ldots, 5\}$) represent (in the same order) the IC scenarios defined in section 4.4.2.2.3; {ind, umi, uma, sub} represent {indoor, urban microcell, urban macrocell, suburban macrocell} environments; and the suffixes $\{1, 5, 10\}$ to these environments represent K_1.

As observed in the figures, an increasing gap occurs between matched filtering results and equalization results when we go to user locations closer to the own BS, which correspond to higher SINR regions. This is especially the valid case for indoor cells, urban microcells, and urban macrocells, where the eye is open for all user locations because white noise (thermal noise and partially inter-cell interference) suppressing CMF is much more affected by intra-cell interference, most of which, however, is suppressed when an equalizer is used and the need for an IC decreases. In other words, in such environments, orthogonality factors at the output of LMMSE equalizers are much higher than those of CMFs.

In the suburban macrocell sizes, for the most distant 30% cellular positions, there is no difference in the performance of receivers. When we go farther to the extreme rural cell sizes, there is almost no difference except at a small number of very close UE positions. These figures clearly show

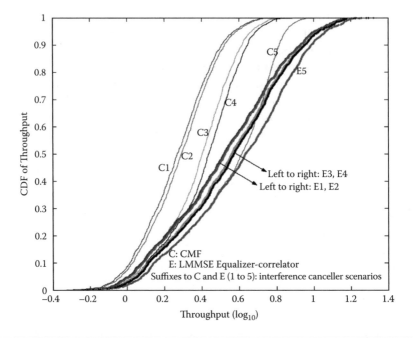

Figure 4.16 Throughput bound CDF of indoor microcell. Number of codes = 10; total HSDPA power portion = 0.6.

the dominance of multi-user interference in small cells where using interference suppressing equalizers becomes meaningful and the dominance of AWGN in the large cells where CMF or RAKE receiver is sufficient.

According to UMTS deployment scenarios, more than 80% of UMTS cells will be pico- or microcells, and hence it will certainly pay off if a UE considers the LMMSE equalizer in order to be scheduled for a high-SINR-demanding HSDPA service. In these settings, the achievable *maximum* $\mathcal{C}$ using equalizers is approximately twice that of CMFs. So, under ideal conditions, CMF has less chance of providing a very high-rate-demanding application.

In Table 4.3, we notice that when an equalizer is used, the median capacity for a UE increases when we move from indoor to urban environments, which is mostly because of the trend of the path-loss exponent. When it is low, inter-cell interference will be high. However, as we further increase the size of the cell, AWGN starts to dominate and median capacity decreases. We also see that w.r.t. CMF, equalizers alone improve the median capacity of pico- and microcells between 60% and 115%. When

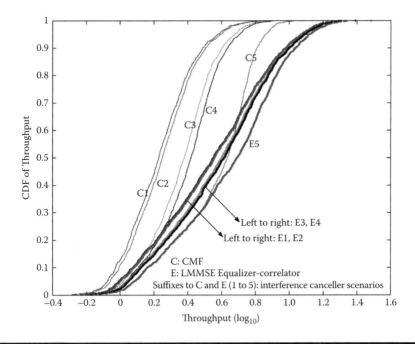

Figure 4.17   Throughput bound CDF of urban microcell. Number of codes = 10; total HSDPA power portion = 0.6.

complete interference cancellation is achieved, these figures increase to 98% and 199%.

Cancelling the pilot tone alone brings very little gain. Moshavi et al., however, claim that it is possible to obtain 11% capacity gain by cancelling the 10% power pilot tone because this much cancelled power can be exploited by the BS to accept a proportional number of new users [36]. This can be valid only if all the UE receivers at the same time cancel the pilot tone, which is not dictated for the moment by the standard. Nevertheless, when equalization is used, it is more worthwhile to subtract known nonorthogonal channels (e.g., the synchronization channels). We, however, will not discuss this aspect any further due to space limitations.

Note that the results obtained are valid when there is no LOS and surrounding cells have identical properties. In reality, we expect higher capacity from picocellular regions because there will be some isolated hot zones like airports and there will be a higher probability of LOS. Furthermore, note that the capacity we are concerned with here is the single-cell capacity. Of course, global system capacity from the adoption of picocells will

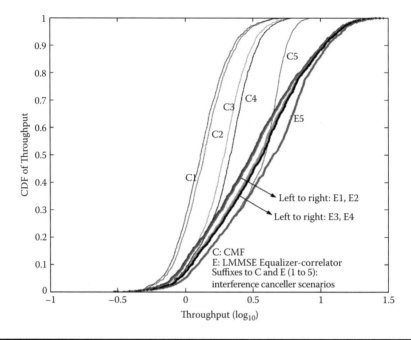

Figure 4.18 Throughput bound CDF of urban macrocell. Number of codes = 10; total HSDPA power portion = 0.6.

be much higher than others because there will be more cells and hence more users will be served.

4.5 Advanced Receivers for Interference Cancellation

Chip-level equalization happens to be a solution of interest for the very specific case of downlink synchronous CDMA. When multi-user interference is of a more general nature (e.g., due to asynchronous or non-orthogonal downlink codes), the solutions for its suppression are equivalently more general and are labeled *multi-user* receivers or detectors. We discuss some well-documented multi-user receivers in this section and later apply them to the specific case of suppression of inter-cell interference in the HSDPA downlink.

4.5.1 Symbol-Rate Signal Model

The signal model in Equation (4.10) is a chip-rate model where $b^{(q)}[i]$ is the ith chip from the qth base station. We can write the equivalent discrete-time

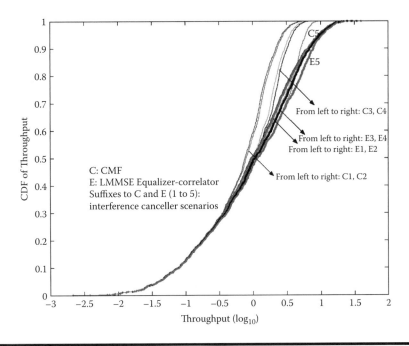

Figure 4.19 Throughput bound CDF of suburban macrocell. Number of codes = 10; total HSDPA power portion = 0.6.

signal model at the symbol-rate as

$$Y[n] = \sum_{q=0}^{Q-1} \overbrace{T(b^{(q)})S^{(q)}[n]C}^{\tilde{G}^{(q)}[n]} A^{(q)}[n] + V[n] \tag{4.36}$$

where K is the number of users assumed without loss of generality the same from all base stations, $Y[n]$ represents the received data block spanning the channel output corresponding to transmission of M symbols at time instant n, $\tilde{G}^{(q)}[n]$ is the symbol-rate channel[12] from base station q composed of the cascade of the propagation channel $T(b^{(q)})$, the diagonal scrambling code matrix $S^{(q)}[n]$, and the block-diagonal channelization code matrix C, assumed to be the same for all base stations. $A^{(q)}[n]$ represents the unit-amplitude MK desired symbols vector. Because the nature of intra-cell and inter-cell interference is the same (non-orthogonal), we assume, in the interest of clarity, only one base station $Q = 1$ and suppress the index q for the purposes of the following discussion.

---

[12] Time-varying because it includes the scrambler.

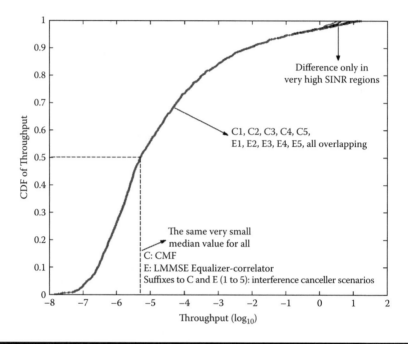

Figure 4.20 Throughput bound CDF of rural cell. Number of codes = 10; total HSDPA power portion = 0.6.

4.5.2 Optimal Receiver

The optimal multi-user detector in terms of minimum symbol error rate (SER) is the Maximum Likelihood Sequence Estimation (MLSE), which is an exhaustive search procedure over the symbol alphabets of all the possible transmitted sequences of all the users with the minimization criterion [22]

$$\hat{A}_{ML} = \arg\min_{A \in \mathcal{X}^{MK}} \left\| Y - \overbrace{T(b)SC}^{\tilde{G}} A \right\|^2 \qquad (4.37)$$

where $\mathcal{X}$ denotes the symbol alphabet.[13] Because this criterion is finite-alphabet constrained, it is NP-hard[14] and perhaps the best that can be done is to use the Viterbi algorithm, which is also exponentially complex with

---

[13] Representing the simple case of same constellation for all users.

[14] A decision problem that is at least as hard as any problem whose solution can be *verified* by polynomial complexity.

Table 4.3 Throughput Bound Median Results

$\mathcal{C}$	C1	C2	C3	C4	C5	E1	E2	E3	E4	E5
ind1	2.46	2.54	2.46	2.54	4.54	3.94	3.99	3.94	3.99	4.87
ind5	1.86	1.94	2.09	2.21	4.13	3.22	3.27	3.33	3.42	4.34
ind10	1.86	1.96	2.53	2.73	4.11	3.31	3.35	3.63	3.73	4.30
umi1	2.18	2.29	2.18	2.29	4.54	4.16	4.20	4.16	4.20	5.30
umi5	1.65	1.74	1.89	2.02	4.34	3.56	3.59	3.71	3.77	4.94
umi10	1.70	1.79	2.41	2.63	4.32	3.59	3.63	4.00	4.16	4.95
uma1	1.78	1.88	1.78	1.88	4.13	3.80	3.86	3.80	3.86	5.16
uma5	1.30	1.37	1.50	1.62	3.97	3.09	3.14	3.26	3.35	4.72
uma10	1.30	1.38	1.95	2.15	3.91	3.10	3.17	3.58	3.74	4.50
sub1	1.11	1.17	1.11	1.17	1.60	1.40	1.42	1.40	1.42	1.61
sub5	0.79	0.82	0.87	0.91	1.19	1.02	1.03	1.05	1.06	1.19
sub10	0.77	0.81	0.97	1.02	1.18	0.98	1.00	1.07	1.09	1.18

the factor MK [44]. Due to these reasons, the ML (Maximum Likelihood) receiver is mostly considered not implementable and its performance serves only as an upper bound. Furthermore, in case of signals from multiple base stations, ML detection of all symbols from all base stations must be considered, making the solution even more undesirable in the downlink problem. Note that Equation (4.37) is at the same time the nonlinear LS estimator.[15]

4.5.3 Decorrelating Receiver

One of the suboptimal but simpler approaches is to relax the finite-alphabet constraint mapping of A from the finite set $\mathcal{X}^{MK}$ into $\mathbb{C}^{MK}$, which turns the nonlinear LS problem in Equation (4.37) into a linear LS problem:

$$\hat{A}_{LS} = \arg \min_{A \in \mathbb{C}^{MK}} \|Y - \tilde{G}A\|^2 \tag{4.38}$$

---

[15] In statistical terms, LS is *disguised* ML when the measurement noise sequence V is zero-mean, *i.i.d.*, and Gaussian.

whose solution is

$$\hat{A}_{LS} = \tilde{F}_{Dec}Y = \left(\overbrace{\tilde{G}^H\tilde{G}}^{R}\right)^{-1}\tilde{G}^HY = R^{-1}X \qquad (4.39)$$

where X and R, respectively, denote the single-user, matched-filter (SUMF) bank output symbol estimates and their cross-correlation matrix.

An equivalent model in terms of linear systems of equations can be written as

$$R\hat{A}_{LS} = X \qquad (4.40)$$

Note that the LS estimator treats the elements of A vector as deterministic unknown parameters having diffuse prior pdfs.

LS estimation (i.e., decorrelation) is the unique least MSE member of the ZF (zero forcing) MUD (multi-user diversity gains, thus called MMSE-ZF receiver) set with the general members expressed as

$$\hat{A}_{ZF} = (T\tilde{G})^{-1}TY$$

with any proper T matrix.

4.5.3.1 Projection Interpretation of Decorrelating Receiver

Projection receiver is another name for the symbol-level MMSE-ZF. It was first presented as the projection receiver for CDMA communications by Schlegel et al. [37] and is based on suppressing both inter-cell and intra-cell interference by projecting the undesired users onto the subspace orthogonal to the one spanned by the desired user's signal vector. We propose different approaches for both exact and approximated interference projection. Considering the symbol-level model,

$$Y[n] = \tilde{G}[n]A[n] + V[n] \qquad (4.41)$$

We call $d[n]$ the symbol to estimate, $\tilde{g}[n]$ the column of $\tilde{G}[n]$ corresponding to that symbol, and $\bar{G}[n]$ all the other columns of $\tilde{G}[n]$. Inter-symbol interference, inter-cell interference, and intra-cell interference are all included in $\bar{G}[n]$. We can write the received signal as

$$Y = \tilde{g}[n]d[n] + \bar{G}[n]\bar{A}[n] + V[n] \qquad (4.42)$$

From a geometrical point of view, the columns of $\bar{G}[n]$ span a certain subspace, the *interference subspace*. $S = \text{span}(\bar{G}[n])$, where $\text{span}(\cdot)$ generates all possible linear combinations of the vectors inside the brackets. We define the projection matrix:

$$\boldsymbol{P}_{\bar{\boldsymbol{G}}[n]} = \bar{\boldsymbol{G}}[n](\bar{\boldsymbol{G}}^H[n]\bar{\boldsymbol{G}}[n])^{-1}\bar{\boldsymbol{G}}^H[n] \qquad (4.43)$$

being the unique orthogonal projection onto $\mathcal{S}$; that is, for any $\boldsymbol{x} \in \mathbb{C}^{\mathcal{D}}$, then $\boldsymbol{P}_{\bar{\boldsymbol{G}}[n]}\boldsymbol{x} \in \mathcal{S}$ and $\|\boldsymbol{x} - \boldsymbol{P}_{\bar{\boldsymbol{G}}[n]}\boldsymbol{x}\|^2$ is minimal. $\mathcal{D}$ is the dimension of the vector $\boldsymbol{x}$. Further, we define

$$\boldsymbol{P}_{\bar{\boldsymbol{G}}[n]}^{\perp} = \boldsymbol{I} - \boldsymbol{P}_{\bar{\boldsymbol{G}}[n]} = \boldsymbol{I} - \bar{\boldsymbol{G}}[n](\bar{\boldsymbol{G}}^H[n]\bar{\boldsymbol{G}}[n])^{-1}\bar{\boldsymbol{G}}[n]^H \qquad (4.44)$$

which is the projection matrix on $\mathcal{S}^{\perp}$ (the orthogonal complement of $\mathcal{S}$), so that for any $\boldsymbol{x} \in \mathbb{C}^{\mathcal{D}}$, then $\boldsymbol{P}_{\bar{\boldsymbol{G}}[n]}^{\perp}\boldsymbol{x} \in \mathcal{S}^{\perp}$.

As the vector obtained in this way has no components in the interference subspace, a simple matched filter ($\tilde{\boldsymbol{g}}^H[n]$) suffices to retrieve the transmitted symbol $d[n]$. Up to a proper scalar scaling factor, the result obtained may be shown to be equivalent to the result obtained by a symbol-level MMSE-ZF equalizer:

$$\hat{d}[n] = \frac{1}{\tilde{\boldsymbol{g}}^H[n]\boldsymbol{P}_{\bar{\boldsymbol{G}}[n]}^{\perp}\tilde{\boldsymbol{g}}[n]}\tilde{\boldsymbol{g}}^H[n]\boldsymbol{P}_{\bar{\boldsymbol{G}}[n]}^{\perp}\boldsymbol{Y} \qquad (4.45)$$

while the classical expression for the MMSE-ZF filter would be

$$\hat{d}^{ZF}[n] = \boldsymbol{e}_n^T(\tilde{\boldsymbol{G}}^H[n]\tilde{\boldsymbol{G}}[n])^{-1}\tilde{\boldsymbol{G}}^H[n]\,\boldsymbol{Y} \qquad (4.46)$$

where $\boldsymbol{e}_n$ is a unit vector.

Notice that both the symbol-level MMSE-ZF receiver and the projection receiver need the matrix $\tilde{\boldsymbol{G}}[n]$ (or $\bar{\boldsymbol{G}}$) to be *tall* to have enough degrees of freedom for the inversion. In general, if the number of interferers is not too large (intra-cell codes as well as inter-cell interfernce), the model in Equation (4.41) allows leveraging on the stacking factor (increasing M) to arbitrarily increase the number of rows in the matrix.

4.5.3.1.1 Successive Projection Algorithms

Most of the computational complexity of the projection receiver lies in the inversion of the Grammian term $\bar{\boldsymbol{G}}^H[n]\bar{\boldsymbol{G}}[n]$, which is a square matrix whose dimension is equal to the total number of interfering columns, J.

To avoid performing the inverse, one simplification can be represented by the idea of projecting the received signal successively on each one of the columns that compose $\bar{\boldsymbol{G}}[n]$, $\bar{\boldsymbol{g}}_1[n]$, until $\bar{\boldsymbol{g}}_J[n]$. Doing it vector by vector, the Grammian terms result in a scalar (the squared norm of the considered vector) and the inversion becomes simply a division by a scalar.

The resulting iterative algorithm is the following:

$$\mathbf{Y}^{(0)} = \mathbf{Y}[n]$$

$$\text{for } i = 1 : J \rightarrow \mathbf{Y}^{(i)} = \mathbf{Y}^{(i-1)} - \frac{1}{\|\bar{\mathbf{g}}_i[n]\|^2} \bar{\mathbf{g}}_i[n] \bar{\mathbf{g}}_i^H[n] \mathbf{Y}^{(i-1)}$$

$$\hat{a}_n = \tilde{\mathbf{g}}^H[n] \mathbf{Y}^{(J)} \qquad (4.47)$$

This algorithm allows a considerable reduction in complexity compared to the full matrix projection. However, projecting vector by vector separately like this is an *approximation* of the full matrix projection exposed above. The two methods would be equivalent *if and only if* the columns $\tilde{\mathbf{g}}_i[n]$ were *orthogonal*. This condition can be achieved by applying a prior orthonormalization process onto the considered vectors, via the Gram–Schmidt procedure or an equivalent one. The one implemented here is a modified version of the Gram–Schmidt procedure, in which the computation of square roots is avoided (square roots have a consistent computational complexity). In fact, for our purposes, the normalization of the basis is not needed. In this way a matrix **U** is computed, whose columns form an orthogonal basis for the interference subspace $\mathcal{S}$, and it is on these vectors that the projections are done. The orthogonalization process is provided as follows:

$$\mathbf{u}_1 = \bar{\mathbf{g}}_1[n]$$

$$\text{for } i = 2 : J \rightarrow \mathbf{u}_i = \bar{\mathbf{g}}_i[n] - \sum_{k=1}^{i-1} c_{k,i} \mathbf{u}_i$$

$$\text{where } c_{k,i} = \langle \mathbf{u}_k, \bar{\mathbf{g}}_i[n] \rangle \qquad (4.48)$$

If one basis vector $\mathbf{u}_i$ is too small (norm lower than a certain value ε), it is discarded. This means that the rank of the interference matrix $\bar{\mathbf{G}}[n]$ is lower than J. Once all the vectors of the basis have been computed, the projection is done on these vectors exactly as in the previous case (Equation (4.47)).

The orthogonalization process allows one to have an *exact* projection algorithm even if performed vector by vector. However, it introduces an additional complexity compared to the direct projection on the columns $\bar{\mathbf{G}}[n]$. So both approaches are reasonable, depending on which trade-off between performances and complexity one wants to achieve. The loss in performance deriving from the first approach depends mainly on two factors:

1. How *sparse* the matrix $\bar{\mathbf{G}}[n]$ is
2. How *tall* the matrix $\bar{\mathbf{G}}[n]$ is

In fact, the more the matrix tends to have such characteristics, the more the inner product between pairs of its columns tends to be small and therefore the condition of orthogonality is approached. Often in practical cases this situation is verified (small number of codes and small number of interfering base stations to cancel) and thus there is no particular need for a prior orthogonalization process because the gap in performance is not enormous.

4.5.3.1.2 Projections on Reduced Subspaces

To reduce the complexity further, one can exploit the fact that not all the interfering vectors $bm\bar{g}_i[n]$ contribute in an equal way to the interference. So the number of vectors to consider can be reduced from the full number J down to a certain subset of *strongest* interferers.

One criterion for the selection of these vectors can be represented by their *inner product* with the column $\tilde{g}[n]$; in this case, we fix a certain threshold ϑ. If the inner product between the columns $\bar{g}_i[n]$ and $\tilde{g}[n]$ is greater than ϑ, the columns will be included in the projection subset, otherwise it will be discarded. The proposed algorithm behaves as follows:

$$\mathbf{Y}^{(0)} = \mathrm{Y}[n]$$
$$\text{for } i = 1 : J$$
$$\quad \text{if } \langle \bar{g}_i[n], \tilde{g}[n] \rangle \geq \vartheta \rightarrow \mathbf{Y}^{(i)} = \mathbf{Y}^{(i-1)} - \frac{1}{\|\bar{g}_i[n]\|^2} \bar{g}_i[n] \bar{g}_i^H[n] \mathbf{Y}^{(i-1)}$$
$$\text{end}$$
$$\hat{a}_n = \tilde{g}[n]^H \mathbf{Y}^{(J)} \tag{4.49}$$

Simulation results show that a proper choice for ϑ can be around 0.2. Furthermore, if the number of selected columns is not very high on average, it is also possible to make the projection on the whole matrix formed by these vectors, which will give better performances.

4.5.4 LMMSE Receiver

Although complete deconvolution is possible for $\tilde{G}$ with the decorrelator $\tilde{F}_{Dec}$, it amplifies the noise term $\mathbf{V}$. A better approach is the LMMSE estimator, which models $\mathbf{A}$ as a random Gaussian vector and solves the cost criterion

$$\tilde{F}_{LMMSE} = \arg_{\tilde{F}} \min_{\mathbf{A} \in \mathbb{C}^{MK}} \mathbb{E}(\tilde{F}\mathbf{Y} - \mathbf{A})(\tilde{F}\mathbf{Y} - \mathbf{A})^H \tag{4.50}$$

with the solution

$$\tilde{F}_{LMMSE} = \left(\tilde{G}^H \tilde{G} + R_{VV}\right)^{-1} \tilde{G}^H \qquad (4.51)$$

which, different from the decorrelator, requires also the noise covariance matrix symbol amplitudes R_{VV}. Note that for vanishing noise, $\tilde{F}_{LMMSE}$ becomes equivalent to the decorrelator. For high noise, on the other hand, it is identical to the single-user matched filter (SUMF).

The equivalent model in terms of linear systems of equations can be written as

$$\underbrace{(R + R_{VV})}_{T} \hat{A}_{LMMSE} = X \qquad (4.52)$$

where T denotes the SUMF bank output covariance matrix. Both decorrelator and LMMSE receiver are very complex due to the fact they require matrix-inversion operations with $O(M^3 K^3)$ complexity. Therefore, reduced rank approximations of the matrix inversion operation have been investigated in the literature with iterative techniques. We elaborate on only the so-called parallel interference cancellation (PIC) family, which is the counterpart of Jacobi iterations for the iterative solutions of linear systems of equations because it works particularly well when user symbols have similar power levels, which is the case for HSDPA. For other state-of-the-art iterative techniques that are not discussed here, such as successive interference cancellation (SIC), which is the counterpart of Gauss–Seidel iterations in matrix algebra or the decorrelating decision feedback equalizer (DFE), see [43] and [11].

4.5.5 Linear Parallel Interference Cancellation (LPIC) Receiver

Conventional LPIC corresponds to using Jacobi iterations for the solutions of linear systems of equations [40]. Splitting the R expression in Equation (4.40) into the two parts as I and $(R - I)$, one can obtain the iterative decorrelation solution as[16]

$$\hat{A}_{LS}^{(i)} = (I - R)\hat{A}_{LS}^{(i-1)} + X \qquad (4.53)$$

The iterations converge provided that the spectral radius $\rho(I - R)$ is less than 2, which is not guaranteed.[17]

---

[16] Similarly splitting T in Equation (4.52) for the LMMSE receiver.

[17] $\rho(X) = \max\{|\lambda|, \lambda \in \Lambda(X)\}$, where $\Lambda(X)$ is the eigenvalue matrix of X.

A better approach is to tackle the problem from the Cayley–Hamilton theorem, which states that every square matrix satisfies its characteristic equation. This principle can be used to find the inverse of an $n \times n$ square matrix by a polynomial expansion as [9]

$$det(\boldsymbol{R} - \lambda \boldsymbol{I}) = 0$$

$$\Rightarrow 1 - c_1\lambda - \cdots - c_{n-1}\lambda^{n-1} - c_n\lambda^n = 0$$

$$\Rightarrow \boldsymbol{I} - c_1\boldsymbol{R} \cdots - c_{n-1}\boldsymbol{R}^{n-1} - c_n\boldsymbol{R}^n = \boldsymbol{0}$$

$$\Rightarrow \boldsymbol{I} = c_1\boldsymbol{R} + \cdots + c_{n-1}\boldsymbol{R}^{n-1} + c_n\boldsymbol{R}^n$$

$$\Rightarrow \boldsymbol{R}^{-1} = c_1\boldsymbol{I} \cdots + c_{n-1}\boldsymbol{R}^{n-2} + c_n\boldsymbol{R}^{n-1}$$

With polynomial expansion it is possible to obtain the decorrelator solution or the LMMSE solution in n iterations. Suboptimal solutions are obtained by stopping at a few iterations, in which case the optimal weights change as well. Although this looks like an attractive solution at first sight, the complexity depends on the weight adaptation. See [26] and [27] for two adaptation schemes, one from the direct derivation from the MMSE expression for a particular number of iterations and one from large system analysis. In this text we are not concerned with weight adaptation, but instead with filter adaptation.

4.5.6 Iterative Receivers Based on Chip Equalizers

The LMMSE chip equalizer-correlator receiver does not exploit subspaces in partially loaded systems. This is in contrast to the symbol-level LMMSE receiver, which as discussed below is time-varying due to the scrambler and hence too complex to implement. A compromise can be found by performing symbol-level multi-stage Wiener filtering (MSWF), which is an iterative solution in which the complexity per iteration becomes comparable to twice that of the RAKE receiver. Because MSWF works best when the input is white, better performance is obtained if the RAKE in each MSWF stage is replaced by a chip equalizer-correlator. One of the main contributions here is to point out that the chip equalizer benefits from a separate optimization in every stage. This is shown through a mix of analysis and simulation results.

The LMMSE receiver is complex for UMTS FDD mobile terminals because it not only requires inversion of a large user cross-correlation matrix, but also needs the code and the amplitude knowledge of all the active users [24]. Furthermore, the LMMSE solution changes every chip period due to aperiodic scrambling. The LMMSE *chip* equalizer-correlator is a suboptimal but much simpler alternative that is derived by modeling the scrambler as a stationary random sequence [20,21]. Another suboptimal multi-user detector that *explicitly* focuses on subtracting the signals of interfering codes is the PIC receiver [43]. It is well known that under very relaxed cell loads, when the number of iterations goes to infinity, PIC might converge to the decorrelating receiver [22]. However, provided that it converges, the convergence rate is still very slow and it requires many stages to obtain a reasonable performance. This is due to the existence of high cross-correlations among users, which in fact is a consequence of the low orthogonality factor obtained initially from the use of a RAKE receiver in the front-end [7,25,29]. In this text, to at least guarantee the convergence in realistic loading factor situations and to increase the speed of convergence, we start the decorrelation operation, that is, the ZF *symbol* equalization from the output of LMMSE chip equalizer-correlator front end receiver whose orthogonality factor is higher than the RAKE receiver. For approximating this matrix inversion operation, we consider the polynomial expansion (PE) technique, which is a better structured equivalent of PIC [26]. Until recently, interference cancellation was considered somewhat reluctantly for the downlink because it unrealistically requires knowing the locations of active codes in the OVSF tree and the amounts of power they carry. Only recently have the merits of inter-cell interference cancellation been acknowledged [5], and efforts at finding viable solutions have more than doubled. The problem of OVSF code indentification can be simplified by an equivalent modeling of the active multi-rate transmission system as a multi-code pseudo-transmission system at any chosen single SF-level L in the OVSF hierarchy. One toy example representing actually the UMTS-TDD case that contains SFs ranging from 1 to 16 is given in Figure 4.21. In this example, the nodes corresponding to the active codes at SF-levels 4 and 8 are demonstrated by black bulbs. Their pseudo-equivalents at SF-level 16 (i.e., $L = 16$) are demonstrated by hatched pattern bulbs.

One can detect the existence or absence of pseudo-codes at the pseudo-level by comparing the powers at their correlator outputs with a noise-floor threshold [23]. These multiple correlations can be implemented with $O(L \log L)$ complexity using fast Walsh-Hadamard transformation (FWHT). Unitary FWHTs (U-FWHT) with proper dimensions can be logically/physically exploited to see/implement the two-way transformations between actual symbol sequences corresponding to the known codes (e.g., HSDPA codes) at various SF-levels and their pseudo-symbol sequence equivalents at a single SF-level. Figure 4.22 demonstrates the two-way transformations between L_2/L_1 consecutive (time-multiplexed) actual symbols a_i at level L_1 and L_2/L_1 parallel (code-multiplexed) pseudo-symbols $\tilde{a}_i$ at a larger SF-level L_2. $P_{\{L_2/L_1\}}/S$ and $S/P_{\{L_2/L_1\}}$ are parallel to serial and serial to parallel converters from/to a bus size L_2/L_1. When actual symbols reside at a higher SF-level, the two transformations have reverse roles.

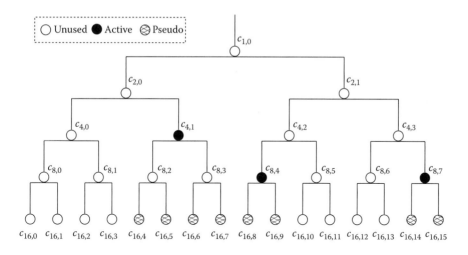

Figure 4.21 Equivalence of active-multirate and pseudo-multicode systems.

4.5.6.1 Polynomial Expansion Receiver

In this section we develop parallel intra-cell interference cancelling (IC) structures based on the polynomial expansion (PE) technique initially proposed in [26]. We exploit the *pseudo-equivalency* concept at the highest *active* SF-level (SF-256), in the UMTS-FDD downlink for applying PE at this level. We ignore the existence of SF-512 because it is rarely used to carry out control commands during an upload operation. The rationale for choosing the *highest active SF*, henceforth called L, is to obtain the highest possible degrees of freedom in determining the PE subspace. If any other level L_x were selected, then an activity on a child code of $c_{x,i}$, $i \in \{0, \dots, L_x - 1\}$, say, at a level $L_y > L_x$ on $c_{y,j}$, $j \in \{(L_y/L_x)i, \dots, (L_y/L_x)(i+1) - 1\}$, would also render mandatory the implicit inclusion of all other child codes of $c_{x,i}$ at level L_y by including $c_{x,i}$ in the PE. This would have an adverse effect of noise amplification.

Pseudo-codes might be used in place of the *unknown actual codes* because the actual symbol estimates and their powers are not necessary as

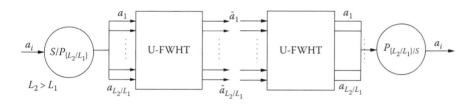

Figure 4.22 Transformations between actual and pseudo-symbols.

long as the pseudo-symbols are treated linearly in interference cancellation. However, knowing or detecting the actual codes is an opportunity for exploiting hard or hyperbolic-tangent nonlinearities or even channel decoding and encoding to refine their symbol estimates [10,18]. In the latter case, one can pass between the symbol blocks of known codes and their pseudo-equivalents at SF-256 by properly dimensioned FWHTs. Through this hybrid treatment, respective nonlinear and linear treatment of known and unknown codes becomes possible.

We model the discrete time received signal over one pseudo-symbol period as

$$\boldsymbol{Y}[n] = \boldsymbol{H}(z)\boldsymbol{S}[n]\boldsymbol{C}\boldsymbol{A}[n] + \boldsymbol{V}[n] = \widetilde{\boldsymbol{G}}(n, z)\boldsymbol{A}[n] + \boldsymbol{V}[n]$$

representing the system at the symbol rate. As shown in Figure 4.23, $\boldsymbol{H}(z) = \sum_{i=0}^{M-1} \boldsymbol{H}[i] z^{-i}$ is the symbol rate $Lm \times L$ channel transfer function, z^{-1} being the symbol period delay operator. The block coefficients $\boldsymbol{H}(i)$ are

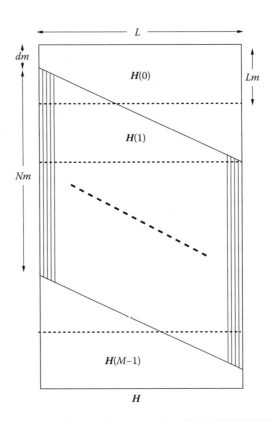

Figure 4.23   Channel impulse response of $H(z)$.

the $M = \lceil \frac{L+N+d-1}{L} \rceil$ parts of the block Toeplitz matrix with $m \times 1$ sized blocks, $\boldsymbol{b}$ being the first column whose top entries might be zero for it comprises the transmission delay d between the BS and the mobile terminal. In this representation, $\boldsymbol{H}[0]$ carries the signal part corresponding to $\boldsymbol{A}[n]$ where there is no user of interest inter-symbol interference (ISI) but only user of interest ICI and MUI. $\boldsymbol{H}(i), (i \in \{1, 2, \dots, M-1\})$, similarly carries the ICI and MUI from $\boldsymbol{A}[n-i]$. The $L \times L$ matrix $\boldsymbol{S}[n]$ is diagonal and contains the scrambler for symbol period n. The column vector $\boldsymbol{A}[n]$ contains the K (pseudo-)symbols and $\boldsymbol{C}$ is the $L \times K$ matrix of the K active codes.

Although it is possible to find an FIR left inverse filter for $\hat{\boldsymbol{G}}(n, z)$ provided that $Lm \geq K$, this is not practical because $\hat{\boldsymbol{G}}(n, z)$ is time-varying due to the aperiodicity of the scrambling. Therefore, we introduce a less complex approximation to this inversion based on the polynomial expansion technique [26]. Instead of basing the receiver directly on the received signal, we first introduce a dimensionality reduction step from Lm to K by equalizing the channels with LMMSE-ZF chip rate equalizers $\boldsymbol{F}(z)$, followed by a bank of correlators. An LMMSE-ZF equalizer is the one among all possible ZF equalizers that minimizes the MSE at its output [14].

Let $\boldsymbol{X}[n]$ be the $K \times 1$ correlator output, which would correspond to the RAKE receiver outputs if channel-matched filters were used instead of channel equalizers. Then,

$$\boldsymbol{X}[n] = \tilde{\boldsymbol{F}}(n, z)\boldsymbol{Y}[n]$$
$$= \boldsymbol{C}^H \boldsymbol{S}^H[n]\boldsymbol{F}(z)(\hat{\boldsymbol{G}}(n, z)\boldsymbol{A}[n] + \boldsymbol{V}[n])$$
$$= \boldsymbol{M}(n, z)\boldsymbol{A}[n] + \tilde{\boldsymbol{F}}(n, z)\boldsymbol{V}[n]$$

where $\boldsymbol{M}(n, z) = \tilde{\boldsymbol{F}}(n, z)\hat{\boldsymbol{G}}(n, z)$ and ZF equalization results in $\boldsymbol{F}(z)\boldsymbol{H}(z) = \boldsymbol{I}$. Hence,

$$\boldsymbol{M}(n, z) = \sum_{i=-\infty}^{\infty} \boldsymbol{M}[n, i]z^{-i} = \begin{bmatrix} \boldsymbol{I} & * \\ * & \boldsymbol{I} \end{bmatrix} \tag{4.54}$$

due to proper normalization of the code energies.

To obtain the estimate of $\boldsymbol{A}[n]$, we initially consider the processing of $\boldsymbol{X}[n]$ by a decorrelator as

$$\widehat{\boldsymbol{A}}[n] = \boldsymbol{M}(n, z)^{-1}\boldsymbol{X}[n]$$
$$= (\boldsymbol{I} - \overline{\boldsymbol{M}}(n, z))^{-1}\boldsymbol{X}[n] \tag{4.55}$$

The correlation matrix $\boldsymbol{M}(n, z)$ has a coefficient $\boldsymbol{M}[n, 0]$ with a dominant unit diagonal in the sense that all other elements of the $\boldsymbol{M}[n, i]$ are much

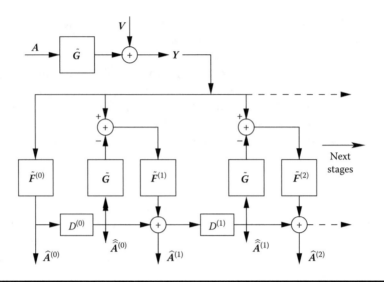

Figure 4.24   Polynomial expansion receiver.

smaller than 1 in magnitude. Hence, the polynomial expansion approach suggests to develop $(I - \overline{M}(n, z))^{-1} = \sum_{i=0}^{\infty} \overline{M}(n, z)^i$ up to some finite order, which leads to the iterative receiver as[18]

$$\widehat{A}^{(-1)} = 0; \quad i \geq 0$$

$$\widehat{A}^{(i)} = X + \overline{M} \widehat{A}^{(i-1)}$$

$$= X + (I - M) \widehat{A}^{(i-1)}$$

$$= \widehat{A}^{(i-1)} + \tilde{F}^i (Y - \tilde{G} \widehat{A}^{(i-1)}) \qquad (4.56)$$

The resultant iterative receiver architecture is given in Figure 4.24 where the numbers in parantheses indicate the iteration indices. A practical receiver would be limited to a few orders, the quality of which depends on the degree of dominance of the static part of the diagonal of $M(n, z)$ given in Equation (4.56) with respect to the ICI carrying dynamic contents of the diagonal elements and MUI carrying off-diagonal elements.

---

[18] Time indices are dropped for brevity.

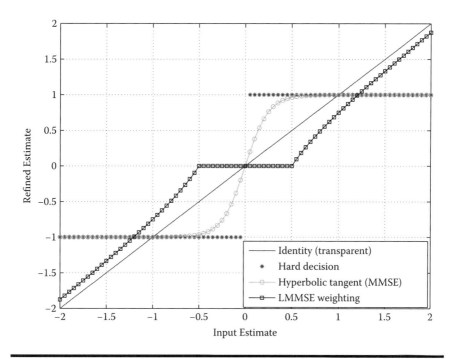

Figure 4.25 **Feedback functionalities for real and imaginary parts of QPSK symbols that have 6-dB SINR.**

In an iterative PE approach, it is advantageous to replace several *local* receiver components obtained from *global* LMMSE-ZF formulation with their LMMSE counterparts. Such modifications should lead to smaller off-diagonal power and hence faster convergence of the iterations to an estimate that is closer to a global MMSE estimate. For example, LMMSE-ZF chip equalizers can be replaced by LMMSE equalizers that, although perturb the orthogonal structure of the received signal from the BS, do not enhance as much the inter-cell interference plus noise [28]. Furthermore, some symbol feedback functionalities $\mathcal{D}$ shown in Figure 4.25 such as LMMSE weighting factors, hard decisions, a variety of soft decisions like hyperbolic-tangent functionality or even channel decoding and encoding blocks can be introduced.

4.5.6.1.1 Filter Adaptation

Figure 4.26 shows the open form of the receiver in Figure 4.24 where we clearly see the chip-level blocks. In case the symbol feedback functionality $\mathcal{D}$ is the identity matrix, we can further obtain a third equivalent architecture

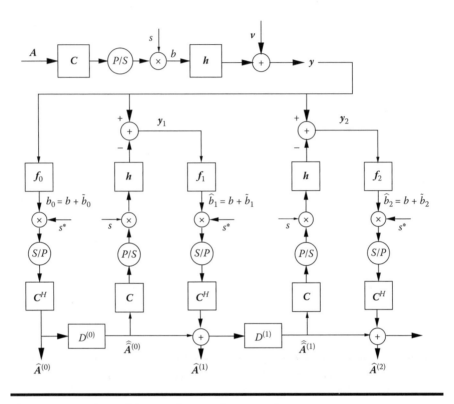

Figure 4.26 Polynomial expansion receiver open format.

given in Figure 4.27 that, different from the previous two, iterates over the *chip* estimates at chip-level filter outputs.

Because the projection operation $S[n]CC^H S^*[n]$ is not a chip-level operation and is not convolutive, it cannot be easily integrated into the filter optimization process. Nevertheless it has two nice properties: (1) the diagonal part is the deterministic value $C_l I$ where C_l is the effective cell loading factor and (2) the expected value of the non-diagonal part is zero. By considering only the diagonal parts of the local projection operations, we reach the multi-stage Wiener (LMMSE) filter adaptation procedure given in the Equation group (4.57) where $\{\mathcal{X}_i, \mathcal{Y}_i, \tilde{B}_i\}$, respectively, denote transfer function between the BS signal and the residual BS signal, transfer function for the intercell interference plus noise, the residual interference plus noise at iteration i.[19] The Wiener (LMMSE)

---

[19] Each bold variable in this section has a (z) suffix, which is dropped for brevity; † stands for z-transform para-conjugate operator meaning matched filter in the time domain.

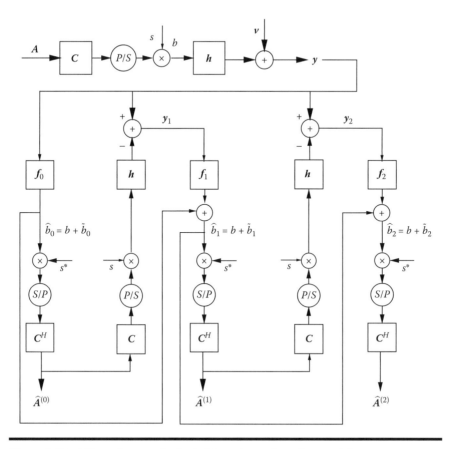

Figure 4.27   PE receiver equivalent chip estimate iterating model.

filter and the unbiased LMMSE filter are denoted by $\boldsymbol{F}_i^{w}$ and $\boldsymbol{F}_i$, respectively.

The LMMSE optimization process output is the complete filter expression of $\boldsymbol{F}_i$, from which we derive its two ingredients $\boldsymbol{S}_{\check{b}_{i-1}\boldsymbol{y}_i}$ and $\boldsymbol{S}_{\boldsymbol{y}_i\boldsymbol{y}_i}$ by *factorization*. The structure of the factorized terms are clear guidelines for understanding that, when unbiased, the chip-level filter $\boldsymbol{F}_i$ intends to estimate and subtract the *residual interference plus noise* term at the preceding iteration, which is expected to be also valid for systems with additional system components such as hard decisions. For example, if we consider the loop among the signals $\hat{b}_0$, $\boldsymbol{y}_1$, and $\hat{b}_1$ that contains the transfer functions $\boldsymbol{F}_1(z)$ and $\boldsymbol{H}(z)$, it estimates the residual signal $\check{b}_0$ and subtracts it from $\hat{b}_0$, which leads to the creation of the new residual signal $\check{b}_1$. The same reasoning holds for subsequent iterations where the amount of interference plus noise variance $\sigma_{b_i}^2$ is expected to decrease with increasing i as long as the spectral radius $\rho(I - C_l F_i H) < 1$.

INITIALIZATION (First Stage)

$$\mathcal{X}_0 = F_0 H - I$$

$$\mathcal{Y}_0 = F_0$$

$$\tilde{B}_0 = \mathcal{X}_0 B + \mathcal{Y}_0 V$$

ITERATIONS (Interference Cancellation Stages)

for $(i > 0)$ and $(i < i_{max})$

$$\mathcal{X}_i = (I - C_l F_i H)\mathcal{X}_{i-1}$$

$$\mathcal{Y}_i = (I - C_l F_i H)\mathcal{Y}_{i-1} + F_i$$

$$\tilde{B}_i = \mathcal{X}_i B + \mathcal{Y}_i V$$

$$\arg_{F_i^w} \min \frac{1}{2\pi j} \oint \frac{dz}{z}\left(\mathcal{X}_i \mathcal{X}_i^\dagger \sigma_b^2 + \mathcal{Y}_i \mathcal{Y}_i^\dagger \sigma_v^2\right) \tag{4.57}$$

$$F_i^w = S_{\tilde{b}_{i-1}y_i} S_{y_i y_i}^{-1}$$

$$S_{\tilde{b}_{i-1}y_i} = C_l \mathcal{X}_{i-1} \mathcal{X}_{i-1}^\dagger H^\dagger \sigma_b^2 - \mathcal{Y}_{i-1}(I - C_l H \mathcal{Y}_{i-1})^\dagger \sigma_v^2$$

$$S_{y_i y_i} = C_l^2 H \mathcal{X}_{i-1} \mathcal{X}_{i-1}^\dagger H^\dagger \sigma_b^2 + (I - C_l H \mathcal{Y}_{i-1})(I - C_l H \mathcal{Y}_{i-1})^\dagger \sigma_v^2$$

$$F_i = \frac{2\pi j F_i^w}{\oint \frac{dz}{z} F_i^w H}: \text{unbiasing operation} \tag{4.58}$$

end

In practice, LMMSE chip equalizer correlator blocks might also be implemented as Generalized RAKE (G-RAKE) receivers in which case, in each stage, filtering with F_i and H will be similar to the filtering part of the RAKE receiver [45]. Hence, each iteration will have twice the complexity of that of the RAKE.

4.5.6.1.2 Impact of Symbol Feedback Nonlinearities on Filter Expressions

When hard decisions or hyperbolic tangent nonlinearities are used on a subset of codes, we see two alternatives to reflect their impact on filtering expressions. The first approach is to simply assume that the associated symbols are perfectly estimated and hence to exclude them after the first stage. In this case, the only required changes are to consider C_l and σ_b^2 as, respectively, the loading factor and the sum chip variance of the remaining codes that are treated linearly. The second approach is to quantify the variances of symbol estimation errors after the nonlinearities at every stage by the scheme and introduce new additive Gaussian noise sources at those points with the obtained variances.

4.5.6.2 Intercell Interference Cancellation Expansion

The polynomial expansion receiver can be modified to include also the inter-cell interference cancellation. The filter adaptations, the changes in signal modeling, and the architecture for cancelling the interference of one neighboring BS are given in Equation group (4.59), Equation group (4.60) and Figure 4.28. The scheme can be easily extended to cover any number of cells by increasing the sizes of vectors and matrices in Equation group (4.60).

INITIALIZATION (First Stage)

$$\mathcal{X}_0 = \mathbf{F}_0 \mathbf{H} - \mathbf{I}$$

$$\mathcal{Y}_0 = \mathbf{F}_0$$

$$\tilde{\mathbf{B}}_0 = \mathcal{X}_0 \mathbf{B} + \mathcal{Y}_0 \mathbf{V}$$

ITERATIONS (Interference Cancellation Stages)

for $(i > 0)$ and $(i < i_{max})$

$$\mathcal{X}_i = (\mathbf{I} - \mathbf{F}_i \mathbf{H} \mathbf{C}_l) \mathcal{X}_{i-1}$$

$$\mathcal{Y}_i = (\mathbf{I} - \mathbf{F}_i \mathbf{H} \mathbf{C}_l) \mathcal{Y}_{i-1} + \mathbf{F}_i$$

$$\tilde{\mathbf{B}}_i = \mathcal{X}_i \mathbf{B} + \mathcal{Y}_i \mathbf{V}$$

$$\arg_{\mathbf{F}_{1,i}^w} \min \frac{1}{2\pi j} \oint \frac{dz}{z} \left(\mathcal{X}_{1,i} \Sigma_b^2 \mathcal{X}_{1,i}^\dagger + \mathcal{Y}_{1,i} \mathcal{Y}_{1,i}^\dagger \sigma_v^2 \right) \qquad (4.59)$$

$$\arg_{\mathbf{F}_{2,i}^w} \min \frac{1}{2\pi j} \oint \frac{dz}{z} \left(\mathcal{X}_{2,i} \Sigma_b^2 \mathcal{X}_{2,i}^\dagger + \mathcal{Y}_{2,i} \mathcal{Y}_{2,i}^\dagger \sigma_v^2 \right)$$

$$\mathbf{F}_{1,i}^w = \mathbf{S}_{\tilde{b}_{1,i-1} y_i} \mathbf{S}_{y_i y_i}^{-1} \qquad \mathbf{F}_{2,i}^w = \mathbf{S}_{\tilde{b}_{2,i-1} y_i} \mathbf{S}_{y_i y_i}^{-1}$$

$$\mathbf{S}_{\tilde{b}_{1,i-1} y_i} = \mathcal{C}_{l_1} \mathcal{X}_{1,i-1} \Sigma_b^2 \mathcal{X}_{1,i-1}^\dagger \mathbf{H}_1^\dagger - \mathcal{Y}_{1,i-1} (\mathbf{I} - \mathcal{C}_{l_1} \mathbf{H}_1 \mathcal{Y}_{1,i-1})^\dagger \sigma_v^2$$

$$\mathbf{S}_{\tilde{b}_{2,i-1} y_i} = \mathcal{C}_{l_2} \mathcal{X}_{2,i-1} \Sigma_b^2 \mathcal{X}_{2,i-1}^\dagger \mathbf{H}_2^\dagger - \mathcal{Y}_{2,i-1} (\mathbf{I} - \mathcal{C}_{l_2} \mathbf{H}_2 \mathcal{Y}_{2,i-1})^\dagger \sigma_v^2$$

$$\mathbf{S}_{y_i y_i} = \mathbf{H} \mathcal{C}_l \mathcal{X}_{i-1} \Sigma_b^2 \mathcal{X}_{i-1}^\dagger \mathcal{C}_l \mathbf{H}^\dagger + (\mathbf{I} - \mathbf{H} \mathcal{C}_l \mathcal{Y}_{i-1}) (\mathbf{I} - \mathbf{H} \mathcal{C}_l \mathcal{Y}_{i-1})^\dagger \sigma_v^2$$

$$\mathbf{F}_{1,i} = \frac{2\pi j \mathbf{F}_{1,i}^w}{\oint \frac{dz}{z} \mathbf{F}_{1,i}^w \mathbf{H}_1} \qquad \mathbf{F}_{2,i} = \frac{2\pi j \mathbf{F}_{2,i}^w}{\oint \frac{dz}{z} \mathbf{F}_{2,i}^w \mathbf{H}_2}$$

end

$$A[n] \longrightarrow \begin{bmatrix} A_1[n] \\ A_2[n] \end{bmatrix} : \text{vector of symbols}$$

$$C \longrightarrow \begin{bmatrix} C_1 & 0 \\ 0 & C_2 \end{bmatrix} : \text{channelization codes}$$

$$S[n] \longrightarrow \begin{bmatrix} S_1[n] & 0 \\ 0 & S_2[n] \end{bmatrix} : \text{scrambling} \qquad (4.60)$$

$$B \longrightarrow \begin{bmatrix} B_1 \\ B_2 \end{bmatrix} : \text{transmitted chip sequences}$$

$$\sigma_b^2 \longrightarrow \Sigma_b^2 = \begin{bmatrix} \sigma_{b_1}^2 & 0 \\ 0 & \sigma_{b_2}^2 \end{bmatrix} : \text{chip-level signal covariance}$$

$$H(z) \longrightarrow \begin{bmatrix} H_1(z) & H_2(z) \end{bmatrix} : \text{chip rate channel}$$

$$F_i(z) \longrightarrow \begin{bmatrix} F_{1,i}(z) \\ F_{2,i}(z) \end{bmatrix} : \text{chip-level equalizers at iteration } i$$

$$\tilde{G}(n, z) \longrightarrow \begin{bmatrix} \tilde{G}_1(n, z) & \tilde{G}_2(n, z) \end{bmatrix} = \begin{bmatrix} H_1(z)S_1[n]C_1 & H_2(z)S_2[n]C_2 \end{bmatrix} :$$
$$\text{symbol rate channel}$$

$$\tilde{F}^{(i)}(n, z) \longrightarrow \begin{bmatrix} \tilde{F}_1^{(i)}(n, z) \\ \tilde{F}_2^{(i)}(n, z) \end{bmatrix} : \text{symbol-level equalizers at iteration } i$$

$$\mathcal{X}_i \longrightarrow \begin{bmatrix} \mathcal{X}_{1,1,i} & \mathcal{X}_{1,2,i} \\ \mathcal{X}_{2,1,i} & \mathcal{X}_{2,2,i} \end{bmatrix} : \text{interference transfer function}$$

$$\mathcal{X}_{1,i} = \begin{bmatrix} \mathcal{X}_{1,1,i} & \mathcal{X}_{1,2,i} \end{bmatrix} : \text{interference transfer function for the first}$$
$$\text{BS signal}$$

$$\mathcal{X}_{2,i} = \begin{bmatrix} \mathcal{X}_{2,1,i} & \mathcal{X}_{2,2,i} \end{bmatrix} : \text{interference transfer function for the second}$$
$$\text{BS signal}$$

$$\mathcal{Y}_i \longrightarrow \begin{bmatrix} \mathcal{Y}_{1,i} \\ \mathcal{Y}_{2,i} \end{bmatrix} : \text{noise transfer function}$$

$$\mathcal{C}_l \longrightarrow \mathcal{C}_l = \begin{bmatrix} \mathcal{C}_{l_1} & 0 \\ 0 & \mathcal{C}_{l_2} \end{bmatrix} : \text{loading factors}$$

$$Y[n] \longrightarrow \begin{bmatrix} H_1(z)S_1[n]C_1 & H_2(z)S_2[n]C_2 \end{bmatrix} \begin{bmatrix} A_1[n] \\ A_2[n] \end{bmatrix} + V[n]$$

$$= \tilde{G}(n, z) A[n] + V[n]$$

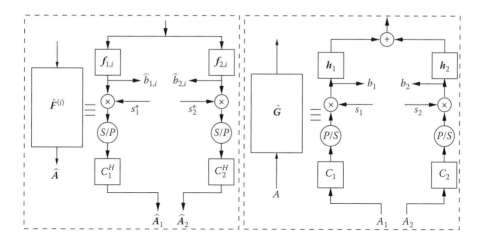

Figure 4.28   Symbol-level transfer function blocks and their chip-level equivalents.

4.5.6.3 Simulations and Conclusions

For the simulations, we consider only intra-cell interference cancellation.

We take a High-Speed Downlink Packet Access (HSDPA) scenario in the UMTS-FDD downlink [5]. We consider five HSDPA codes at SF-16 assigned to the UE, each consuming 8% of the base station power. The PCPICH pilot tone at SF-256 consumes 10% power. There is the PCCPICH code at SF-256 that consumes 4% power. To effectively model all remaining multirate user codes that we do not know, we place 46 pseudo-codes at level 256, each having 1% power. So in total, five HSDSCH codes at SF-16 being equivalent to 80 pseudo-codes at SF-256, the system is effectively 50% loaded with 128 (pseudo-)codes at SF-256; that is, $C_l = 0.5$. Although, in practice, the pseudo-codes should be detected by a method explained in the text, for the moment we assume that they are known. We also assume perfect knowledge of the channel. An oversampling factor of 2 and one receive antenna are used.[20] Static propagation channel parameters are randomly generated from the ITU Vehicular-A power delay profile. Pulse shape is the UMTS-standard, root-raised cosine with a roll-off factor of 0.22. Therefore, the propagation channel, pulse shape cascade (i.e., the overall channel) has a length of 19 chips at a 3.84 Mchips/s transmission rate. Symbols are QPSK. $\hat{I}_{or}/I_{oc}$ denotes the received base station power to inter-cell interference plus noise power ratio. We took the average SINR result of five HSDPA codes over 100 realizations of one UMTS slot (160 symbol period) transmissions.

---

[20] The order of filtering and rechanneling operations have an impact on the noise term in case of polyphase filtering, which we neglect for the moment.

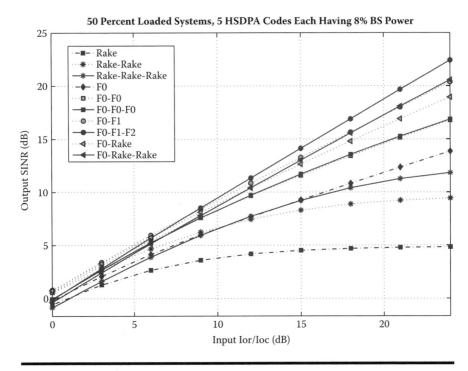

Figure 4.29 SINR versus $\hat{I}_{or}/I_{oc}$ linear decisions results, Vehicular A channel, $N = 19$. 50% loaded systems; 5 HSDPA codes, each having 8% BS power.

Figure 4.29 shows the performances of the PE scheme with various chip-level filter usages and iterations from one to three. The legends indicate the used filters with iteration order. For example, F0-F1-F2 means optimized filters are used in all stages; the F0-Rake-Rake hybrid scheme means the first stage filter is an LMMSE chip equalizer and the subsequent two are Rake receivers; Rake-Rake-Rake corresponds to the *conventional* linear PIC. Many other variants different from the ones shown can also be used. As expected, the RAKE receiver performs the worst. The conventional linear PIC with only RAKE receivers starts diverging after the first iteration, especially for $\hat{I}_{or}/I_{oc}$ values below 10 dB. This is consistent with the literature because it is well known that, for guaranteeing the convergence of the LPIC, the loading factor should be lower than 17% [15].[21] The scheme that uses only F0 does not improve significantly after the second iteration. Using RAKE receivers after F0 performs very well. As expected, adapting the filters at all

---

[21] In the random CDMA, flat fading case.

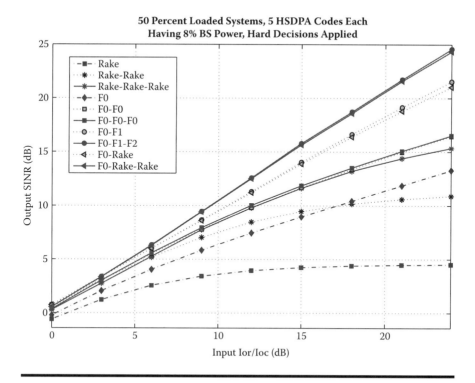

Figure 4.30 SINR versus $\hat{I}_{or}/I_{oc}$ hard decisions results, Vehicular A channel, $N = 19$. 50% loaded systems; 5 HSDPA codes, each having 8% BS power; hard decision applied.

iterations performs the best. Such a scheme obtains almost the same performance as F0-Rake-Rake in one less iteration, that is, with configuration F0-F1. At low $\hat{I}_{or}/I_{oc}$ values, which reflect the cell-edge situations, the performance of the first iteration is better than the second one. This is due to the fact that in low SNR regions, the gain from the interference reduction is not sufficient to compensate for the loss from noise amplification, because the iterative scheme is still a decorrelation. One might also attribute this to the well-known *ping-pong* effect for LPIC [33].

Figure 4.30 shows the performances when we apply hard decisions on the five HSDPA codes, which have an effective loading impact $C_{HSDPA} = \frac{5}{16}$. With the assumption of correct decisions we subtract C_{HSDPA} from the overall cell load of 0.5 and apply the $C_l = \frac{3}{16}$ value in the filter adaptation process in Equation (4.58). In this case, using RAKE receivers after first-stage equalization catches up with the optimized filters after three stages. We also observe that conventional PIC also starts getting into a convergence trend.

It is not, however, explicit from the $\mathcal{X}_i$ and $\mathcal{Y}_i$ expressions why things should improve despite the fact that the $\mathcal{C}_l$ value decreases, resulting in lower iteration gain in chip estimation. Due to this fact, one would at first sight expect an inability of the interference reduction to compensate for the amplified noise. This is, however, not the case, because almost all ingredients of the additional noise term coming from the previous iteration are in the subspace belonging to the codes whose symbols are estimated linearly, whereas the final SINR performance metric is computed on codes such as HSDPA codes, which are treated by hard decisions. In the full linear treatment, however, the additional noise that traverses the iterations with amplification is in the whole signal space. Therefore, when hard decisions are applied, there is an implicit reduction in additional noise by a factor $\frac{\mathcal{C}_l}{\mathcal{C}_l+\mathcal{C}_{HSDPA}}$. These interpretations seem to be in conflict with the chip equalizer adaptation expressions where we ignored the non-diagonal part of the projection operation $\mathbf{S}[n]\mathbf{CC}^H\mathbf{S}^*[n]$ in order to avoid considering dependence on codes. For the interpretations of performances at symbol levels, however, one must look from a different perspective, taking into account the code knowledge.

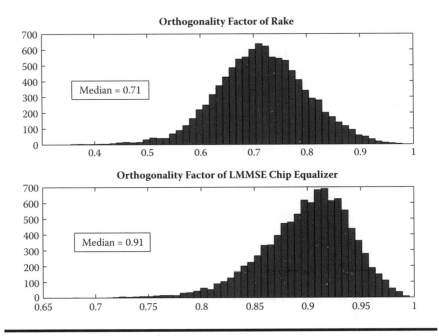

Figure 4.31 Orthogonality factor histograms of two-phase CMF (upper) and two-phase LMMSE chip equalizer (lower) in Vehicular A channel with $\hat{I}_{or}/I_{oc} = 10$ dB.

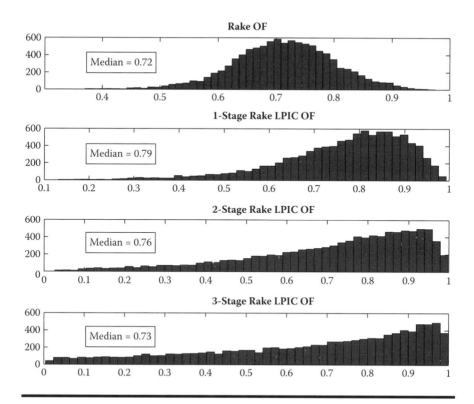

Figure 4.32 Orthogonality factor histogram of conventional LPIC with two-phase CMF in the Vehicular A channel.

Comparing Figure 4.29 and Figure 4.30 we observe that at medium and high $\hat{I}_{or}/I_{oc}$ working regions, hard decisions increase the obtained SINR by 1 to 3 dB. At low $\hat{I}_{or}/I_{oc}$ regions, there is no gain, which is understandable because in those regions hard decisions are not reliable.

We next look at the orthogonality factor (OF) histograms of the considered receivers by randomly generating 10^4 static channels from the Vehicular A power delay profile and an $\hat{I}_{or}/I_{oc}$ value of 10 dB. Figure 4.31 shows the histograms for the CMF and LMMSE equalizer. We see that, besides giving worse median OF, CMF might also give OFs less than 0.4. In Figure 4.32, Figure 4.33 and Figure 4.34 we, respectively, see the trend of OFs obtained from all CMF usage, CMF usage after first stage equalization, and all chip-level LMMSE equalizer usage in LPIC iterations. To obtain these, we first compute the $\|\mathcal{X}_i\|^2$ and pass to OF as $\alpha_i = \frac{1}{1+\|\mathcal{X}_i\|^2}$ since all the filters are unbiased. The histograms in Figure 4.32 clearly demonstrate the problem with conventional LPIC. From a median value perspective, the OF

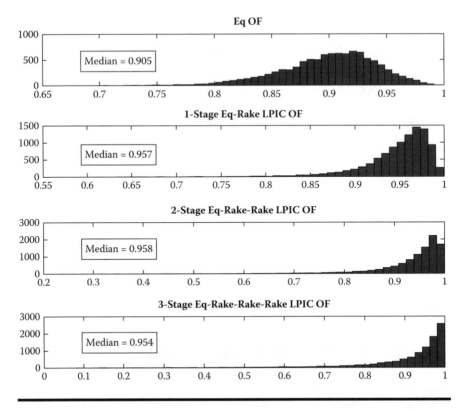

Figure 4.33 Orthogonality factor histogram of LPIC with first stage two-phase LMMSE chip equalizer followed by two-phase CMFs in the Vehicular A channel with $\hat{I}_{or}/I_{oc} = 10$ dB.

first improves at the first stage and then starts degrading. An even more important concern is the widening of the OF range. After four iterations, there are even channel cases where OF is close to zero. The histograms in Figure 4.33 demonstrate the importance of LMMSE chip equalization as a starting point. Although there are still very few corner cases leading to small OFs, the overall performance is at an acceptable level. Finally, the histograms in Figure 4.33 clearly indicate the desirability of using optimized chip equalizers at all stages. Not only the median value but also the worst-case OF improves with each iteration: 0.86 at first and fourth stages, respectively. In brief, we can say that when the mobile knows multiple codes, as in the HSDPA service, applying RAKE receivers after a first-stage equalization stage is a proper choice. In the case of only one code, however, it is beneficial to adapt filters at each stage.

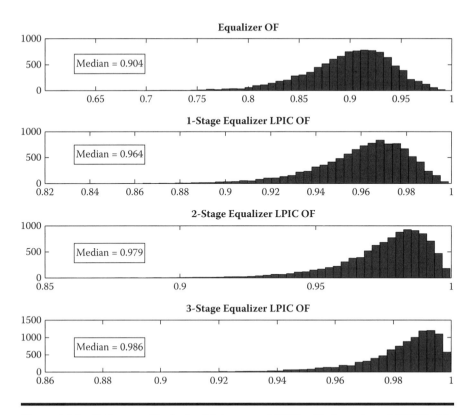

Figure 4.34 Orthogonality factor histogram of LPIC with two-phase LMMSE chip equalizers in all stages in the Vehicular A channel with $\hat{I}_{or}/I_{oc} = 10$ dB.

References

[1] 3GPP. Physical Layer Aspects of HSDPA. Technical Report TS 25.848. [Online]. Available: http://www.3gpp.org/ftp/Specs.

[2] 3GPP. Physical Layer Procedures (FDD). Technical Report TS 25.214. [Online]. Available: http://www.3gpp.org/ftp/Specs.

[3] 3GPP. QoS Concept and Architecture. Technical Report TS 23.107, [Online]. Available: http://www.3gpp.org/ftp/Specs.

[4] 3GPP. Technical Specifications. Technical Report. [Online]. Available: http://www.3gpp.org.

[5] 3GPP. User Equipment (UE) Radio Transmission and Reception (FDD) — Release 8. Technical Report TS 25.101, 3GPP, [Online]. Available: http://www.3gpp.org/ftp/Specs.

[6] 3GPP. UTRA High Speed Downlink Packet Access; Overall UTRAN Description. Technical Report TS25.855. [Online]. Available: http://www.3gpp.org/ftp/Specs.

[7] R.M. Buehrer, S.P. Nicoloso, and S. Gollamudi. Linear versus nonlinear interference cancellation. *J. Commun. Networks*, 1(2): 118–133, June 1999.

[8] C. Cozzo, G.E. Bottomley, and A.S. Khayrallah. Rake receiver finger placement for realistic channels. In *Proc. of the Wireless Communications and Networking Conference*, 2004, pp. 316–321.

[9] H.P. Decell. An application of the Cayley-Hamilton theorem to generalized matrix inversion. *Jr. SIAM Rev.*, 7(4): 526–528, October 1965.

[10] D. Divsalar, M.K. Simon, and D. Raphelli. Improved parallel interference cancellation. *IEEE Trans. Commun. Theory*, 46: 258–268, February 1998.

[11] A. Duel-Hallen. Decorrelating decision-feedback multiuser detector for synchronous code-division multiple-access channel. *IEEE Trans. Commun.*, 41(2): 285–290, February 1993.

[12] P. Frenger, S. Parkvall, and E. Dahlman. The evolution of WCDMA towards higher speed downlink packet data access. In *Proc. Vehicular Technology Conf.*, October 2001.

[13] I. Ghauri and D.T.M. Slock. Linear receivers for the DS-CDMA downlink exploiting orthogonality of spreading sequences. In *Proc. 32nd Asilomar Conf. on Signals, Systems & Computers*, 1: 650–654, November 1998, Pacific Grove, CA.

[14] I. Ghauri and D.T.M. Slock. MMSE-ZF receiver and blind adaptation for multirate CDMA. In *Proc. Vehicular Technology Conf.*, September 1999.

[15] A. Grant and C. Schlegel. Convergence of linear interference cancellation multiuser receivers. *IEEE Trans. Commun.*, 49(10): 1824–1834, October 2001.

[16] L.J. Greenstein, Y.S. Yeh, V. Erceg, and M.V. Clark. A new path gain/delay-spread propagation model for digital cellular channels. *IEEE Trans. Vehicular Technol.*, 46(2): 477–485, May 1997.

[17] H. Holma and A. Toskala. *WCDMA for UMTS*. John Wiley & Sons, 2000.

[18] R. Irmer, A. Nahler, and G. Fettweis. On the impact of soft decision functions on the performance of multistage parallel interference cancelers for CDMA systems. In *Proc. Vehicular Technol. Conf.*, Rhodes, Greece, May 2001.

[19] J.F. Kurose and K.W. Rose. *Computer Networking: A Top Down Approach Featuring the Internet*. Addison-Wesley, 2003.

[20] M. Lenardi and D.T.M. Slock. SINR maximizing equalizer receiver for DS-CDMA. In *Proc. of the EUSIPCO Conf.*, Tampere, Finland, September 2000.

[21] M. Lenardi and D.T.M. Slock. A RAKE structured SINR maximizing mobile receiver for the WCDMA downlink. In *Proc. Asilomar Conf. on Signals, Systems and Computers*, Pacific Grove, CA, November 2001.

[22] R. Lupas and S. Verd'u. Linear multiuser detectors for synchronous code-division multiple-access channels. *IEEE Trans. Information Theory*, IT-35: 123–136, January 1989.

[23] M.F. Madkour, S.C. Gupta, and Y.E. Wang. Successive interference cancellation algorithms for downlink W-CDMA communications. *IEEE Trans. Wireless Commun.*, 1(1): 169–177, January 2002.

[24] U. Madhow and M.L. Honig. MMSE interference suppression for direct-sequence spreadspectrum CDMA. *IEEE Trans. Commun. Theory*, 42: 3178–3188, December 1994.

[25] N.B. Mehta, L.J. Greenstein, T.M. Willis, and Z. Kostic. Analysis and results for the orthogonality factor in WCDMA downlinks. In *Proc. of the Vehicular Technol. Conf.*, May 2002.

[26] S. Moshavi, E. Kanterakis, and D.L. Schilling. Multistage linear receivers for DS-CDMA systems. *Int. J. Wireless Information Networks*, 3(1): 1–17, 1996.

[27] R.R. Muller and S. Verdú. Design and analysis of low complexity interference mitigation on vector channels. *IEEE J. Selected Areas in Commun.*, 19(8): 1429–1441, August 2001.

[28] C. Papadias and D.T.M. Slock. Fractionally spaced equalization of linear polyphase channels and related blind techniques based on multichannel linear prediction. *IEEE Trans. Signal Processing*, 47(3): 641–654, March 1999.

[29] K.I. Pedersen and P.E. Mogensen. The downlink orthogonality factors influence on WCDMA system performance. In *Proc. Vehicular Technology Conf.*, September 2002.

[30] H. Poor and S. Verd'u. Probability of error in MMSE multiuser detection. *IEEE Trans. Information Theory*, 43(3): 858–871, May 1997.

[31] J.G. Proakis. *Digital Communications*, 3rd edition. McGraw-Hill, 1995.

[32] T.S. Rappaport. *Wireless Communications—Principles and Practice*. Prentice Hall, 1996.

[33] L.K. Rasmussen and I.J. Oppermann. Ping-pong effects in linear parallel interference cancellation for CDMA. *IEEE Trans. Wireless Commun.*, 2(2): 357–363, March 2003.

[34] COST Telecom Secretariat, European Commission. COST231 Final Report. Digital Mobile Radio towards Future Generation Systems. Technical report, Brussels, Belgium, 1999.

[35] J.S. Sadowsky, D. Yellin, S. Moshavi, and Y. Perets. Cancellation accuracy in CDMA pilot interference cancellation. In *Proc. Vehicular Technol. Conf.*, April 2003.

[36] J.S. Sadowsky, D. Yellin, S. Moshavi, and Y. Perets. Capacity gains from pilot cancellation in CDMA networks. In *Proc. Wireless Commun. and Networking Conf.*, March 2003.

[37] C. Schlegel, S. Roy, P. Alexander, and Z. Xiang. Multiuser projection receivers. *IEEE J. Selected Areas in Commun.*, 14(8): 1610–1618, October 1996.

[38] D.T.M. Slock and I. Ghauri. A blind maximum SINR receiver for the DS-CDMA forward link. In *Proc. ICASSP'2000*, Istanbul, Turkey, June 2000.

[39] W. Stallings. *Data and Computer Communications, 7th edition*. Prentice Hall, May 2003.

[40] G. Strang. *Introduction to Linear Algebra, 3rd edition*. Wellesley-Cambridge Press, June 1998.

[41] Telecommunication Industry Association. Mobile Station Base-Station Compatibility Standard for Dual-Mode Wideband Spread Spectrum Cellular System. Technical Report TIA/EIA/IS-95, 1993.

[42] D. Tse and P. Viswanath. *Fundamentals of Wireless Communication*. Cambridge University Press, May 2005.

[43] M.K. Varanasi and B. Aazhang. Multistage detection in asynchronous code-division multiple-access communications. *IEEE Trans. Commun. Theory*, 38: 509–519, April 1990.

[44] S. Verdú. *Multiuser Detection*. Cambridge University Press, 1998.

[45] Y.P.E. Wang and G.E. Bottomley. Generalized rake reception for cancelling interference from multiple base stations. In *Proc. Vehicular Tech. Conf.*, Boston, MA, September 2000.

[46] K. Zayana and B. Guisnet. Measurements and modelisation of shadowing crosscorrelations between two base stations. In *Proc. ICUPC*, Rome, Italy, October 1998.

[47] J. Zhang, E.K.P. Chong, and D.N.C. Tse. Output MAI distributions of linear MMSE multiuser receivers in DS-CDMA systems. *IEEE Trans. Information Theory*, 47(3): 1128–1144, March 2001.

Chapter 5

Packet Scheduling Principles and Algorithms for Downlink

Kamal Deep Singh, Gerardo Rubino, and César Viho

Contents

5.1 Introduction

This chapter presents the packet scheduling principles and different scheduling algorithms for High-Speed Downlink Packet Access (HSDPA). HSDPA is an enhancement of UMTS (Universal Mobile Telecommunications

173

System) networks that supports data rates of several megabits per second (Mbps), making it suitable for data applications ranging from file transfer to multimedia streaming. Despite the fairly high data rates that HSDPA offers, the shared downlink radio channel used in HSDPA is a challenging environment for delay- and loss-sensitive applications like video streaming. Thus, resource management algorithms are needed to ensure good quality-of-service.

MAC (medium access control) layer packet scheduling is one of the salient points of HSDPA to perform resource management (i.e., bandwidth allocation between terminals), taking into account the radio channel conditions of all users. In some proposals, additional factors such as fairness between users or cell throughput or quality-of-service (QoS) parameters are also considered in the scheduling mechanism.

The focus of this chapter is on downlink-centric services in the 3G network. For this reason, other services such as VoIP (Voice over Internet Protocol), which are also dependent on uplink, are not discussed in detail, although some discussion about them is provided at the end. Moreover, the services considered in this chapter, such as video streaming, have relatively flexible delay requirements as compared to VoIP. In addition, it should be noted that the emphasis of this chapter is only on MAC layer scheduling and other UTRAN (UMTS Terrestrial Radio Access Network) functions, shown in Figure 5.1, that affect the QoS are out of scope of this chapter.

This chapter is structured as follows. First the key concepts related to HSDPA scheduling are presented. Then, different packet scheduling

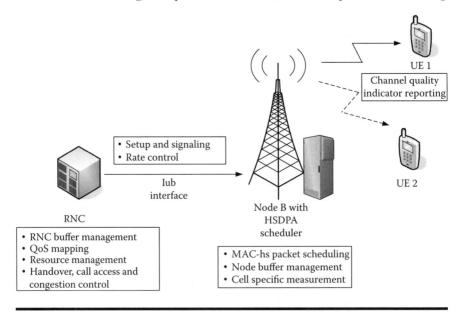

Figure 5.1 Overview of functional split between RNC, Node B, and UE.

algorithms are discussed and their performance analyses is done. The chapter concludes after a short discussion of some points that are not considered in this chapter.

5.2 Packet Scheduling

The HSDPA scheduler is the key to resource management in the UTRAN downlink because it decides which user is to be scheduled in each transmission time interval (TTI). In general, the scheduling decision does not take into account only the channel conditions, but additional factors such as fairness between users, cell throughput, and QoS parameters are typically considered in the scheduling mechanism. The choice of a scheduler usually involves some trade-offs between these factors.

In HSDPA, a new transport channel called high-speed downlink shared channel (HS-DSCH) has been introduced. HS-DSCH is supported by an auxiliary channel called high-speed shared control channel (HS-SCCH). Node B uses HS-SCCH to send the parameters related to demodulation and decoding and sends this information two slots in advance. The value of TTI in HSDPA has been reduced to 2 ms, which is equal to three slots, as compared to the 10, 20, 40, and 80-ms intervals supported by UMTS Release 99. In the uplink, high speed dedicated physical control channel (HS-DPCCH) is used to feed back ACK/NACK and to allow for fast monitoring of the radio channel conditions of all users: Every TTI of 2 ms, a UE (User Equipment) can send a channel quality indicator (CQI) to Node B over this control channel. The relationship between these channels is shown in Figure 5.2.

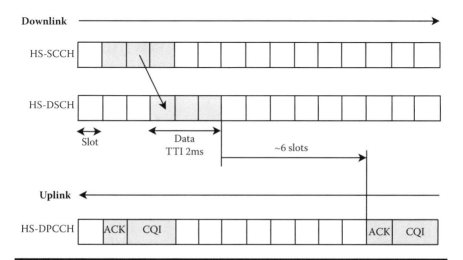

Figure 5.2   HSDPA channels and timing relationship.

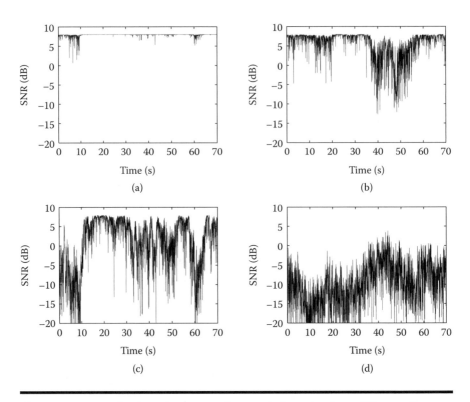

Figure 5.3 Example of received power by UEs (Ped A, 3 km/h) at different distances from Node B: (a) 50 m, (b) 200 m, (c) 350 m, and (d) 500 m.

CQI feedback makes it possible to optimize the transmission by means of channel-adaptive schemes. The quality of reception can vary due to variable radio channel conditions over time. There are many factors such as distance of UE from Node B, transmission power, fading, noise that can affect the reception quality. For example, in Figure 5.3 it can be seen that the channel conditions,[1] especially for the farther users, may change rapidly. To improve the chances of correct reception, the robustness of the transmitted signal is adapted to the channel conditions and this is called Adaptation, Modulation and Coding (AMC).

With AMC, HSDPA can adaptively choose the modulation and coding for a given time slot as a function of current channel quality. While choosing modulation and coding, the trade-off is between robustness and data rate. This approach, combined with the packet scheduling and CQI, has certain advantages. For instance, the users experiencing good channel quality can

---

[1] The curves in the figure were obtained through the simulation [1] of a PHY-layer model.

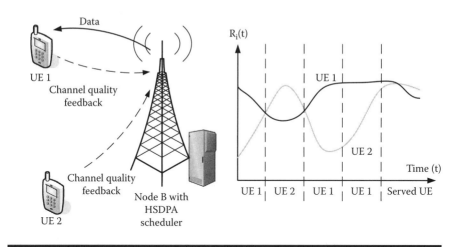

Figure 5.4   Example of channel-aware HSDPA scheduling. Here the user with the best channel conditions, in the given TTI, is scheduled.

benefit from high data rate and, in the presence of several users, the packet scheduling algorithms can benefit from user diversity and, thus, schedule users during constructive fades. In the case of several users, the resource management is performed by the scheduler that may adapt to fluctuating radio conditions. Every TTI, the scheduler chooses the next user to be served based on the channel conditions of *all* the users. This channel-adaptive scheduling can be used, for instance, to maximize the global cell throughput by scheduling the user with best channel condition, as shown in Figure 5.4. Note that a given UE can experience strong fluctuations in bandwidth over time, due to such channel-dependent scheduling policies.

5.2.1 Quality-of-Service-Aware Scheduling

Some applications, such as VoIP or multimedia streaming, have requirements that cannot be satisfied with *Best-Effort* service when there are not enough network resources available. This can occur when the network is congested or when other services, those without stringent requirements, compete aggressively for available resources. For these applications, the network infrastructure must be adapted to give them special treatment in terms of minimum bandwidth, delay, loss tolerance, jitter, etc. This special treatment makes it possible to satisfy the specific requirements of different services and ensures good QoS.

Different QoS parameters or UMTS bearer attributes such as traffic class and traffic handling priority are defined in [2]. However, the values of these parameters are not available at Node B where the HSDPA scheduler is implemented. In fact, Node B can also be from a manufacturer that is

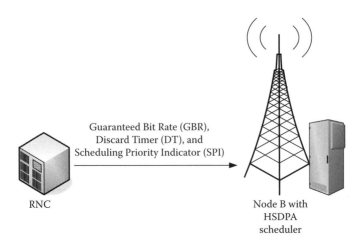

Guaranteed Bit Rate (GBR),
Discard Timer (DT), and
Scheduling Priority Indicator (SPI)

RNC

Node B with
HSDPA
scheduler

Figure 5.5   RNC sets QoS parameters for HSDPA scheduler present in Node B.

different from that of the radio network controller (RNC). Thus, an interface between the RNC and Node B is standardized and the RNC is responsible for setting QoS parameters in Node B through the Iub interface [3], as shown in Figure 5.5. Some of the relevant parameters are guaranteed bit rate (GBR), the minimum guaranteed rate at which the user data will be scheduled; scheduling priority indicator (SPI), the priority with respect to other flows or users; and discard timer (DT) [3], the value in seconds or milliseconds such that any packet delayed more than this value must be discarded by Node B. Note that the RNC has to map all the QoS parameters defined in Release 99 to these parameters in order to provide QoS.

After these parameters are configured by the RNC, it is then up to the HSDPA packet scheduler algorithm to use these values for serving different users. 3GPP does not specify the exact behavior of the scheduler as a function to these parameters and, hence, this is kept open for the implementation. The choice of values for these parameters or mapping of different UMTS QoS class attributes to these parameters by the RNC is also operator dependent. In addition, the operator must decide how to handle the user traffic when it exceeds GBR or when the user data rate is variable, even when the average bit rate remains equal to the GBR.

In some cases, the RNC need not set the DT at Node B. For example, in the case of multimedia streaming, where delays on the order of some seconds are tolerated, it may disable the DT at Node B. Nevertheless, it can still use a DT at the RNC such that, on timeout, it can drop full IP packets before segmenting them into RLC-PDUs (Protocol Data Units) and before sending the segments to Node B. On the other hand, when VoIP service is used, the delay requirements are very strict and are on the order of 100 ms. In this case, the RNC should avoid queuing the IP packets and instead should send them to Node B as soon as possible. In addition, it should configure the DT at Node B, (e.g., to a value of 100 ms) to avoid queue buildup at Node B. That way, any queue buildup will be avoided because Node-B queues will always be emptied after timeout because either the enqueued data will be transmitted or it will be discarded. This will work even when a flow control algorithm is used between the RNC and Node B, because after the queue is emptied at Node B, it will automatically ask for more data from the RNC.

Regarding the decision to set SPI values, usually the higher priority will be set for VoIP flows, lower priority for streaming flows, and lowest priority for the flows belonging to non-real-time traffic.

5.3 Packet Scheduling Algorithms

Numerous MAC-layer schedulers have been proposed and studied in the literature. Here we discuss some of the schedulers that have been proposed for HSDPA and are representative of different design trade-offs and constraints. We divide the set of schedulers into two parts and call them Best-Effort schedulers and QoS schedulers. Best-Effort schedulers do not provide QoS guarantees, whereas QoS schedulers can provide QoS guarantees. Further explanation about them is given in the following text.

The simulations in this section are done using a packet-level simulator. To simulate HSDPA, this chapter uses the Enhanced UMTS Radio Access Network Extensions for ns-2 (EURANE) extensions [1] to ns-2 [4]. The simulation settings for UTRAN are as follows. The Okamura-Hata distance loss model and correlation model [1,5] for shadow fading are used with 10 W of Node-B transmission power resources for HS-PDSCH, 17 dBi of antenna gain, 30 dBm of intra-cell interference, and −70 dBm of inter-cell interference. UM mode is used in the RLC layer, UE category 1-6 is assumed, and thus the maximum number of available codes for HS-DSCH is 5 and the maximum instantaneous bit rate is 3.6 Mbps.

The multipath environment used is Pedestrian-A with a user speed of 3 kmph. The cell radius is assumed to be 500 m. In the simulations, the distance of users from Node B is randomly picked to be $500\sqrt{u}$ meters, where u is a uniformly distributed variable between 0 and 1.0. Note that this approach ensures uniform distribution over a circle of radius 500 m, whereas directly picking the values of user distance equal to a uniform random variable between 0 and 500 m, instead, does not ensure uniform distribution. Moreover, once the distance is chosen for a user, it is assumed that the user moves in a circle without changing the distance from Node B to keep the simulations tractable with respect to user distance. The simulations are run 20 times for each simulation configuration to obtain good confidence intervals. The parameters that change are explicitly specified in the following text, and other detailed parameters are the same as the default settings used in [1].

5.3.1 Best-Effort Schedulers

Best-Effort schedulers do not provide any QoS guarantees to different flows. Mapping of some flows to Best Effort can be done, for example, by mapping Interactive class and Background class flows to Best Effort. The QoS parameters, such as GBR are set to zero. Because Best-Effort flows do not have strict requirements and QoS guarantees are not provided to them, the main objective of the scheduler algorithm in this case is to optimize the global throughput and ensure that resources are divided fairly among different users. A summary of some Best-Effort schedulers is provided below. Please refer to Table 5.1 while reading it.

The Round Robin (RR) scheduler, which is probably the simplest one in terms of implementation, gives the time slot to users in a cyclic manner. Note that RR is fair with respect to system resources, or time slots, but in general the capacity is *not* shared equally among the UEs, as shown in Figure 5.6 that shows the cumulative distribution function (CDF) of user throughput when ten FTP users are in the system. This is explained by the fact that each user may experience different channel conditions. Moreover, RR policy is not optimal with respect to maximizing the global throughput, as is shown in Figure 5.7, because this policy does not utilize channel condition feedback.

The most aggressive policy to utilize the channel feedback is called Maximum C/I scheduling.[2] This is the same algorithm as illustrated in Figure 5.4. The Max C/I scheduler gives the channel to a user having the best channel conditions, which in turn is the user that can support the maximum instantaneous data rate in the current TTI. This algorithm is also summarized in Table 5.1. Note that $R_i(t)$, used in Table 5.1, is the instantaneous supportable data rate at which user i can be served at time t, depending on CQI and $BLER_{target}$, which is typically 10%.

The Max C/I scheduler provides the highest *cell* (global) throughput because it always serves the users that can support the highest data rates. This is shown in Figure 5.7. On the other hand, this scheme is very unfair because a user nearest to Node B can get all the resources, and the users farther away will be starved. This can be seen in Figure 5.6(a), which shows the CDF of user throughput. It can be seen that some users get very good throughput, up to 1500 kbps, because they are scheduled most of the time due to their good channel quality, whereas some other users get very bad throughput in the range of 10 kbps to 30 kbps because of their relatively poorer channel conditions.

Unlike the Max C/I (Carrier-to-Interference Ratio) scheduler, the Proportionally Fair (PF) scheduling algorithm [6–8] looks at the relative channel

---

[2] Max C/I denotes the signal-to-noise ratio that is used to estimate the optimal value of the instantaneous supportable data rate, $R_i(t)$, with a given block error rate target ($BLER_{target}$).

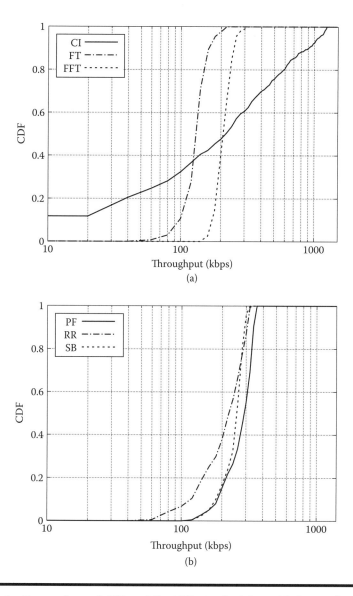

Figure 5.6 Comparison of different Best-Effort schedulers. (a) Comparison of CI, FT, and FFT schedulers and (b) comparison of PF, RR, and SB schedulers. CDF of user throughput over a time window of 1 minute is plotted. The bin size used for calculation of CDF is 20 kbps. There are ten FTP users in the system.

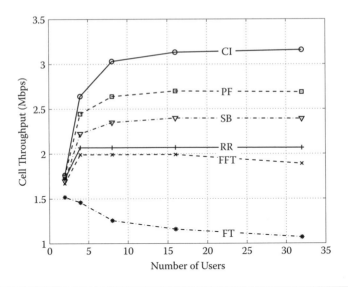

Figure 5.7   Cell throughput.

conditions of a user with respect to its own past channel conditions. PF assigns the current time slot at time t to a user i^* having maximum ratio of supportable instantaneous rate $R_i(t)$ and its own exponentially weighted moving average throughput $\lambda_i(t)$, which is calculated as follows:

$$\lambda_i(t) = \left(1 - \frac{1}{\tau}\right) \cdot \lambda_i(t - \Delta t) + \frac{s_i}{\tau} \cdot R_i(t) \qquad (5.1)$$

with $\tau > 1$, Δt is equal to the length of the TTI, $s_i = 1$ when user i is served in the current time slot, and $s_i = 0$ otherwise. It should be noted that, with respect to Equation (5.1), different implementation variants are possible in order to deal with a few special cases such as when the user queue is empty or the data present in the queue of the scheduled user queue is less than that corresponding to $R_i(t)$. In the implementation considered in this chapter, when user i is being scheduled and the data in this queue is less than that corresponding to $R_i(t)$, we replace $R_i(t)$ in Equation (5.1) with the rate corresponding to the actual amount of data in the queue. Moreover, when a user's queue is empty, and thus is not being scheduled in the current time slot, $\lambda_i(t)$ is not updated and $\lambda_i(t) = \lambda_i(t - \Delta t)$.

The PF scheduler offers a good trade-off between cell throughput (see Figure 5.7) and fairness (see Figure 5.6(b)), as it gives the channel to the user having "relatively good" channel conditions and also results in a fair distribution of resources. This scheduler provides the so-called proportional fairness criterion defined in [9]. A Generic PF (GPF) scheduler [9,10] is also shown in Table 5.1. The GPF scheduler uses the ratio of $[R_i(t)]^\eta$ and $[\lambda_i(t)]^\gamma$

Table 5.1   Most Studied Best-Effort Scheduling algorithms

Scheduler	Algorithm (Note that ties are broken randomly):	Scheduled User i^* Is Given by:
Round Robin (RR)	Schedule users in cyclic order	Cyclic manner
Fair Throughput (FT)	Equalize throughput of all users	$\arg\min_i \{\lambda_i(t)\}$
Max C/I	Choose user with best instantaneous supported data rate	$\arg\max_i \{R_i(t)\}$
Proportionally Fair (PF)	Choose user with best transmission rate relative to user's own average throughput	$\arg\max_i \left\{ \frac{R_i(t)}{\lambda_i(t)} \right\}$
Generic Proportionally Fair (GPF)	Similar to PF algorithm with additional exponents γ and η	$\arg\max_i \left\{ \frac{[R_i(t)]^\eta}{[\lambda_i(t)]^\gamma} \right\}$
Score Based (SB)	Choose user with best rank for channel quality based on history	$\arg\min_i \{ S_i(t) \}$
Fast Fair Throughput (FFT)	Equalize throughput of all users and exploit channel conditions	$\arg\max_i \left\{ \frac{R_i(t)}{\lambda_i(t)} \frac{\max_j(\overline{R_j(t)})}{\overline{R_i(t)}} \right\}$

Note: Here, for a given user i at time t, $\lambda_i(t)$ is the exponentially weighted moving average throughput, $R_i(t)$ is the instantaneous supportable data rate, $\overline{R_i(t)}$ is the exponential weighted moving average of $R_i(t)$, $\max_j(\overline{R_j(t)})$ is the maximum among different users, $S_i(t)$ is the rank of the current channel condition with respect to past W values for the same user. In addition, W, γ, η are implementation parameters.

for deciding the user to be scheduled. When the exponents γ and η are close to 1.0, then the GPF scheduler behaves similarly to the PF scheduler. Whereas, when γ is increased more than 1.0, or η is decreased less than 1.0, then GPF starts behaving in an "extra" fair manner by favoring users with bad channel conditions. On the other hand, when γ is decreased less than 1.0 or η is increased more than 1.0, then it starts behaving like the Max C/I scheduler. This can be seen in Figure 5.8 that shows the CDF of user throuputs with different values of γ and η.

The work in [11] argues that the PF scheduler results in fair distribution only under ideal conditions, when users experience symmetric fading statistics, and becomes unfair in realistic conditions. A Score-Based (SB) scheduler is proposed that ranks the current channel condition of a user with respect to the past channel conditions over a window of size W. For example, the current and past W values $R_i(t)$ to $R_i(t - W + 1)$ for a user i, are ranked from 1 to W, depending on their values. Similarly, the current

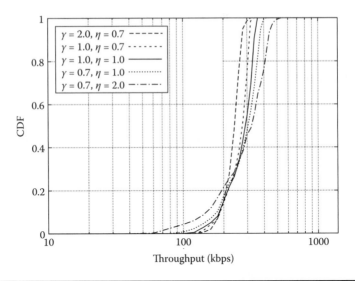

Figure 5.8 Generic PF scheduler with different values of γ and η.

channel conditions of other users are ranked based on their respective past channel conditions. After that, users are sorted according to the rank of their current channel condition and then the user having the best rank is scheduled in the current TTI. Note that ties are broken randomly and similarly, while in ranking, if two values related to channel conditions are the same for a given user, then either of them is ranked better than the other with equal probability. It is reported in [11] that SB scheduling is able to distribute the time slots equally even when asymmetric channel variations are encountered. Figure 5.6 shows the CDF of user throughput obtained with SB and from Figure 5.7 it can be seen that equal distribution of time slots is achieved at the cost of slightly decreased overall cell throughput when compared with the PF scheduler. However, the cell throughput obtained with SB is still higher than that obtained with RR as shown in Figure 5.7.

In fact, the implementation of SB scheduling used in this chapter is slightly different and better than in [11]. It is suggested in [11] that in the case of a tie, while comparing the current channel condition with one of the past values, either of them is ranked better than the other with equal probability and comparison is moved to next value in the window. However, during the simulations done for this chapter, it was found that this approach can lead to throughput starvation when most of the past values are the same. For example, when past W values are the same as the current CQI representing current channel conditions, then after W coin tosses, it is expected that the unbiased coin will produce one of the two outcomes $\frac{W}{2}$ times. Thus, the rank comes to be around $\frac{W}{2}$ most of the time. This is a relatively bad rank; and when the number of users in the system increases, then the chances are high that at least one user will have a better rank than $\frac{W}{2}$. This means that the chances of a user having similar past W or numerous CQI values getting scheduled will be much less.

This particular case of a user having numerous similar past CQI values is more frequent with users close to Node B because they experience very good channel conditions and most of the time the CQI value will be the maximum value possible and thus will be same. In the implementation used in this chapter, instead of tossing an unbiased coin each time, when a value equal to the past value is encountered, the implementation proceeds as follows: Denoting by W_s the number of past CQI values in the window that are equal to current CQI, a random integer between 0 and W_s having a uniform distribution, is drawn and added to the rank of current CQI obtained after comparing it with the unequal past CQI values in the window. It was observed during the simulations, the results of which are shown in Figures 5.6(b) and 5.7, that this approach worked and the users near Node B were no longer starved.

There exist other schedulers that use different criteria for fairness. A scheduler called Fair Throughput (FT) equalizes the throughputs of all users by scheduling the user with minimum average throughput. With FT, all users in the system, at a given time, obtain equal throughput. In Figure 5.6(a), the CDF is not exactly vertical because although the throughputs are the same at a particular time, even then they can still vary from time to time. FT has a drawback in that it does not exploit channel condition feedback. Moreover, it should be noted that with the FT scheduler, some users experiencing bad channel quality can cause deterioration of throughput for other users as well. This leads to a low value of overall cell throughput, as shown in Figure 5.7. This is because the scheduler will start providing more resources to the users with bad quality, which in turn would mean low average throughput for all the users. Thus, if FT is used, then it will have to rely on resource management procedures to drop some users experiencing bad quality in order to avoid deteriorating the throughput for other users. Another version of the FT scheduler is Fast Fair Throughput (FFT). FFT, proposed in [12], aims to provide fair throughput distribution as well as exploit channel condition feedback. As shown in Table 5.1, this distribution is achieved by multiplying the term in PF by the ratio of $max_j(\overline{R_j(t)})$ to $\overline{R_i(t)}$, where $\overline{R_i(t)}$ is the exponential weighted moving average of $R_i(t)$ and $max_j(\overline{R_j(t)})$ is the maximum $\overline{R_j(t)}$ among different users.

5.3.2 QoS Schedulers

QoS guarantees will be required by users of services such as video streaming that have requirements in terms of minimum bit rate and maximum packet delay. User satisfaction is the main factor to consider. Satisfaction criteria are different for different services, each having its own QoS

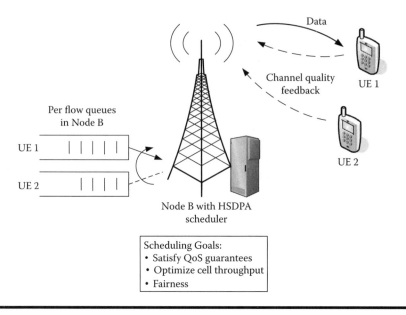

Figure 5.9   General goals of QoS-aware packet scheduling algorithms.

requirements. Thus, as shown in Figure 5.9, packet scheduling should aim to fulfill these QoS requirements instead of just focusing on cell throughput and fairness.

QoS classes and parameters have already been defined by 3GPP standards [2]. However, these standards do not define the scheduler implementation details, and thus provide flexibility to implementers and Node B vendors to choose their own scheduler in order to map these QoS classes and parameters.

Several schedulers that can offer QoS assurances in HSDPA have been proposed and studied in the literature [13]. Pedersen et al. [14] discuss different options for providing QoS using different QoS-aware schedulers. This section discusses various QoS schedulers. For quantitative analysis, different simulations are performed with a focus on video streaming services. The video traffic is generated by taking a video of 1-minute duration. The uncompressed video is a raw YUV video formed by concatenating three well-known Quarter Common Intermediate Format (QCIF) reference video sequences ("Foreman," "Mother and Daughter," and "News") in order to have a video of 1 minute and also to have a video with variable complexity. The video is then compressed using x264 codec [15] with H.264 baseline profile level 1.2 with no Bi-directional (B) frames, 15 frames per second, and strict 100-kbps CBR with a maximum 10% of bit rate variations. The EvalVid [16] methodology is used to perform EURANE simulations with the packet traces. In the following text, some QOS schedulers are discussed. Please refer to Table 5.2 while reading the discussion.

5.3.2.1 Weighted Delay-Based Schedulers

A scheduler known as Modified Largest Weighted Delay First (MLWDF) is proposed in [17]. MLWDF chooses the user $i^* = \arg\max_i\{a_i w_i(t)\frac{R_i(t)}{\lambda_i(t)}\}$, where $w_i(t)$ can be the delay of head-of-line packet in the user queue, user queue length, or $\frac{Q_i(t)}{GBR_i}$, where $Q_i(t)$ is the number of tokens corresponding to the user token bucket algorithm, which is explained later. This choice ragarding the value of $w_i(t)$ depends on different versions of MLWDF and whether QoS guarantee is in terms of minimum delay or guaranteed bit rate (GBR_i) for user i.

MLWDF can provide two notions of QoS. First, it can achieve QoS requirements of the form $\Pr\{w_i(t) > D_i\} \le \delta_i$, where $w_i(t)$ is the delay of the head-of-line packet in the user queue, D_i is the target delay, and δ_i is the probability of QoS requirement violation. Second, it should be noted that this QoS requirement in terms of delay can easily be rewritten in another form to provide minimum rate guarantees. Moreover, MLWDF can support mixed types of QoS guarantees in the system such that some of them are provided delay guarantees and others are provided rate guarantees. In the case when a minimum rate guarantee is provided, MLWDF uses a token

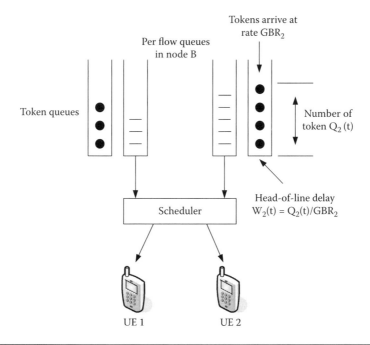

Figure 5.10   Using token queues with the MLWDF scheduling algorithm.

bucket control algorithm, as shown in Figure 5.10, where for each user, i tokens arrive at a constant rate equal to GBR_i. The MLWDF rule is used to schedule the users, and $w_i(t)$ is replaced by the delay of the head-of-line token in the token queue that, in turn, is equal to $\frac{Q_i(t)}{GBR_i}$ because tokens arrive at constant rate GBR_i. When a user is scheduled in the current TTI, a number of tokens equal to the amount of data sent for that user are removed from his token bucket.

The parameter a_i is an implementation parameter that reflects the strictness of QoS guarantees and in case of delay guarantees; [18] suggests the value of a_i to be $-log(\delta_i)/D_i$. In the case of rate guarantees, a_i is *not* equal to $-log(\delta_i)/GBR_i$ because GBR_i is not exactly equivalent to D_i. Any specific value of a_i, when rate guarantees are provided, is not suggested in [18], although it is recommended to set a higher value of a_i to provide stricter assurance of minimum rate guarantee for a given user i.

Figure 5.11 shows the benefits of using a QoS-aware scheduler, MLWDF, over Best-Effort schedulers when some users in the cell require QoS guarantees. A mixed scenario is considered in the simulation with some video streaming users and FTP users in the background. To decide on the maximum number of video streaming users that can be supported, a criterion is used such that 90% of the video streaming users should be satisfied across different simulation runs. The value of GBR_i is set to 110 kbps. This is slightly higher than the compressed video bit rate of 100 kbps. The value of a_i is set to 1.0 for all users. Figure 5.11 shows the percentage of satisfied users with respect to an increasing number of video streaming users. It should be noted that, in this figure, a video streaming user is assumed to be satisfied if PSNR (peak-signal-to-noise ratio) of the received video as compared to original uncompressed video is higher than 30 dB. The PSNR is calculated by considering the whole video as one image and then comparing it with the original uncompressed video. It should be noted that deterioration in video quality is caused due to lost packets and delayed packets that are discarded.

In Figure 5.11(a), it can be seen that in case of no Best-Effort flows, MLWDF can support a significantly greater number of users as compared to Best-Effort schedulers. In the case of Best-Effort load, generated using ten FTP flows, the difference in terms of the maximum number of video streaming users supported as compared to that of Best-Effort schedulers is even higher as shown in Figure 5.11(b). In fact, no Best-Effort scheduler is able to satisfy 90% of users in any case when ten FTP users are present in background, as shown in Figure 5.11(b). This is because the Best-Effort scheduler cannot guarantee minimum throughput, and this leads to packet losses that, in turn, degrade the video quality. On the other hand, a QoS-aware scheduler can provide a minimum rate guarantee for up to 18 video streaming users in the case of no Best-Effort load and 16 video streaming users in case of Best-Effort load.

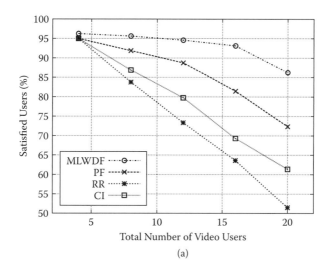

(a)

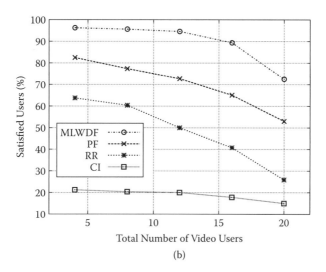

(b)

Figure 5.11 **Percentage of satisfied video users with increasing number of video users. (a) No background traffic; and (b) ten FTP users as background traffic. A video user is assumed to be satisfied if video PSNR > 30. Drop Tail queue management is used at RNC with 60 packets of queue limit. Discard timer is not set, and a credit-based flow control is used between RNC and Node B.**

While studying MLWDF, another interesting scheduling function was reported in [18] and called the Exponential Rule (EXP). The EXP scheduler chooses the user

$$i^* = \arg\max_i \left\{ a_i \frac{R_i(t)}{\lambda_i(t)} \exp\left(\frac{a_i w_i(t) - \frac{1}{N}\sum_i a_i w_i(t)}{b + \sqrt{\frac{1}{N}\sum_i a_i w_i(t)}} \right) \right\}$$

where N is the total number of users and b is another implementation parameter. This function was further studied in [19]. The idea of the EXP scheduler is to keep the values of $a_i w_i(t)$, for different users, close to each other. This is because when the value of $a_i w_i(t)$ for user i is more than $\frac{1}{N}\sum_i a_i w_i(t)$ by a value that is of the order of $\sqrt{\frac{1}{N}\sum_i a_i w_i(t)}$, then the priority of that user starts to increase [18]. This in turn increases the probability of that user being scheduled. It was also noted in [18] that the term $\frac{1}{N}\sum_i a_i w_i(t)$ can be removed from the numerator in the EXP scheduler equation because it is common for all the users and is kept just for the purpose of illustration. Thus, only the term $\sqrt{\frac{1}{N}\sum_i a_i w_i(t)}$ present in the denominator matters in the scheduling decision.

Intuitively, it can be remarked that there are two differences between EXP and MLWDF. The exponential weight in EXP scheduler always remains more than 1.0, with $a_i \geq 0$ and $w_i(t) \geq 0$, after the common term $1/N \sum_i a_i w_i(t)$ is removed from the numerator in the EXP scheduler equation. Whereas in the case of MLWDF, the weight is equal to $w_i(t)$, which can take values close to or equal to 0 and, in this case, a user will not be scheduled even with a very high value of $R_i(t)$. Moreover, instead of absolute deterioration of QoS guarantees, EXP reacts to relative deterioration of QoS guarantees. The latter is because EXP tries to bring the values of $a_i w_i(t)$ for all users close to $\frac{1}{N}\sum_i a_i w_i(t)$, which in turn means that if QoS is bad for some users then it will become bad for all users. These two schedulers are further compared in the later sections.

Another point that can be made about both schedulers is that the scheduler function is not defined for Best-Effort users as all users get either delay or minimum throughput guarantees. Thus, when RNC does not set the values for GBR_i and Discard Timer, the behavior of MLWDF and EXP implementation is not defined. However, this can be easily overcome. For example, if RNC sets $GBR_i = 0$ for some flows, the scheduler implementation in Node B can set GBR_i to a small value, that is, more than 0 (for example, equal to 10 kbps). This means that the Best-Effort flows, even though they may no longer be called Best-Effort, will get at least 10 kbps of throughput. They can always get more throughput, depending on the channel conditions, as any additional capacity will be divided among all the users. Further it should be noted that even non-real-time flows can always be assigned some QoS guarantees, depending on the QoS mapping defined by the operator.

5.3.2.2 Utility-Based Schedulers

As we discuss more QoS schedulers, it will be seen that most of these schedulers take the slot assignment decision and pick the user i^* such that

$$i^* = \arg\max_i \left\{ B_i(t) \frac{R_i(t)}{\lambda_i(t)} \right\} \tag{5.2}$$

where $B_i(t)$ represents a function that increases or decreases a user's priority, depending on his QoS. Note that a user's priority in turn affects his chances of being scheduled in the current TTI. In the case of MLWDF, $B_i(t) = a_i w_i(t)$; and in the case of EXP,

$$B_i(t) = a_i \exp \left(\frac{a_i w_i(t) - \frac{1}{N} \sum_i a_i w_i(t)}{b + \sqrt{\frac{1}{N} \sum_i a_i w_i(t)}} \right)$$

To design QoS schedulers, it was shown in [20] that, given a user utility $U_i(\lambda_i)$, it is possible to obtain a "good" scheduler that picks the user i^* such that

$$i^* = \arg\max_i \left\{ R_i(t) \frac{\partial U_i(\lambda_i)}{\partial \lambda_i} \right\} \tag{5.3}$$

Note in the above formulation that the corresponding utility functions for RR, Max C/I, and PF scheduling are $U(\lambda) = 1$, $U(\lambda) = \lambda$, and $U(\lambda) = \log \lambda$, respectively. However, it should be remarked that defining the required utility function is a difficult task. A Rate-Guarantee (RG) scheduler is designed in [21] using Equation (5.3). A utility function of the form $U(\lambda) = \log \lambda + 1 - \exp(-\beta \cdot (\lambda - GBR))$ is assumed, with $\beta > 0$, resulting in the RG scheduler with

$$B_i(t) = \begin{cases} 1 + \lambda_i(t) \cdot \beta \cdot \exp(-\beta \cdot (\lambda_i(t) - GBR_i)) & \forall i \in \mathcal{Q} \\ 1 & \forall i \in \mathcal{B} \end{cases} \tag{5.4}$$

which corresponds to the case where the minimum throughput, GBR_i, is required by a QoS user i; $\mathcal{Q}$ and $\mathcal{B}$ denote the set of QoS and Best-Effort (BE) users, respectively. Note that the exponential function ensures that whenever a user with minimum throughput guarantee GBR_i is getting a rate lower than GBR_i, then the exponential function will increase rapidly, which in turn will increase the chances of that user getting scheduled. On the other hand, when the user throughput is greater than GBR_i, then the value of the exponential function will decrease slowly. The implementation parameter β controls the aggressiveness of the exponential function, to increase and decrease, depending on the value of the difference between the current user throughput and GBR_i.

Similarly, other schedulers based on delay guarantees can be designed [20]. In the literature, there exist several studies [14,20,22] that use RG to provide rate guarantees to QoS users. Unlike MLWDF and EXP, the RG scheduler has an option to schedule Best-Effort users that, in turn, are the users without any QoS guarantees. With RG, QoS users are scheduled according to their QoS requirements and any remaining throughput capacity is distributed in a proportionally fair manner to Best-Effort users. Whether a user flow is assigned to the Best-Effort class is decided by RNC that, in turn, sets the QoS parameters like GBR to 0 for those flows.

It should be noted that when the minimum throughput guarantee GBR_i of a QoS user is getting satisfied, and thus $\lambda_i(t) = GBR_i$, then $B_i(t) = 1 + \beta \cdot GBR_i$, which is a value that is higher than $B_i(t) = 1$ for Best-Effort users. The study in [22] thus replaces the term $\lambda_i(t) \cdot \beta$, before the exponential term in Equation (5.4), with α, for QoS users, and replaces $B_i(t) = 1$ with $B_i(t) = 1 + \alpha$ for Best-Effort users. This implies that the value of $B_i(t)$ for Best-Effort users becomes the same as that for QoS users when their QoS guarantees are being satisfied. However, this variant is not recommended for implementation, as it was reported in [23] that it shows disparity among different QoS users when they have different values of GBR_i, as discussed below.

The work in [23,24] studied the RG scheduler and it was found that both the original RG scheduler in [20] and its variant in [22] suffer from the deterioration of rate guarantees of the QoS users as the number of "active" BE users $n_{be} = |\mathcal{B}|$ increases. This is because as n_{be} increases, the value of λ_i for BE users decreases. This in turn increases the term $B_i(t)/\lambda_i$ in Equation (5.2) for BE users as $B_i(t) = 1$ remains constant. This means that BE users will start taking more resources from QoS users as their number increases. Moreover, the above schedulers, especially the variant in [22], are reported to be biased toward some users, depending on their values of rate guarantees. To solve these issues, the work in [24] proposes an alternate scheduler called Required Activity Detection (RAD), which is not a utility-based scheduler and is discussed in the next section, and the work in [23] proposes a normalized version of the RG scheduler called Normalized Rate Guarantee (NRG).

NRG is another utility-based scheduler that is based on RG. It overcomes the drawbacks of RG and is a normalized version of the RG scheduler. Unlike RG, which uses the absolute difference between λ and GBR, NRG uses a normalized term $(\lambda - GBR)/GBR$ in order to increase or decrease the exponential function based on the percentage drop in the throughput rather than using the absolute difference. Another modification is done to ensure that the increase in BE load should not deteriorate the quality of the QoS flows, and the increased BE load should be shared among BE users only. Thus, the goal is to segregate BE users from QoS users, as explained below.

NRG assumes that the utility for the QoS users is such that $U_Q(\lambda) = GBR \cdot (\log(\lambda) + 1 - \exp(-\beta \cdot \frac{\lambda - GBR}{GBR}))$. For Best-Effort users, NRG considers, similarly to the RG scheduler, that the rate guarantees are "always satisfied," and the utility for BE users is such that $U_B(\lambda) = \frac{k_{be}}{n_{be}} \log(\lambda)$. The value of k_{be}, which is an implementation parameter, determines the proportion of resources allocated to the BE users and the n_{be} term, similar to the concept used in [24], in the denominator, will ensure that if the number of BE users increases, then it will not increase the load on the resources of QoS users. Furthermore, a value $n_{be}^{(min)}$ can be chosen such that $B_i(t) = k_{be}/n_{be}^{(min)}$ if $n_{be} < n_{be}^{(min)}$ to avoid the division by zero and to slightly improve the QoS when the number of BE users goes low.

Using the above utilities and Equation (5.3) results in the NRG scheduler with

$$
B_i(t) = \begin{cases} GBR_i + \lambda_i(t)\beta \cdot \exp\left(-\beta \cdot \frac{(\lambda_i(t) - GBR_i)}{GBR_i}\right) & \forall i \in \mathcal{Q}, \\ \frac{k_{be}}{n_{be}} & \forall i \in \mathcal{B} \end{cases} \tag{5.5}
$$

The parameter β determines the aggressiveness of the scheduler to the percentage drop in the user throughput and $\beta = 6.0$ is used in [23]. The typical values of k_{be} considered in this chapter are between 550 and 1500 when GBR_i and $\lambda_i(t)$ are in kbps unit.

It should be noted that the term $B_i(t)/\lambda_i(t)$ in the scheduler Equation (5.2), with $B_i(t)$ defined by NRG in Equation (5.5), becomes equal for all the QoS users when they achieve their minimum throughput or when the percentage drop in their throughput is the same. Thus, it is called a Normalized Rate Guarantee scheduler because both the terms $(\lambda_i(t) - GBR_i)/GBR_i$ for QoS users and $B_i(t)$ for BE users are normalized. The results reported in [23] show that NRG is able to avoid the deterioration in rate guarantees of the QoS users when the number of Best-Effort users are increased in the system; whereas the increase in Best-Effort load significantly deteriorates the QoS when RG is used. In addition, NRG was shown [23] to be unbiased toward different values of GBR_i.

Figure 5.12 compares the performance of NRG with the performance of a Proportionally Fair (PF) scheduler, a Best-Effort scheduler. Note that the goal of a QoS scheduler is to, first, satisfy the QoS guarantees and then maximize the cell throughput. Thus, we focus on the Best-Effort throughput with respect to the number of QoS users in the system. The setting is similar to that used in a previous section ("Weighted Delay-Based Schedulers") except that Drop Tail is not used and instead a relaxed Discard Timer value of 5.0 s is used. Moreover, a user is assumed to be satisfied if the amount of discarded packets due to Discard Timer is less than 5%.

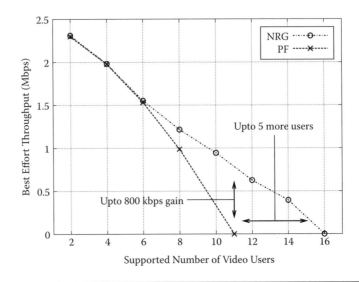

Figure 5.12 QoS-aware scheduler results in the gains in terms of overall throughput and maximum number of QoS users supported in the system. In the simulations, a relaxed Discard Timer value of 5.0 s is used instead of using Drop Tail. The credit-based rate control algorithm between RNC and Node B is not used.

In the case of NRG, ten FTP users are used as Best-Effort users, and the number of video users is increased gradually until 90% of the video users remain satisfied across 20 simulation runs and overall throughput obtained by Best-Effort users is noted. Slight optimization is used with NRG such that when the number of video users is less than ten, then the value of k_{be} in the NRG scheduler function is taken to be 1000 with λ and *GBR* in kbps. But, when the number of video users is more than ten, the value of k_{be} decreased to 500. On the other hand, in the case of PF, the number of video users as well as the number of Best-Effort users varies. It should be noted that for a given number of video users and in the case of the PF scheduler, when the Best-Effort load is increased gradually, it reaches a point where just 90% of the video users are satisfied across 20 simulation runs. This throughput value is noted, for PF, with respect to the number of video users, as it is the maximum Best-Effort throughput possible that can still ensure 90% video user satisfaction.

From Figure 5.12 it is clear that NRG outperforms PF when the number of video users in the system increases. The gain is in terms of up to 800 kbps of additional overall throughput for Best-Effort flows and up to five more video users that can be supported in the system. The gain over PF is due to the fact that after some point, the PF scheduler cannot accept more Best-Effort users as it would not be able to satisfy the QoS users. But NRG can accept any amount of Best-Effort load because it first satisfies the QoS

Table 5.2 QoS Scheduling Algorithms

Scheduler	QoS Guarantee	Choose User $i^* = \arg\max_i \left\{ \frac{B_i(t)R_i(t)}{\lambda_i(t)} \right\}$	
		$B_i(t)$ for QoS Users	$B_i(t)$ for BE Users
MLWDF	Delay/Rate	$a_i w_i(t)$	Not defined
EXP	Delay/Rate	$a_i \exp\left(\frac{a_i w_i(t) - \frac{1}{N}\sum_i a_i w_i(t)}{b + \sqrt{\frac{1}{N}\sum_i a_i w_i(t)}} \right)$	Not defined
RG	Rate	$1 + \lambda_i(t)\beta \exp\left(-\beta(\lambda_i(t) - GBR_i)\right)$	1
RAD	Rate	$\begin{cases} \frac{GBR_i}{\lambda_{sch}^{(i)}(t)} & \text{if } \rho(t) < 1, \\[2ex] \frac{GBR_i/\lambda_{sch}^{(i)}(t)}{\rho(t)} & \text{if } \rho(t) \geq 1 \end{cases}$ where $\rho(t) = \sum_i \frac{GBR_i}{\lambda_{sch}^{(i)}(t)}$	$\begin{cases} \frac{1-\rho(t)}{n_{be}} & \text{if } \rho(t) < 1, \\[2ex] 0 & \text{if } \rho(t) \geq 1 \end{cases}$ where $\rho(t) = \sum_i \frac{GBR_i}{\lambda_{sch}^{(i)}(t)}$
NRG	Rate	$GBR_i + \lambda_i(t)\beta \exp\left(\frac{-\beta(\lambda_i(t)-GBR_i)}{GBR_i} \right)$	$\frac{k_{be}}{n_{be}}$

Note: Here, for a given user i at time t, $\lambda_i(t)$ is the exponentially weighted moving average throughput, $R_i(t)$ is the instantaneous supportable data rate, and $w_i(t)$ is either head-of-line packet delay or {Number of Tokens}/GBR_i as shown in Figure 5.10. $\lambda_{sch}^{(i)}$ is the effective scheduling throughput based on only the time slots when user i is actually scheduled. N and n_{be} are the total number of users and Best-Effort users, respectively. The parameters a_i, b, β, and k_{be} are implementation parameters.

users and then divides the remaining capacity, if any, among the Best-Effort users, which is the goal of NRG.

5.3.2.3 Required Activity Detection Scheduler

The idea of a RAD scheduler is that when user i is being served at a data rate of $\lambda_{sch}^{(i)}(t)$ in the time slot when that user is scheduled, then that user needs to be scheduled for $GBR_i/\lambda_{sch}^{(i)}(t)$ time slots per unit time. This term is called the required activity of the user. Thus, this term is taken to be $B_i(t)$ in Equation (5.2) because then it acts as weighted PF, and over the long term it tends to divide the total time slots among different users in proportion to $B_i(t)$. This, is turn, ensures a minimum bit rate of GBR_i for user i, as shown by the results in [24].

RAD is a QoS-aware scheduler with different definitions of $B_i(t)$ during congestion and during no congestion. Congestion is defined with respect to the total load $\rho(t) = \sum_i \frac{GBR_i}{\lambda_{sch}^{(i)}(t)}$. When $\rho(t) < 1$, the system is not congested as there are resources available that can satisfy QoS guarantees at that time.

In this case, $B_i(t)$ is taken as

$$B_i(t) = \begin{cases} \dfrac{GBR_i}{\lambda_{scb}^{(i)}(t)} & \forall i \in \mathcal{Q}, \\[3mm] \dfrac{1 - \rho(t)}{n_{be}} & \forall i \in \mathcal{B} \end{cases} \tag{5.6}$$

On the other hand, when the system is congested and $\rho(t) \geq 1$, then

$$B_i(t) = \begin{cases} \dfrac{GBR_i/\lambda_{scb}^{(i)}(t)}{\rho(t)} & \forall i \in \mathcal{Q}, \\[3mm] 0 & \forall i \in \mathcal{B} \end{cases} \tag{5.7}$$

where n_{be} is the number of Best-Effort users and $\lambda_{scb}^{(i)}(t)$ is the effective scheduling throughput based only on TTIs when user i is actually scheduled. It is calculated as follows:

$$\lambda_{scb}^{(i)}(t) = \left(1 - \frac{s_i}{\tau_{scb}}\right) \lambda_{scb}^{(i)}(t - \Delta t) + \frac{s_i}{\tau_{scb}} R_i(t)$$

with $s_i = 1$

when user i is scheduled in the current time slot, and $s_i = 0$ otherwise. It should be noted that $\lambda_{scb}^{(i)}(t)$ is computed in a slightly different way as compared to $\lambda_i(t)$ in Equation (5.1) because of the different value of τ_{scb} and; in addition, when user i is not scheduled in the current time slot, unlike $\lambda_i(t)$ that decreases, $\lambda_{scb}^{(i)}(t)$ is exactly set to the previous value $\lambda_{scb}^{(i)}(t - \Delta t)$. The performance of the RAD scheduler is studied in the next section.

5.3.2.4 Comparison among the Different QoS-Aware Schedulers

Before comparing different QoS aware schedulers, it should be noted that a fair comparison is very difficult because of numerous implementation parameters and, more importantly, due to different configuration options offered by the different QoS-aware schedulers. In fact, we can actually compare the advantages of different QoS-aware schedulers discussed in this chapter with respect to the different configuration options that are offered. The advantage of MLWDF and EXP is that they can support delay-based as well as rate-based QoS guarantees, simultaneously. The positive point about utility-based schedulers is that if one can design the utility of a QoS service, then one can design the corresponding optimal scheduler. Regarding the RAD scheduler, its advantage lies in the fact that unlike other QoS-aware schedulers that require the tuning of numerous implementation parameters, it does not have any implementation parameters to tune and only requires the value of *GBR*. In addition, unlike MLWDF and EXP, the NRG and RAD schedulers have the option to first satisfy the QoS guarantees and then divide the remaining capacity among the Best-Effort users, such as those with $GBR = 0$, without affecting the QoS users.

To further compare different QoS-aware schedulers, two types of settings are considered. Because the focus of this chapter is on downlink-centric services such as video streaming, Video-on-Demand, FTP, and others, we assume that these services will have a varied range of delay requirements. Thus, RNC can set different values of the Discard Timer, such as a strict timer value of 1.5 s or lower and up to much more relaxed values like 5.0 s or no Discard Timer at all. This section thus compares the performance of different schedulers in two scenarios: one in which the Discard Timer value is set to 1.5 s and the other in which it is set to 5.0 s. We look at the percentage of packets belonging to the QoS users that are dropped by the Discard Timer. To provide a fair comparison of these schedulers, their numerous implementation parameters are tuned, after a lot of simulation runs, such that the Best-Effort throughput becomes equal to that of other schedulers under the same conditions. Most of the time, these implementation parameters are re-tuned whenever the configuration changes, for example, when the number of video users is changed. Similar to previous sections, the results reported in this section are the average values obtained after 20 independent simulation runs. In case of MLWDF and EXP schedulers that do not define the function for Best-Effort users, a low rate guarantee of 10 kbps is used and parameter a_i is moved to match the required overall Best-Effort throughput. It is observed that a low rate guarantee for Best-Effort users results in better overall throughput as compared to when a higher rate guarantee, such as 28 kbps [19] or 64 kbps, is used with MLWDF and EXP for Best-Effort users. In the case of NRG, the value of β is taken as 6.0 and the parameter k_{be} is used for tuning. Because there are no parameters that can be tuned in the case of RAD, the value of GBR is moved such that for a video of 100 kbps, the values of GBR are chosen that are increasingly higher than 100 kbps. Other simulation parameters are the same as those in previous sections.

It can be seen in Figure 5.13 and Figure 5.14 that in the case of MLWDF, it is not possible to match the Best-Effort throughput with other QoS schedulers without significantly increasing the discard rate of QoS users. This is a disadvantage of MLWDF, which uses a weighted delay approach, whereas other QoS-aware schedulers divide the remaining capacity using a Proportional Fair-like algorithm. This is true even for the EXP scheduler, and the difference derives from the weighted delay approach of MLWDF that varies the value of $B_i(t)$ from 0 to some value that in turn performs poorer as compared to PF-like schedulers in terms of overall throughput. In contrast to MLWDF, the lowest possible value of the exponential function in EXP is 1, which in turn means a_i for $B_i(t)$. Thus, from Figures 5.13 and 5.14, it can be seen that MLWDF results in the loss of up to 400 kbps of overall throughput for similar and sometimes a worse percentage of discarded packets.

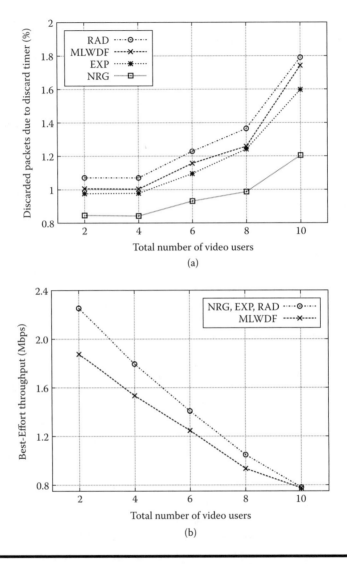

Figure 5.13 Comparison of different QoS schedulers when Discard Timer value is 1.5 s. All schedulers are tuned such that the BE throughput is the same with the exception of MLWDF. For MLWDF, higher throughput is not possible without significantly increasing the average loss rate. (a) Average loss rate of video flows and (b) Best-Effort throughput.

Another point to observe is that when the Discard Timer value is strict and equal to 1.5 s, the NRG performs best, as it results in the lowest percentage of discarded packets with the same, or better in the case of MLWDF, overall throughput. This is because the exponential function used in NRG behaves aggressively toward the drops in throughput values for QoS users;

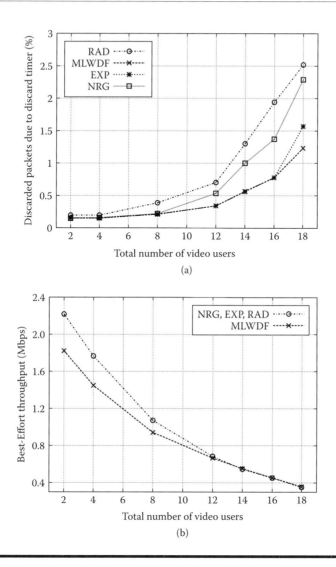

Figure 5.14   Comparison of different QoS schedulers with relatively relaxed Discard Timer value of 5.0 s: (a) Average loss rate of video flows and (b) Best-Effort throughput.

whereas, other schedulers have a relatively more relaxed approach when throughput drops beyond *GBR*. On the other hand, when a relaxed Discard Timer value is used, such as 5.0 s, then the EXP scheduler performs best with the exception of MLWDF performing slightly better when the number of video users is close to the maximum number of video users that can ever be supported. The RAD scheduler performs slightly worse than other schedulers because it has the most relaxed approach to satisfy throughput guarantees and, in addition, it starves the Best-Effort users when QoS guarantees are not getting satisfied, which means slightly lower throughput for Best-Effort users. This in turn increases the discard rate when the *GBR* value is tuned to match the Best-Effort throughput of other schedulers, even when *GBR* is always kept above 110 kbps.

In addition, it is found that although MLWDF has a slight disadvantage in terms of lower overall throughput, it still supports the most or an equal number of QoS users, when compared with other schedulers, in all the cases when Best-Effort traffic is not present. This is because when the system is 100% loaded by QoS users, the Best-Effort throughput is minimal anyway and does not make any difference. It is observed that for MLWDF, NRG, EXP, and RAD, the maximum number of video users supported in two cases are 12, 12, 9, and 9 when the Discard Timer is 1.5 s; and 18, 16, 17, and 13 when the Discard Timer is 5.0 s, respectively. This special case, of maximum number of QoS users supported, corresponds to the case when no background traffic is present.

5.4 Discussion

This chapter considered some of the existing scheduling algorithms with a focus on downlink-centric services. Services such as VoIP that are equally uplink centric were not discussed. Moreover, it was assumed that only one user is scheduled in one TTI, and code multiplexing, which allows multiple users to be scheduled in same TTI, was not considered. It should be noted that the use of code multiplexing can increase the intra-cell interference and thus it is not always relevant in the case of downlink-centric services. However, in the case of services such as VoIP, it becomes essential to use code multiplexing. This is because of the inherent delay characteristics of VoIP data flow, which means end-to-end delay requirements on the order of 200 ms and data rates as low as 16 kbps. So, at any given time, it is highly likely that the amount of data available for transmission will be less than that which can be sent in one TTI. Thus, most of the time, with VoIP, there will be additional resources available for more than one user in each TTI. The absence of code multiplexing can mean inefficient utilization of radio resources, and thus code multiplexing is highly recommended in the case of VoIP service. When considering code multiplexing and other enhancements, a total capacity of up to 120 VoIP users was reported in [25] over HSPA with 3GPP Release 7.

The role of the scheduling priority indicator (SPI) was not discussed in analyzing the different schedulers. One simple approach is to schedule users with strict priority, depending on SPI, such that lower priority users are not scheduled until there is data available in the queues of higher priority users. However, this approach will lead to frequent starvation of lower priority users. In this chapter, a "soft priority" approach was used while scheduling users such that lower priority users, for example Best-Effort users, are not completely segregated from QoS users. Even when some backlog data is available for QoS users, Best-Effort users can still be scheduled, depending on the scheduler algorithm and the current channel conditions. In the context of the schedulers discussed in this chapter, the value of SPI will be used to determine the implementation parameters of the respective schedulers.

Moreover, it should be noted that this chapter focused mainly on scheduling algorithms present in Node B. Some other resource management mechanisms (such as call admission control, buffer management algorithms at RNC, congestion control, handover management, and QoS mapping), although outside the scope of this chapter, are very important for overall QoS satisfaction of the users. These mechanisms, combined with the scheduling algorithm, will determine the overall impact on user-perceived quality-of-experience (QoE).

While assessing the impact of several schedulers on the QoS satisfaction of the users, we used objective metrics such as PSNR and loss rate. However, it is well known that objective metrics such as these do not necessarily correlate well with user-perceived quality when a network is involved. See, for example, [26]. On the other hand, subjective evaluation of video quality is very costly, and one good way to obtain a quality metric as correlated as possible with human visual perception is to use some recently proposed techniques such as those proposed in [27–29].

5.5 Conclusion

This chapter focused on HSDPA scheduling principles and different scheduling algorithms. The key point of HSDPA is its fast and adaptive packet scheduling. This chapter discussed some existing HSDPA packet schedulers. These schedulers were evaluated with a focus on download-centric services in a 3G network. For evaluating the performance, a mixed traffic scenario was considered.

In terms of the maximum number of users that can be served at one time, it was found that QoS-aware schedulers performed better. For the low-load conditions, Best-Effort schedulers could satisfy different users; but as the load increased, only QoS schedulers were able to guarantee a minimum quality to 90% of the video users. The trade-off was the lower per-user throughput that the BE users were getting in comparison. Nevertheless, the QoS-aware schedulers were still dividing the remaining capacity among the BE users in a fair manner.

References

[1] EURANE, http://www.ti-wmc.nl/eurane/.

[2] 3GPP TR 23.107 v 5.13.0, Technical Specification Group Services and System Aspect; Quality of Service (QoS) Concept and Architecture, Release 5, 2005.

[3] 3GPP TS 25.433 v 5.16.0, UTRAN Iub Interface Node B Application Part (NBAP) Signalling, October 2006.

[4] S. McCanne and S. Floyd, ns Network Simulator, http://www.isi.edu/nsnam/ns/.

[5] M. Gudmundson, Correlation model for shadow fading in mobile radio systems, *Electronics Lett.*, 27(23): 2145–2146, 1991.

[6] A. Jalali, R. Padovani, and R. Pankaj, Data throughput of CDMA-HDR A high efficiency-high data rate personal communication wireless system, in *Proc. IEEE VTC 2000 Spring*, 3: 1663–1667, 2000.

[7] S. Borst, User-level performance of channel-aware scheduling algorithms in wireless data networks, *IEEE/ACM Trans. Networking*, 13(3): 636–647, 2005.

[8] T.E. Kolding, Link and system performance aspects of proportional fair scheduling in WCDMA/HSDPA, in *Proc. IEEE VTC 2003—Fall*, 3: 1717–1722, October 2003.

[9] F. Kelly, Charging and rate control for elastic traffic, *Eur. Trans. Telecommun.* 8: 33–37, 1997.

[10] 3GPP2, 1xEV-DV Evaluation Methodology (v14), C30-20030616-043R1, April 2003.

[11] T. Bonald, A score-based opportunistic scheduler for fading radio channels, in *European Wireless 2004*, Barcelona, Spain, February 2004.

[12] G. Barriac and J. Holtzman, Introducing delay sensitivity into the proportional fair algorithm for CDMA downlink scheduling, in *IEEE Seventh Int. Symp. Spread Spectrum Techniques and Applications, 2002*, September. 2002.

[13] P. Ameigeiras, Packet Scheduling and Quality of Service in HSDPA, Ph.D. dissertation, Institute of Electronic Systems, Aalborg University, Denmark, October 2003.

[14] K.I. Pedersen, P.E. Mogensen, and T.E. Kolding, QoS considerations for HSDPA and performance results for different services, in *Proc. IEEE VTC 2006-Fall*, September 2006.

[15] x264, http://www.videolan.org/developers/x264.html.

[16] J. Klaue, B. Rathke, and A. Wolisz, EvalVid—A framework for video transmission and quality evaluation, in *Proc. 13th Int. Conf. Modelling Techniques and Tools for Computer Performance Evaluation*, Urbana, IL, September 2003.

[17] M. Andrews, K. Kumaran, K. Ramanan, A. Stolyar, P. Whiting, and R. Vijayakumar, Providing quality of service over a shared wireless link, *IEEE Commun. Mag.*, 39(2): 150–154, February 2001.

[18] M. Andrews, K. Kumaran, K. Ramanan, A. Stolyar, R. Vijayakumar, and P. Whiting, CDMA Data QoS Scheduling on the Forward Link with Variable Channel Condtions, *Bell Labs Technical Memo.*, April 2000.

[19] S. Shakkottai and A. Stolyar, A Study of Scheduling Algorithms for a Mixture of Real- and Non-Real-Time Data in HDR, *Bell Labs Technical Memo.*, August 2000.

[20] P. Hosein, QoS control for WCDMA high speed packet data, in *IEEE Proc. Int. Workshop on Mobile and Wireless Network Commun.*, September 2002, pp. 169–173.

[21] P. Hosein, A TCP-friendly congestion control algorithm for 1xEV-DV forward link packet data, in *The 57th IEEE Semiannual Vehicular Technol. Conf., 2003. VTC 2003—Spring*, April 2003, pp. 621–625.

[22] M. Lundevall, B. Olin, J. Olsson, N. Wiberg, S. Wanstedt, J. Eriksson, and F. Eng, Streaming applications over HSDPA in mixed service scenarios, in *Proc. IEEE VTC 2004–Fall*, 2: 841–845, September 2004.

[23] K.D. Singh and D.Ros, Normalized rate guarantee scheduler for high speed downlink packet access, in *IEEE GLOBECOM'07*, Washington, D.C., November 2007.

[24] T.E. Kolding, QoS-aware proportional fair packet scheduling with required activity detection, in *Proc. IEEE VTC 2006—Fall*, September 2006.

[25] H. Holma, M. Kuusela, E. Malkamaki, K. Ranta-aho, and C. Tao, VOIP over HSPA with 3GPP Release 7, in *IEEE 17th Int. Symp. on Personal, Indoor and Mobile Radio Commun.*, September 2006.

[26] R. Wu, T. Ferguson, and B. Qiu, Digital video quality using quantitative quality metrics, in *4th Int. Conf. Signal Processing*, Beijing, China, 1998, pp. 1013–1016.

[27] S. Mohamed and G. Rubino, A study of real-time packet video quality using random neural networks, *IEEE Trans. Circuits and Systems for Video Technol.*, 12(12): 1071–1083, December 2002.

[28] G. Rubino, M. Varela, and S. Mohamed, Performance evaluation of real-time speech through a packet network: A random neural networks-based approach, *Performance Evaluation*, 57(2): 141–162, May 2004.

[29] G. Rubino, P. Tirilly, and M. Varela, Evaluating users' satisfaction in packet networks using random neural networks, in *Proc. ICANN'06*, Athens, Greece, September 2006.

Chapter 6

Packet Scheduling and Buffer Management in HSDPA

Khalid Al-Begain, Suleiman Y. Yerima, and Belal AbuHaija

Contents

6.1 Introduction

Providing high-speed data has always been an important goal of the wireless community. The 3rd Generation Partnership Project (3GPP) has introduced the High-Speed Downlink Packet Access (HSDPA) as a step forward in this direction. HSDPA evolved from WCDMA (Wideband Code Division Multiple Access) utilizing a number of existing technologies. Several techniques have been employed to compensate for the changing link conditions. The main theme is based on link adaptation by modifying the transmission parameters of the system to adapt to the instantaneous transmission conditions. Among many, HSDPA employs fixed spreading factor, adaptive modulation and coding (AMC), fast scheduling, and physical layer retransmission by applying Hybrid Automatic Repeat Request (HARQ) to provide high-speed downlink packet access by means of High-Speed Downlink Shared Channel (HS-DSCH). All this implies that substantial changes have been made to the Node B to enhance Release 99 WCDMA with packet scheduling embedded in a new MAC sub-layer known as the MAC-hs (Medium Access Control) (MAC-high speed). Because this necessitates data buffering at the air interface, which poses a bottleneck to end-to-end communication, buffer management in the Node B MAC-hs is essential and critical for Quality-of-Service (QoS) provision.

In Node B, an efficient packet scheduling mechanism is crucial to HSDPA performance. Hence, this chapter discusses various packet scheduling algorithms that have been proposed for HSDPA. Packet scheduling algorithms that support multimedia traffic with diverse concurrent classes of flows being transmitted to the same end user are discussed. Furthermore, new approaches for such end-user multimedia sessions based on integrated packet scheduling with buffer management are also presented. In these approaches, the packet scheduling functionality selects a user for downlink transmission based on a given scheduling discipline (i.e., inter-user prioritization), while the buffer management scheme determines the class of flow to be transmitted from the users' multiplexed flows (i.e., inter-class prioritization).

The chapter begins with a detailed description of the HSDPA physical and MAC layers as given in the 3GPP standards, followed by a discussion of the HSDPA packet scheduling algorithms. Finally, the integrated buffer management and packet scheduling solutions for HSDPA multimedia sessions with concurrent diverse flows are presented.

6.2 HSDPA Physical Layer

On the radio system level, several WCDMA functionalities are adapted to enable HSDPA. Three of the fundamental properties of WCDMA have been disabled: soft handover, variable spreading factor (SF), and fast power control. In HSDPA, only hard handover is allowed. The SF is fixed at SF16 under one scrambling code [1]. Figure 6.1 depicts the spreading factor and the number of channelization codes available for HSDPA. Fifteen channel codes are used for data transmission. One code is used for signaling with an SF of 128. AMC techniques have been implemented. Depending on the channel condition, the scheduler can have a choice between QPSK (quadrature phase-shift keying) and 16QAM (quadrature amplitude modulation). In QPSK, each signaling element represents 2 bits; however, in 16QAM, each signaling element represents 4 bits, which in theory can double the amount of data carried in a frame. Turbo coding is employed in HSDPA, ranging from $R = 1/4$ to $R = 3/4$, depending on the channel condition. Spreading is applied to the physical channel after coding and modulation. Spreading, in effect, increases the bandwidth of the signal. The number of chips per data symbol is called the spreading factor. A scrambling operation is applied to the signal, and this operation is a means to separate base stations [2].

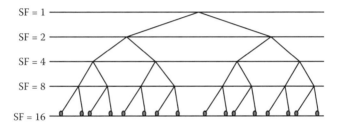

Figure 6.1 Spreading factor.

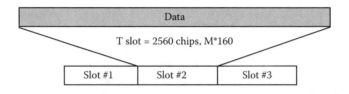

Figure 6.2   Subframe structure for the HS-PDSCH.

Three new physical channels have been defined for HSDPA: the HS-PDSCH (High-Speed Physical Downlink Shared Channel) used to carry data on the downlink direction, the HS-SCCH (High-Speed Shared Control Channel) used to carry signaling in the downlink direction, and the HS-DPCCH (High-Speed Dedicated Physical Control Channel) used to carry signaling in the uplink direction. An HS-PDSCH corresponds to one channel code of fixed spreading factor from the set of 15 channelization codes available for HS-DSCH transmission. Multi-code transmission is allowed, which translates to a UE (user equipment) being assigned multiple codes in the same HS-PDSCH transmission time interval (TTI = 2 ms), depending on the UE capability. The subframe and slot structure of HS-PDSCH are shown in Figure 6.2.

HS-PDSCH can use QPSK or 16QAM modulation symbols, where M is the number of bits per modulation symbols; that is, $M = 2$ for QPSK and $M = 4$ for 16QAM. The 10-ms frame of WCDMA has been divided into five subframes, 2 ms each for fast scheduling and retransmission. The HS-SCCH is a fixed-rate (60 kbps, SF = 128) downlink physical channel used to carry downlink signaling related to HS-DSCH transmission. Figure 6.3 shows the subframe structure of the HS-SCCH.

HS-SCCH is transmitted two time slots ahead of HS-PDSCH. It carries critical signaling information such as the channelization code set, modulation scheme, and transport block size to the UE. HS-SCCH also carries the HARQ process number, redundancy version, and UE identity, among other information [3].

HS-DPCCH is used to carry the signaling information in the uplink direction. The channel has SF = 256, 10 bits per slot, and 30 bits for the subframe

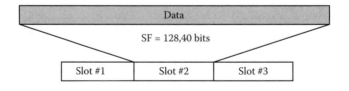

Figure 6.3   Subframe structure for the HS-SCCH.

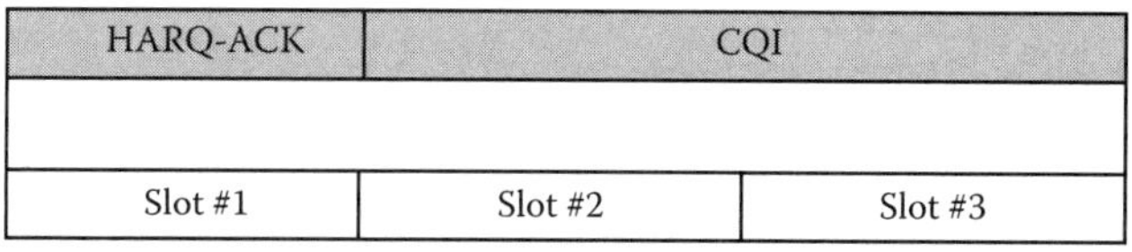

HARQ-ACK	CQI	
Slot #1	Slot #2	Slot #3

Figure 6.4 Subframe structure for the HS-DPCCH.

(15 kbps). There is, at most, one HS-DPCCH per UE. HS-DPCCH carries the ARQ ACK, in one slot and the Channel Quality Indicator (CQI) in the other two slots. Figure 6.4 shows the subframe structure for the HS-DPCCH.

The CQI represented by 5 bits for a value is between 1 and 30; based on this feedback from UE, the scheduler in Node B decides on the modulation and transport block size to send in the next TTI transmission if there is available data for the UE in the buffer [4]. HS-DPCCH is transmitted 7.5 time slots (5 ms) after the reception of the HS-PDSCH.

6.3 HSDPA MAC Architecture

To support HS-DSCH, a functional split in the MAC is required. This new entity is called MAC-hs and it is deployed closer to the radio link in Node B for fast retransmission. In each cell that supports HS-DSCH, there is one MAC-hs entity, responsible for management of physical resources allocated to HSDPA. HSDPA did not affect the architecture of the upper layers. MAC-hs is composed of four functional entities as depicted in Figure 6.5 [5]. These entities are described below.

6.3.1 Flow Control

A flow control function is needed to govern the exchange of data between MAC-hs and MAC-d. There are two cases for implementing flow control functionality. In the first case, an SRNC (serving radio network controller) is connected to Node B through the Iub interface. MAC-d entity is located in the SRNC and connected to MAC-hs through the Iub interface. The SRNC in the second case is connected to the CRNC (controlling RNC) through the Iur interface and, in turn, CRNC is connected to Node B through the Iub interface. In the first case, the flow control is directly exchanging data between MAC-d (located in the SRNC) and MAC-hs (located in Node B). In the second case, there are two flow controls, one between MAC-d (located in SRNC) and MAC-c/sh (located in CRNC) and the second one between MAC-c/sh and MAC-hs. The UTRAN side of the MAC architecture is depicted in Figure 6.6.

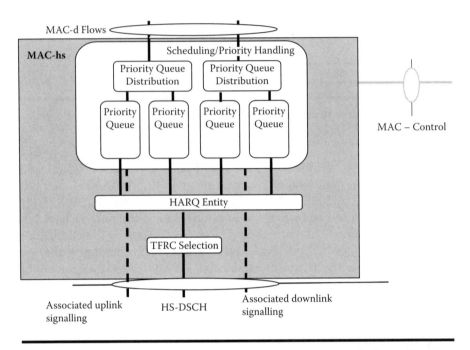

Figure 6.5 MAC-hs UTRAN side. (*Source:* From 3GPP, Technical Specifications Group RAN, Medium Access Control (MAC) Protocol Specifications, 3GPP TS 25.321 version 8.4.0.)

There are two control messages available in the 3GPP standards for exchanging data between MAC-hs and MAC-d. HS-DSCH Capacity Request is shown in Figure 6.7. Entity A and Entity B can be either SRNC and Node B, respectively, if the flow control is directly between the SRNC and Node B, or (SRNC, DRNC) and (CRNC, Node B) if the flow control is handled separately on Iur and Iub. The capacity request message is sent from SRNC/DRNC to Node B to request that data be transferred to Node B and to indicate the buffer size of the user.

The HS-DSCH Capacity Allocation signaling message is illustrated in Figure 6.8. This message indicates to SRNC/DRNC the allocated buffer space in Node B for a specific user.

SRNC indicates the amount of data available for a user-given Common Transport Channel Priority Indicator (CmCH-PI). The range of the priority is between 0 and 15, with 0 being lowest priority and 15 being the highest priority. SRNC also indicates the buffer size of the user—that is, the amount of data in the SRNC buffer. In the capacity allocation control frame, the amount of data or credits that SRNC is allowed to send is specified by Node B. In case of congestion on the Iub interface, the credits are reduced equally for all users [6].

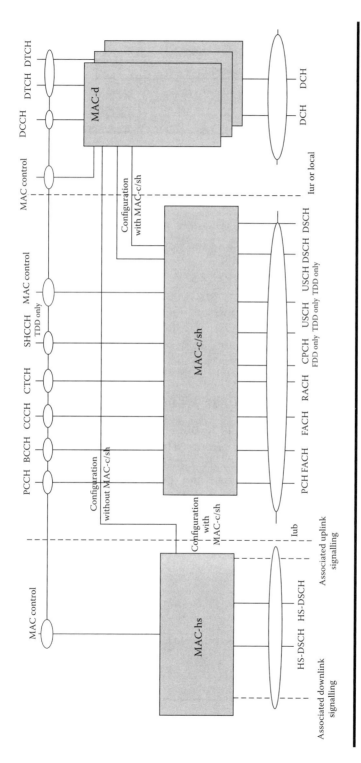

Figure 6.6 UTRAN side of the MAC architecture. (*Source:* From 3GPP, Technical Specifications Group RAN, Medium Access Control (MAC) Protocol Specifications, 3GPP TS 25.321 version 8.4.0.)

Figure 6.7   HS-DSCH capacity request.

Flow control is intended to limit layer 2 signaling latency and reduce discarding and retransmission of data as a result of HSDPA congestion. Flow control is provided independently by MAC-d flow for a given MAC-hs entity. As a matter of fact, one of the design goals of MAC-hs is to limit signaling to the upper layers.

Studies on HSDPA performance (see, for example, [8–10]) indicate that Iub flow control has a significant impact on MAC-hs packet scheduler performance and the resulting end-to-end QoS (Quality of Service). In [11], a flow control algorithm to support transfer of multiplexed diverse flows in an end-user multimedia session is proposed for the integrated buffer management and scheduling approach that is discussed later.

6.3.2 *Scheduling and Priority Handling*

This function manages HS-DSCH resources between HARQ entities and the data flows according to their priority. There is one priority queue entity (HARQ entity) per user, and there are eight different queues in each entity. For each MAC-hs PDU (protocol data unit) being serviced, the priority entity determines the queue ID and the transmission sequence number (TSN is explained when we discuss the frame format of MAC-hs PDU). Based on the HS-DPCCH report in the uplink signaling, MAC-hs has the choice of new transmission or retransmission of lost PDU. 3GPP standards do not specify

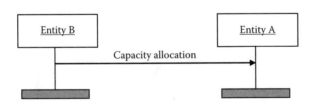

Figure 6.8   HS-DSCH capacity allocation.

the scheduling technique to use; however, several scheduling techniques that have been researched are described later in this chapter.

6.3.3 HARQ

There is one HARQ entity for each user. Each entity can support multiple instances of HARQ processes. There is one HARQ process created per HS-DSCH per TTI. There are two main schemes of implementing HARQ, *Chase Combining* and *Incremental Redundancy*. Chase Combining is retransmission of the same data block. In Incremental Redundancy, each retransmission includes additional redundancy bits from the channel encoder (HARQ type 2). HARQ type 3 is also an incremental redundancy scheme but with a difference that each transmission is self-decoded. There are two types of protocols that are used with HARQ: Selective Repeat (SR) and Stop-and-Wait (SAW). The first protocol repeats the blocks that are in error and is usually insensitive to delay. However, the disadvantage of this scheme is that the UE memory requirements very highly depends on the maximum number of blocks in the transmission sequence. The other disadvantage is that HARQ requires the receiver to reliably determine the sequence number of the transmission; therefore, stronger code must be used for the sequence number. The SAW protocol is simple and requires very little overhead. It keeps working on the current block until it is received successfully. This protocol will save signaling bandwidth and reduce the amount of memory required by the UE. However, such a simplistic approach comes at the expense of waiting of 7.5 time slots (5 ms) for the acknowledgment to arrive at the transmitter. This wait will reduce the system capacity. Therefore, we run several instances in parallel with the SAW to eliminate this problem and fully utilize the system resources [1].

6.3.4 Transport Format and Resource Combination (TFRC)

Depending on the reported channel quality condition, TFRC is determined. There are two values for modulation (QPSK and 16QAM) and several coding values ranging from 1/4 to 3/4. The combination of the modulation and coding is adjusted based on the channel quality conditions as previously mentioned. TFRC largely depends on the UE capabilities, modulation, and coding rate (Table 6.1) [5, 7].

6.3.5 MAC-hs Protocol Data Unit

MAC-hs PDU is a bit string but not necessarily in multiples of 8 bits. It consists of a MAC-hs header and one or more MAC-hs SDUs (service data

Table 6.1   Theoretical Bit Rates with 15 Multi-codes

TFRC	Modulation	Effective Code Rate	Maximum Throughput (Mbps)
1	QPSK	$\frac{1}{4}$	1.8
2	QPSK	$\frac{2}{4}$	3.6
3	QPSK	$\frac{3}{4}$	5.3
4	16QAM	$\frac{2}{4}$	7.2
5	16QAM	$\frac{3}{4}$	10.7

units). The MAC-hs header is of variable size. Each MAC-hs SDU equals one MAC-d PDU. A maximum of one MAC-hs PDU can be transmitted in a TTI per UE. The MAC-hs SDUs belong to the same queue. MAC-hs consists of the following fields (Figure 6.9):

- Version Flag (VF): This is a 1-bit flag providing extension capabilities to the PDU format. This bit is always set to 0 and the value 1 is reserved for future use.
- Queue Identifier (Queue ID): This is a 3-bit field that provides identifications to the reordering queue in the receiver, in order to support independent buffer handling of data belonging to different reordering queues.
- Transmission Sequence Number (TSN): This field is used to support in-sequence delivery to higher layers. The TSN identifies the transmission sequence on HS-DSCH. This field is 6 bits.
- Size Index Identifier (SID): This field identifies the size of a set of consecutive MAC-d PDUs. The MAC-d PDU size for a given SID is configured by higher layers and is independent for each queue. This field is 3 bits long.

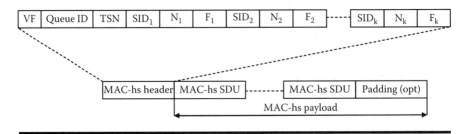

Figure 6.9   MAC-hs PDU structure.

- Number of MAC-d PDUs (N): This is the number of consecutive MAC-d PDUs with equal size. This field is 7 bits long. In FDD (frequency division duplex) mode, the maximum number transmitted in one TTI is assumed to be 70.
- Flag (F): This is a 1-bit field. If the value is 0, then the field is followed by an additional SID; if the value is 1, then the field is followed by a MAC-d PDU. The maximum number of repeated $F = 0$ in a TTI is assumed to be 7.

3GPP standards do not specify a certain scheduling mechanism for HSDPA. There are three QoS parameters specified in the Iub: the guaranteed bit rate (GBR), the Scheduling Priority Indicator (SPI) of values 0 to 15 as mentioned earlier, and the discard timer (DT), which specifies how long the packet should be buffered in Node B (MAC-hs) before it is discarded. The mapping of these parameters and how they are used in MAC-hs scheduler has been left to the operators. The scheduler can schedule all UEs within a cell based on the SPI from the higher layer (in the RNC). Alternatively, the UEs can be scheduled based on other metrics specified according to a given scheduling algorithm in the MAC-hs; the most popular algorithms are discussed later. The scheduler determines the HARQ entity and the queue to be serviced. It sets the TSN, starting from value 0 for each Queue ID. The scheduler also increments the value by one for each new transmission, but not for retransmission. It indicates the Queue ID and TSN to the HARQ entity for each MAC-hs PDU to be transmitted. The scheduler may decide to discard any out-of-date MAC-hs SDU. The scheduler also determines the redundancy version for the transmitted and retransmitted frame. The next section discusses well-known packet scheduling algorithms that can be implemented for scheduling the UEs in the MAC-hs packet scheduler.

6.4 HSDPA Packet Scheduling Algorithms

The MAC-hs packet scheduler is responsible for accommodating, within the limited and variable resources, HSDPA users with different QoS requirements and those experiencing different air interface conditions [12]. There are several possible solutions to such a scheduling problem, and this section discusses various algorithms proposed by researchers for HSDPA MAC-hs packet scheduling, including techniques that support multiple classes of flows per user session.

With the purpose of enhancing the cell throughput (capacity), the HSDPA scheduling algorithm can take advantage of the instantaneous channel variations and temporarily raise the priority of the users. Because the users' channel quality varies asynchronously, the time-shared nature of HS-DSCH introduces a form of selection diversity with important benefits for spectral efficiency [13]. However, this could mean that users more distant from Node B (at cell edge), and therefore requesting lower data rates, could be starved for service. Consequently, the scheduling algorithms must balance the conflicting goals of maximizing throughput, while at the same time ensuring some degree of fairness to all users requesting a service [14].

Opportunistic scheduling methods or *channel-aware schedulers* are those that exploit multi-user (selection) diversity, such that with a large number of users in the cell, a gain in total cell throughput can be achieved by giving transmission priority to users with favorable channel conditions over those with poor channel conditions. On the other hand, *blind* scheduling methods do not take into consideration the radio conditions of the users. According to [15], with opportunistic scheduling, the QoS of all users can be improved over (blind) scheduling schemes that do not take channel conditions into account. Gutierrez [13] classifies the well-known HSDPA packet scheduling algorithms according to the pace of the scheduling process into *fast* and *slow* as described below.

6.4.1 Fast Scheduling Methods

Fast scheduling methods are those that base the scheduling decisions on the recent UE channel quality measurement so that the instantaneous variations in the user's supportable data rate can be tracked. With fast scheduling methods, the scheduling decisions are executed on a TTI basis. Where there is a sufficient number of time multiplexed users, fast scheduling methods can exploit multi-user selection diversity to provide significant capacity gain. Examples of fast scheduling methods include Maximum C/I, Proportional Fair, and Fast Fair Throughput.

6.4.1.1 Maximum C/I Scheduling

The Maximum Carrier-to-Interference Ratio scheduler (Max-C/I), which is also referred to as the maximum throughput scheduler, is designed to maximize HSDPA cell throughput. The Max-C/I packet scheduler always schedules the user with the best instantaneous channel quality. Because the Max-C/I scheduler monopolizes the cell resources for users with favorable channel conditions, there may be a number of users at the cell edge that may never be scheduled. However, because fast fading dynamics have a larger range than the average radio propagation conditions, it is possible for users with poor average radio conditions to still have access to the shared channel. Max-C/I is inherently unfair in scheduling resources and suffers from coverage limitations, which are its main drawbacks.

6.4.1.2 Proportional Fair Scheduling

This scheduling algorithm serves the user with the best *relative* channel quality according to the following scheduling metric:

$$P_i = \frac{r_i(t)}{R_i} \quad i = 1, \ldots, K$$

where P_i denotes the user priority, $r_i(t)$ is the instantaneous data rate that can be supported by the user in the current TTI if served by the packet scheduler, and R_i is the average throughput experienced by user i. The denominator in the scheduling metric allows a user that is being allocated little scheduling resources to increase its priority over time. R_i can be updated every TTI prior to selecting a user by employing the following first-order recursion:

$$R_i(n) = (1 - \beta) R_i(n - 1) + \beta r_i(n - 1)$$

where $r_i (n - 1)$ is the requested rate of user i in the previous TTI $(n - 1)$ if the user i was selected for transmission in TTI $(n - 1)$, and 0 otherwise. Because β is the time constant of the (exponential smoothing) filter, β^{-1} equals the equivalent averaging period in TTIs. Thus, Kolding [16] recommends that the averaging length should be set long enough to ensure that the process averages out the fast fading variations, but short enough to still reflect medium-term conditions such as shadow fading. Note that the throughput calculation for a user is only done for periods of time when the user has buffered data in the MAC-hs. This is important for the stability of QoS-aware packet scheduling methods, which will otherwise try to compensate for inactive users with no data to transmit [17].

The Proportional Fair (PF) algorithm is presented in [18] and also analyzed in several other papers (see, for example, [19, 20]). The PF scheduler provides a trade-off between fairness and achievable cell throughput, thus providing significant coverage extension over the Max-C/I (maximum throughput) scheduler. It aims to serve users under very favorable instantaneous radio conditions relative to their average ones, thus taking advantage of the temporal variations of the fast fading channel. This relation is popularly interpreted as "scheduling users on top of their fades."

6.4.1.3 Fast Fair Throughput Scheduling

Fast fair throughput scheduling is also known as Proportional Fair Throughput (PF-T) in [21]. Like the Proportional Fair method, this scheduling method is aimed at distributing the cell throughput fairly among the users while taking advantage of the short-term fading variations of the radio channel. In [22], a fast fair throughput algorithm that modifies the PF algorithm to provide equalization of user throughput is proposed by Barriac as follows:

$$P_i = \alpha \frac{r_i(t)}{R_i} \quad i = 1, \ldots, K$$

where α is the term that weights the PF algorithm to equalize the user throughputs and is given by

$$\alpha = \frac{\max j\{\overline{Rj}\}}{Rj}$$

where $\max j\{\overline{Rj}\}$ is a constant that indicates the maximum average supportable data rate from all users, and $\overline{Rj}$ is the average supportable data rate of user i.

6.4.2 Slow Scheduling Methods

The slow scheduling methods are less aggressive than the aforementioned fast scheduling methods. They include scheduling algorithms that base their scheduling decisions on the average user's signal quality (e.g., the average C/I scheduler) or those that schedule without any consideration for user channel quality (e.g., "blind" Round Robin scheduler).

6.4.2.1 The Round Robin Scheduler

Round Robin scheduling is a "blind" scheduling technique that serves users in a cyclic manner without taking their radio channel conditions into consideration. Round Robin is also "fair time" scheduling; this means that the same power is allocated to all users such that a higher throughput is experienced by users with better channel conditions. This kind of scheduler is noted for its simplicity and assurance of a fair resource distribution among the users in the cell. Because Round Robin scheduling decision is oblivious to channel conditions, it typically suffers from low average throughput.

6.4.2.2 The Average C/I Scheduler

This scheduler selects, in every TTI, the user with the largest average C/I that has buffered data ready for transmission. An averaging window length of, for example, 20 to 100 ms is used such that only the user with maximum slow-averaged channel quality/throughput is served [13, 21].

6.4.2.3 The Fair Throughput Scheduler

The fair throughput scheduling approach aims at ensuring that all simultaneously queued users receive the same average throughput, which means that users in bad channel conditions receive relatively more allocation of HS-DSCH resources. Thus, in every TTI, the user with the lowest average throughput is scheduled. This scheduler is considered a slow scheduling method because it does not employ instantaneous channel quality information.

6.4.3 Delay Differentiated Packet Scheduling Schemes

The previously discussed basic channel-aware and non-channel-aware scheduling algorithms do not explicitly take packet delays into account in their scheduling metrics. To provide QoS differentiation and to meet the delay requirements of real-time services, such as real-time streaming audio/video, conversational voice, and interactive video, researchers have proposed various scheduling algorithms that take packet delay into account for HSDPA MAC-hs scheduling. A few of these are discussed here and include Channel-Dependent Earliest Due Deadline (CD-EDD) [23], the Exponential Rule (ER) [22], and the Modified Largest Weighted Delay First (M-LWDF) [15]. These algorithms extend the opportunistic PF scheduling algorithm to include queuing delay metrics, thus allowing for delay differentiation between user packet flows.

6.4.3.1 Channel-Dependent Earliest Due Deadline

The CD-EDD algorithm is a combination of PF scheduling and an earliest deadline due (EDD) component resulting in the scheduling metric [23]:

$$P_i = w_i \frac{r_i(t)}{R_i} \frac{W_i(t)}{T_i - W_i(t)} \quad i = 1, \ldots, \ K$$

where w_i is a weighting factor for the user i, and $W_i(t)$ denotes the waiting time of the packet at the head of the queue (i.e., the HOL (head of line) packet). T_i is the maximum allowable delay for ith user, while $T_i - W_i(t)$ indicates the time until the deadline is reached. Thus, as time until deadline draws closer (i.e., when HOL packet delay approaches T_i), the EDD term dominates the scheduling metric, thereby giving the user a higher scheduling priority. On the other hand, when HOL packet delay is low, the user gets a low scheduling priority due to the EDD term.

6.4.3.2 Exponential Rule

The exponential rule (ER) algorithm strives to equalize the weighted delays of the queues of all flows if their differences happen to be large; but, for regular situations, the PF algorithm dominates. It has the following scheduling metric [22]:

$$P_i = w_i \frac{r_i(t)}{R_i} \exp\left(\frac{w_i\, W_i(t) - \overline{w\, W(t)}}{1 + \sqrt{\overline{w\, W(t)}}}\right) \quad i = 1, \ldots, K$$

w_i is given by $w_i = -\log(\delta_i)/T_i$ with δ_i being the largest probability with which the scheduler may violate its deadline. $\overline{w\, W(t)}$ is defined as

$$\frac{1}{K}\sum_k w_i\, W_i(t)$$

6.4.3.3 Modified Largest Weighted Delay First

The modified largest weighted delay first (M-LWDF) algorithm not only takes advantage of the multi-user diversity available in the shared channel available through the PF algorithm, but also increases the priority of flows with HOL packets close to their deadline violation. The scheduling metric is as follows. In each time slot, serve the queue j for which

$$w_i\, W_i(t)(r_i(t)/R_i)$$

is maximal, where $W_i(t)$ is the HOL packet delay for queue i, $r_i(t)/R_i$ is the PF term, and w_i is given by $w_i = -\log(\delta_i)/T_i$.

Thus, the greater the user i current packet delay, the channel quality relative to its average level, and the higher the QoS requirement (set through weights w_i), the greater the chance of that user being scheduled. One key feature of the M-LWDF algorithm is that a scheduling decision depends on both current channel conditions and the states of the queues. M-LWDF has proved to be throughput optimal in [24]; furthermore, it remains throughput optimal if for all or some users, the delay $W_i(t)$ is replaced by the queue length (amount of data) $Q_i(t)$ [15]. Because the M-LWDF algorithm only needs to time stamp the arriving data packets of all users, or keep track of the current queue length, it is quite easy to implement.

6.4.4 Scheduling with Inter-class Prioritization for End-User Multiplexed Diverse Flows

The above considered algorithms, while being able to provide delay differentiation, also possess metrics for *per-user* scheduling decisions only; the scheduling metrics do not include provisions for class differentiation between flows when an end user receives multiple flows with diverse QoS requirements in the same session. To address this problem, proposals based on modifications of the existing algorithms have been investigated.

For example, Golaup et al. [25] have modified the M-LWDF algorithm to allow for inter-class prioritization in addition to inter-user prioritization, thus supporting class differentiation between flows of the same user. The proposed algorithm is termed the Largest Average Weighted Delay First (L-AWDF) [26].

6.4.4.1 The Largest Average Weighted Delay First (L-AWDF)

The basic idea behind L-AWDF is that it maintains a PF factor for fairness with respect to channel conditions while giving transmission priority to those users whose scheduled packets are nearing the maximum allowable delays for their class through a relative delay factor; this is constructed in such a way as to allow traffic class performances to be influenced by different weighting factors. Note that in this algorithm, the priority of each class of traffic/flow is expressed through the specification of a *maximum allowable delay* and a selected *weighting factor*. The prioritization function for L-AWDF is defined as follows [25]:

$$P_i = \frac{R_i(t)}{\lambda_i(t)} \cdot \sum_{j=1}^{n} \left(\alpha j \cdot \frac{D_{i,\, j}(t)}{T_{i,\, j}} \right), \quad \sum_{j=1}^{n} \alpha_j = 1$$

where α_j is the weighting factor for traffic class j, and $D_{i,j}(t)$, $T_{i,j}$ are the observed head-of-line and maximum permissible delays, respectively, for traffic class j at user i. P_i denotes the user priority, $R_i(t)$ is the instantaneous data rate that can be supported by the user in the current TTI if served by the packet scheduler, and $\lambda_i(t)$ is the average throughput experienced by user i. Performance analyses of L-AWDF algorithm can be found in [25–27].

6.5 Buffer Management-Based Scheduling Approaches

The 3GPP HSDPA specifications do not define specific buffer management schemes for Node B operation, therefore allowing for network performance optimization through operator-specific implementation. However, employing buffer management schemes in the MAC-hs not only improves resource utilization at the air interface and efficiency at upper protocol layers, but also allows for customized QoS control, especially for multiple diverse flows transmitted to an end user in the same HSDPA session. Furthermore, with MAC-hs buffer management, end-to-end traffic performance gains can be achieved because of the queued flows at the air interface, which present a bandwidth bottleneck and unpredictable time-varying channel quality.

As mentioned previously, the packet scheduler's function is to schedule all UEs within the cell as well as to service priority queues, that is, priority handling (see Figure 6.5). The scheduler schedules MAC-hs SDUs (generated from one or more queued MAC-d PDUs) based on information from upper layers. One UE may be associated with one or more MAC-d flows (priority queues). Each MAC-d flow contains HS-DSCH MAC-d PDUs for one or more priority queues. Hence, the buffer management algorithm (BMA) can be viewed as a subfunction of the MAC-hs scheduler that oversees the responsibility of handling priorities for multiple flows (priority queues) associated with one UE in the MAC-hs. With this approach, classical packet scheduling algorithms such as Round Robin, Max-C/I, PF, and M-LWDF can be applied to select the user eligible to receive transmission in the current TTI (inter-user transmission scheduling) while the BMA decides between the multiple flows/priority queues associated with the selected user (inter-class transmission prioritization) according to the BMA transmission prioritization policy. This approach works very well because a robust and efficient BMA can be designed to operate offline, while the packet scheduler performs the online task of scheduling the users in the cell.

The BMA performs buffer admission control (BAC) on arriving MAC-d PDUs according to the priority scheme's BAC mechanism. The BMA also determines the MAC-d flow (identifies the UE priority Queue ID), from which the MAC-d PDUs will be assembled into MAC-hs SDUs for transmission to the UE on a given HARQ entity. If priority switching is enabled [in the case of a dynamic buffer management scheme such as D-TSP (dynamic time-space priority) discussed later], the priority switching algorithm is consulted to identify the correct MAC-d flow (Queue ID) for next transmission. The scheduler then schedules the MAC-d flow on a selected HARQ process, if the scheduling algorithm has allocated the next transmission opportunity to the UE associated with the MAC-d flow. Thus, the BMA as a subentity of the MAC-hs scheduler, is executed on each UE's MAC-d flows, that is, the *priority handling* for UE's with multiple flows per session [e.g., multiplexed RT (real-time) and NRT (non-real-time) flows concurrent in the multimedia connection]. The entire scheduling process (see Figure 6.10) for multi-flow sessions therefore can be viewed as occurring in a logical hierarchy with the BMA performing buffer admission control and inter-class (inter-flow) scheduling, while MAC-hs scheduling performs the inter-UE scheduling according to any of the selected scheduling disciplines (e.g., Max-C/I, PF, or M-LWDF).

6.5.1 Integrated Scheduling and Buffer Management for HSDPA Multimedia Traffic

The queuing of packets in the Node B MAC-hs provides an opportunity to incorporate buffer management into the MAC-hs packet scheduling functionality in order to improve end-to-end traffic performance and resource

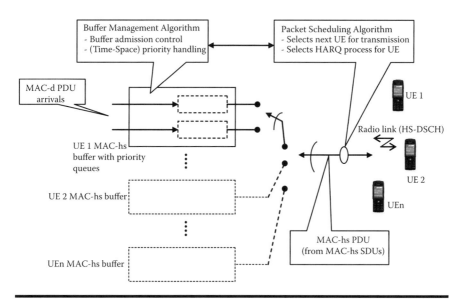

Figure 6.10 Integrated scheduling and buffer management in HSDPA. Logical hierarchy of UE₁ multiplexed RT and NRT MAC-d flows in Node B MAC-hs.

utilization. With higher data rates and improved downlink packet-switched services support enabled by HSDPA, multimedia applications, i.e., those with concurrent diverse "media" or data flows with different QoS requirements being multiplexed in the same end-user session, are a possible communication scenario for example, voice/video chatting with simultaneous data/file download. To schedule such end-user multimedia traffic in the MAC-hs, inter-user scheduling that also allows for inter-class transmission prioritization is necessary because the diverse media belong to different traffic classes that have different QoS requirements. The real-time voice/video components of the multimedia flow are partially loss tolerant while requiring low delay/jitter. On the other hand, the NRT components of the multimedia flow are delay tolerant but sensitive to packet loss. Thus, with the conflicting QoS requirements of the multiplexed components of the multimedia flow belonging to the same user, integrating a buffer management scheme with packet scheduling algorithms provides an effective mechanism for QoS control and performance optimization. The incorporated buffer management scheme must, of course, be able to prioritize the RT class for minimum queuing delay/jitter in the MAC-hs buffer, while at the same time minimizing packet loss for the NRT class. A buffer management scheme known as the time-space priority (TSP) has been proposed for inter-class priority handling and QoS control between multimedia session components (RT and NRT classes), in conjunction with inter-UE scheduling as shown in Figure 6.10. Thus, TSP (as an inter-class BMA) provides per-session buffer management for the priority queues/flows associated with the multimedia user in the MAC-hs, by allowing space priority for the NRT flow while according time (transmission) priority to the RT flow.

6.5.1.1 Time-Space Priority Buffer Management

TSP [28] buffer management is a priority queuing scheme designed for joint QoS control of concurrent RT and NRT flows in an end-user multimedia session. Unlike most priority queuing that provides either delay or loss differentiation, TSP combines both loss and delay differentiation in a single (logical) queue, thus yielding transmission (time) priority for RT packets and space priority for NRT packets.

A threshold R is used to partition the queue, as shown in Figure 6.11, such that NRT packets are accorded a larger buffer space because of their loss sensitivity. R is a soft threshold that limits the maximum number of RT packets admitted into the queue at any given time to R. Due to delay/jitter sensitivity of RT packets such as voice/video packets, they are queued on a first-come-first-served basis in front of the delay-tolerant NRT packets, which are also queued in a first-come-first-served manner behind the RT packets. However, the NRT packets enjoy unlimited access to the buffer such that the maximum number of admitted NRT packets at any given time can vary from $N - R$ to N, where N is the total queue capacity. This allows for more efficient buffer space utilization and further minimization of NRT packet loss.

By virtue of TSP queuing, when the packet scheduler selects the UE with multimedia traffic for transmission in the current TTI, the transmission order of the multimedia flow components in TSP is RT packets first, and then NRT packets are allowed to be transmitted when there are no RT packets present in the queue/buffer. This implies static prioritization of real-time flow/class over the non-real-time flow/class. The restriction of admitted RT packets

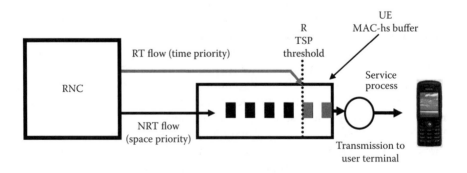

Figure 6.11 **Time-Space Priority queuing for end-user multimedia traffic priority queuing and QoS control in the Node B MAC-hs packet scheduling entity.**

into the queue also provides a mechanism to ensure that the non-time-flow does not get starved of transmission bandwidth especially at high real-time streams arrival intensity.

Due to loss sensitivity, the arrival of NRT packets at the RNC necessitates the use of RLC (Radio Link Control) Acknowledged Mode (AM) for onward transmission over the Iub interface to the Node B. RLC AM packets require feedback from a peer RLC entity in the UE, which is typically sent via a STATUS message. NRT packets that are lost or discarded from the Node B MAC-hs due to buffer overflow or excessive queuing delays can only be recovered with the RNC-based RLC (selective repeat) ARQ retransmissions. Because the ARQ mechanisms operate between the RNC and recipient UE, retransmissions increase the RLC round-trip-time, resulting in overall end-to-end delay of the NRT packets, which manifests in severe degradation of end-to-end-throughput for TCP-based NRT flow within a multimedia session. This is undesirable because the majority of NRT traffic utilizes TCP as the transport protocol. Furthermore, retransmissions lead to waste of Iub resources and Node B buffer space, as well as air interface bandwidth. The aforementioned motivates the utilization of Iub flow control to regulate MAC-hs queues of the multimedia sessions to allow efficient buffer and transmission resource utilization. An enhanced TSP buffer management scheme (E-TSP) proposed and evaluated in [11, 29] uses the Iub credit-based flow control mechanism discussed earlier to regulate the flow of the NRT component of the multimedia traffic in order to enhance end-to-end performance.

The E-TSP scheme is depicted in Figure 6.12. In addition to the TSP threshold R, it has additional thresholds **L** and **H** and a *credit allocation algorithm* designed to enable more efficient utilization of buffer space and air interface resources with minimal Iub signaling load. The allocation algorithm utilizes NBAP (Node B Application Part) signaling [6] as shown in the bottom part of Figure 6.12, to issue credits to the RNC that determine the number of arriving packet data units (PDUs) to the Node B for each flow in the user's session.

The number of credits per transmission interval is determined as follows [11]. Let the total credits per TTI for a particular multi-flow/multimedia user be given by

$$C_{Total} = C_{NRT} + C_{RT}$$

where

$$C_{RT} = (\lambda_{RT}/PDU\_size) \cdot TTI$$

C_{RT} is the number of credits per TTI for the RT flow in the multimedia session, while PDU_size denotes the size of the PDU in bits. λ_{RT} is the minimum guaranteed bit rate of the RT flow (in bps), a parameter that can

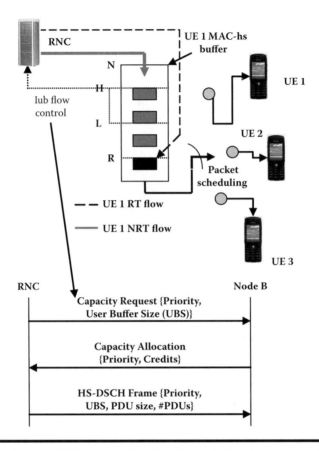

Figure 6.12 Enhanced Time-Space Priority (E-TSP) buffer management for per-UE multimedia traffic QoS control.

be obtained from the QoS attributes of the flow during bearer negotiation. For the NRT flow, the number of credits C_{NRT} is given by

$$C_{NRT} = \min\{C_{NRTmax}, UBS_{NRT}\}$$

where UBS_{NRT} is the user's NRT buffer occupancy in the RNC and C_{NRTmax} is the maximum NRT grants per TTI, which depends on the HSDPA channel load, scheduling policy, and the recipient UE radio conditions. C_{NRTmax} is calculated from

$$C_{NRTmax} = (\lambda'_{NRT}/PDU\_size) \cdot TTI, \ \boldsymbol{N}_T < \boldsymbol{L}$$

$$k \cdot (\lambda'_{NRT}/PDU\_size) \cdot TTI, \ \boldsymbol{L} \leq \boldsymbol{N}_T \leq \boldsymbol{H}$$

$$0, \ \boldsymbol{N}_T > \boldsymbol{H}$$

where **L** and **H** are the flow control thresholds in Figure 6.12, and N_T is the total number of RT and NRT PDUs in the queue. $0 < k < 1$ is a factor for overflow control and λ'_{NRT} is an estimate of the user's NRT data rate allocated by the packet scheduler in the MAC-hs. The estimate is obtained using an exponentially weighted moving average filter according to

$$\lambda'_{NRT} = \alpha \cdot \lambda'_{NRT-1} + (1 - \alpha) \cdot \lambda_{NRT}$$

where λ_{NRT} is the instantaneous NRT bit rate. With the expression for C_{NRTmax}, NRT grant allocation is made dependent of load and user channel quality, which is appropriate because of the elastic nature of the NRT flow. Because averages are used in the grant calculation, the space between **H** and **N** absorbs instantaneous burst arrivals.

6.5.1.2 Buffer Management with Dynamic Inter-Class Prioritization

Although the TSP and E-TSP schemes can be utilized as the BMA to allow inter-class QoS control for integrated scheduling and buffer management solution in the MAC-hs, the static transmission prioritization of real-time class in multimedia flow may not always yield the optimum QoS. Within the BMA, *dynamic* transmission prioritization will allow the delay flexibility and partial loss tolerance of the RT streams to be exploited to optimize bandwidth allocation to the NRT streams. This will improve end-to-end performance as well as further mitigate potential NRT streams starvation in the multimedia session. In [30], dynamic time-space priority (D-TSP) is proposed as an enhancement to E-TSP to enable optimized QoS control of the RT and NRT streams of the multimedia session. The end-to-end performance evaluation of D-TSP using extensive system-level HSDPA simulations can also be found in [30].

D-TSP as shown in Figure 6.13 incorporates dynamic switching of time (transmission) priority between the RT and NRT streams present in the multimedia user's MAC-hs buffer.

For a given transmission opportunity assigned to the UE with multimedia traffic by the (inter-UE) packet scheduling algorithm, when there is no danger of HOL queuing delay of multimedia real-time packets exceeding a given delay budget, the transmission priority is switched to the NRT flow. If real-time HOL delay is greater than or equal to the delay budget or no non-real-time packets are present in the D-TSP queue, then transmission priority remains with the real-time flow. The delay budget can be expressed in terms of the number of queued real-time packets via a parameter **k**, where [30]

$$\text{Delay budget} = \boldsymbol{k} \times (\text{RT packet inter-arrival time})$$

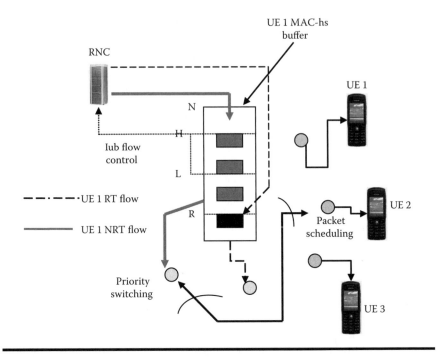

Figure 6.13   Enhanced TSP with dynamic priority switching (D-TSP).

Thus, $k = 2$ with a real-time packet inter-arrival time of 20 ms is equivalent to a delay budget of 40 ms.

Let *MAX_delay* represent the maximum allowable queuing delay to enable end-to-end QoS delay guarantee for the RT streams in the multimedia flow. A discard timer (DT) is set on arrival of real-time packets (PDUs) to the MAC-hs buffer. DT is configured to time-out after a period of *MAX_delay*, triggering the dropping of HOL RT packet(s) queued for up to *MAX_delay* seconds. DT is cancelled on transmission of RT packet(s). We can therefore express the time priority switching strategy as follows:

IF RT packets < **k** AND RT HOL delay < *MAX_delay* AND
NRT packets > 0
Time Priority = NRT flow
Generate Transport Block from NRT PDUs
ELSE
Time Priority = RT flow
Generate Transport Block from RT PDUs

D-TSP provides an effective buffer management algorithm for an integrated scheduling and buffer management solution in HSDPA because it ensures that the NRT stream is allocated the optimal bandwidth at the air interface within the allowable delay tolerance of the companion RT stream in the multimedia session of the user, while the allocation of transmission slots between the multimedia user and other users in the cell is done via an implemented inter-user packet scheduling algorithm such as PF, M-LWDF, or Max-C/I, depending on operator requirements.

6.6 Summary

This chapter provided detailed coverage of physical and MAC layer mechanisms in the 3GPP standards that enable High-Speed Downlink Packet Access (HSDPA) in WCDMA UMTS systems. This provided a basis for our discussion of packet scheduling and buffer management solutions in HSDPA. Packet scheduling is embedded in a new MAC-hs entity at Node B close to the air interface. The MAC-hs also incorporates credit-based flow control algorithms to facilitate HS-DSCH congestion control. The MAC-hs entity in an HSDPA Node B is also equipped with priority handling functionality to manage resources between HARQ entities, which could consist of multiple queues for flows associated with a single user. Thus, the HSDPA standards have existing mechanisms to facilitate buffer management for prioritizing multiple flows with different traffic class requirements in an ongoing, end-user multimedia session. This allows integrated scheduling and buffer management solutions to be employed for enhanced end-to-end QoS performance of real-time and non-real-time classes of flows in the same end-user multimedia sessions.

In HSDPA, opportunistic or channel-aware packet schedulers exploit multi-user diversity to increase the gain in total cell throughput usually at the expense of fairness. Nevertheless, the conflicting goals of maximizing throughput while ensuring some degree of fairness of users requesting service is a major design consideration of HSDPA scheduling algorithms. While basic schedulers such as the Maximum C/I, Round Robin, and Proportional Fair do not explicitly account for packet delays in their scheduling metrics, some proposed algorithms such as M-LWDF and CD-EDD provide delay differentiation to allow the delay requirements of real-time services to be met.

To allow packet scheduling with inter-class prioritization between classes or media in a multimedia session of the same user, algorithms such as the L-AWDF have been proposed. L-AWDF allows service differentiation through an expressed maximum allowable delay and a weighting factor for each class of the user's multimedia traffic. An alternative approach to scheduling with inter-class prioritization, which was also discussed, is the integrated scheduling and buffer management approach. This uses a buffer management scheme for inter-class prioritization within the same user multimedia flows while the packet scheduling discipline selects the eligible user for transmission. We presented a TSP buffer management scheme as an effective per-user algorithm that can be used jointly with existing inter-user packet scheduling disciplines for multimedia sessions with real-time and non-real-time classes of flows. An enhanced version (E-TSP) leverages the HSDPA flow control to enhance end-to-end performance. A dynamic version (D-TSP) that allows adaptive inter-class priority switching between the real-time and non-real-time flows, thereby optimizing the QoS control between the end-user multimedia session flows, was also described.

References

[1] 3GPP, Technical Specifications Group RAN, Physical Layer Aspects of UTRA High Speed Down Link Packet Access, 3GPP TS 25.848 version 4.0.0.

[2] 3GPP, Technical Specifications Group RAN, Spreading and Modulation (FDD), 3GPP TS 25.13 version 8.3.0.

[3] 3GPP, Technical Specifications Group RAN, Multiplexing and Channel Coding, 3GPP TS 25.212 version 8.4.0.

[4] 3GPP, Technical Specifications Group RAN, Physical Layer Procedures (FDD), 3GPP TS 25.14 version 8.4.0.

[5] 3GPP, Technical Specifications Group RAN, Medium Access Control (MAC) Protocol Specifications, 3GPP TS 25.321 version 8.4.0.

[6] 3GPP, Technical Specifications Group RAN, High Speed DownLink Packet Access Iub/Iur Protocol Aspects, 3GPP TS 25.877 version 5.1.0.

[7] 3GPP, Technical Specifications Group RAN, High Speed DownLink Packet Access (HSDPA) Overall Description, 3GPP TS 25.308 version 8.4.0.

[8] M.C. Necker and A. Weber, Impact of Iub flow control on HSDPA system performance, in *Proc. Personal Indoor and Mobile Radio Communications (PIMRC 2005)*. Berlin, Germany. September 2005.

[9] P.J. Legg, Optimised Iub flow control for UMTS HSDPA, in *Proc. IEEE Vehicular Technol. Conf. (VTC 2005Spring)*. Stockholm, Sweden. June 2005.

[10] T.L. Weerawardane, A. Timm-Giel, C. Gorg, and T. Reim, Impact of the transport network layer flow control for HSDPA performance, in *Proc. IEE 2006 Conf.* Colombo, Sri Lanka. September 2006.

[11] S.Y. Yerima and K. Al-Begain, End-to-end QoS improvement of HSDPA end-user multi-flow traffic using RAN buffer management, in *Proc. IEEE Int. Conf. New Technologies, Mobility and Security, NTMS 2008*. Tangier, Morocco. November 2008.

[12] D. Soldani, M. Li, and R. Cuny, *QoS and QoE Management in UMTS Cellular Systems,* John Wiley & Sons Ltd. 2006.

[13] P.A. Gutierrez, Packet Scheduling and Quality of Service in HSDPA, Ph.D. thesis, Institute of Electronic Systems, Aalborg University, Denmark. October 2003.

[14] R.C. Elliott and W.A. Krzymien, Scheduling algorithms for the cdma2000 packet data evolution, in *Proc. IEEE 56th Vehicular Technol. Conf. (VTC 2002-Fall)*, 1: 304–310, 2002.

[15] M. Andrews et al., Providing quality of service over a shared wireless link, *IEEE Commun. Mag.*, 39(2): 150–154, February 2001.

[16] T.E. Kolding, Link and system performance aspects of proportional fair packet scheduling in WCDMA/HSDPA, in *Proc. IEEE Vehicular Technol. Conf.*, September 2003, pp. 1717–1722.

[17] H. Holma and A. Toskala, *HSDPA/HSUPA for UMTS*, John Wiley & Sons, 2006.

[18] J.M. Holtzman, CDMA forward link waterfilling power control, in *Vehicular Technol. Conf., 2000. VTC—Spring*, 3: 1663–1667, 2000.

[19] J.M. Holtzman, Asymptotic analysis of proportional fair algorithm, in *Proc. Vehicular Technol. Conf. (VTC)*, May 2000, pp. 1663–1667.

[20] F. Kelly, Charging and rate control for elastic traffic, *Eur. Trans. Telecommun.*, 8: 33–37, 1997.

[21] T.E. Kolding, F. Frederiksen, and P.E. Mogensen, Performance aspects of WCDMA systems with high speed downlink packet access (HSDPA), in *Proc. IEEE 56th Vehicular Technol. Conf. (VTC 2002-Fall)*, 1: 477–481, 2002.

[22] G. Barriac and J. Holtzman, Introducing delay sensitivity into the proportional fair algorithm for CDMA downlink scheduling, in *IEEE Proc. Int. Symp. Spread Spectrum Techniques and Applications*, September 2002, pp. 652–656.

[23] A.K.F. Khattab and K.M.F. Elsayed, Channel-quality dependent earliest due fair scheduling schemes for wireless multimedia networks, in *Proc. MSWiM 2004*. Venice, Italy. October 2004.

[24] M. Andrews et al., CDMA Data QoS Scheduling on the Forward Link with Variable Channel Conditions, Bell Labs Technical Memo, April 2000.

[25] A. Golaup, O. Holland, and A.H. Aghvami, Concept and optimization of an effective packet scheduling algorithm for multimedia traffic over HSDPA, in *IEEE Int. Symp. Personal, Indoor and Mobile Radio Commun*. Berlin, Germany, 2005.

[26] O. Holland, A. Golaup, and H. Aghvami, Efficient packet scheduling for HSDPA allowing inter-class prioritization, *IET Electronics Lett.*, 42(18): pp. 1045–1046, August 2006.

[27] A. Golaup, O. Holland, and A.H. Aghvami, Issues in scheduling multimedia traffic over high speed downlink access link of UMTS, in *Proc. IST Mobile and Wireless Commun. Summit*. Dresden, Germany. June 2005.

[28] K. Al-Begain, A. Dudin, and V. Mushko, Novel queuing model for multimedia over downlink in 3.5G wireless networks, *J. Commun. Software Syst.*, 2(2): 68–80, 2006.

[29] S.Y. Yerima and K. Al-Begain, An enhanced buffer management scheme for multimedia traffic in HSDPA, in *Proc. IEEE Int. Conf. Next Gen. Mobile Applications, Services and Technol. (NGMAST '07)*. Cardiff, United Kingdom. September 2007.

[30] S.Y. Yerima and K. Al-Begain, A dynamic buffer management scheme for end-to-end QoS enhancement of multi-flow services in HSDPA, in *Proc. IEEE Int. Conf. Next Gen. Mobile Applications, Services and Technol. (NGMAST '08)*. Cardiff, United Kingdom. September 2008.

Chapter 7

HSPA Radio Access Network Design

Jamil Yusuf Khan and Xinzhi Yan

Contents

7.1 Introduction

Radio access network (RAN) is one of the key elements of a mobile communication network. The RAN plays a key role in offering multimedia services to users as well as maintaining QoS (quality of service) for different services. The design of RAN parameters is very important in maintaining the appropriate QoS for different services, particularly in a packet-switched wireless network where resource-sharing techniques are applied to maximize network resource utilization and to minimize operating costs. The UMTS (Universal Mobile Telecommunications System) network uses a standard mobile telephone network architecture where various functionalities are grouped and distributed over logical subnetworks that are connected through different logical interfaces. Functionally, all UMTS network elements are grouped within the RAN known as UTRAN (UMTS Terrestrial Radio Access Network) and the core network (CN), which connects the UTRAN to external networks such as PSTNs (public switched telephone networks) and the Internet. Figure 7.1 shows the generic UMTS architecture, depicting the basic reference points and interfaces [1]. The separation of subnetworks allows ease of interconnection and independence of signaling and data transparent networks. Currently, both the legacy WCDMA (Wideband Code Division Multiple Access) network and the HSPA (High-Speed Packet Access) network share the same CN but have different UTRAN architectures, functions, and resource management algorithms. UMTS networks, which are currently offering both legacy and HSDPA (High-Speed Downlink Packet Access)/HSPA services, have upgraded their UTRAN functionalities based on Release 5/6 or higher 3GPP (3rd Generation Partnership Project) standards. The new standard supports both legacy services as well as advanced packet based HSPA services. Introduction of HSDPA and HSUPA (High-Speed Uplink Packet Access) services have increased the packet-switched traffic volume in the UTRAN and in the CN. The UTRAN architecture is currently evolving toward a high data rate and high QoS network. Recently, the E-UTRAN (Evolved UTRAN) architecture was introduced; it was designed to support advanced packet-switched services using a flat network architecture to accommodate new services as well as to offer high QoS to all services.

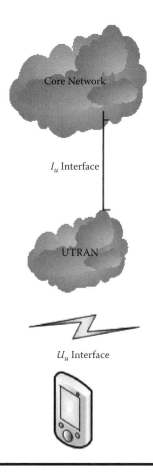

Figure 7.1   UMTS architecture.

The UTRAN consists of two main network elements that form the radio network subsystem (RNS). Those elements are Node-B, which is the base station, and the RNC (radio network controller), the key network controller. The user equipment (UE) connects to the UTRAN using the U_u interface (radio link) as shown in Figure 7.1. Elements of the UE consist of a mobile equipment (ME), which is a radio terminal, and the UMTS subscriber identity module (USIM). The UTRAN architecture of the HSPA network is defined by 3GPP Release 6 and onward standards, which are quite different from the legacy Release 99 standard. The HSDPA standard was introduced in Release 5 of the 3GPP standard in 2002, followed by the introduction of the HSUPA standard in Release 6 in 2004. The HSUPA and HSDPA standards are unified, and the combined standard is known as the HSPA standard. Since the introduction of the HSPA standard, the role of the UTRAN has changed significantly due to the support of diverse types of packet-switched traffic. The HSPA standard was developed mainly to support high data rate packet-switched traffic with low latency and high QoS requirements. With the introduction of the E-UTRAN architecture, the expected target transmission packet delay on the UE to RNC link in an HSPA network should be maintained at less than 10 ms [2]. To support the low latency and high QoS (packet loss, jitter, etc.) requirements, the UTRAN architecture is gradually evolving by incorporating various advanced functionalities. In Release 7 of the 3GPP standard, several HSPA enhancements were introduced, to include providing major improvements to end-user QoS performance, support of high data rate services, inclusion of specialized services such as VoIP (Voice over Internet Protocol), and increasing network efficiency [3]. To further enhance the QoS of HSPA services, the 3GPP is also proposing a new flat architecture similar to the LTE (Long Term Evolution) architecture introduced in Release 8 [4]. Improvements in the QoS and latency of HSPA connections can be achieved by developing enhanced resource management algorithms for the UTRAN; these are beyond the scope of 3GPP standards and have been left for researchers to develop.

To achieve high QoS and low latency for HSPA connections it will be necessary to develop efficient radio resource management algorithms for various interfaces and network entities of the UTRAN. The 3GPP standards do not specify any specific algorithms; rather, it is left to vendors to develop appropriate algorithms that offer high QoS for different services. This chapter concentrates on the development and performance analysis of radio resource management algorithms for the I_{ub} interface, which is one of the key interfaces of the UTRAN. The QoS profile of the I_{ub} link can significantly influence both the uplink and downlink as well as control traffic. Before presenting the resource management algorithms, various key elements of the UTRAN are introduced in this chapter. The chapter is organized into sections as follows:

- "UTRAN Architecture" overviews the UTRAN network and interface architectures. This section also reviews the functional requirements of the UTRAN.
- "UTRAN Protocol Structure" describes several key functional procedures and protocol architecture of the UTRAN.
- "E-UTRAN Architecture" briefly introduces the Evolved UTRAN architecture and lists some of the initial specifications.
- "UTRAN Design: I_{ub} Link Resource Management Algorithms" reviews different UTRAN resource management algorithms, mainly concentrating on the I_{ub} interface, which is responsible for connecting the Node-B and the RNC. This section also very briefly reviews some algorithms developed by different researchers.

- "Air Interface and I_{ub} Parameter Interdependencies" examines the HSDPA air interface and the I_{ub} link parameter interdependencies to evaluate the performance of the joint resource management algorithm.
- The "Summary" provides conclusions.

7.2 UTRAN Architecture

The UTRAN consists of a number of RNSs connecting UEs to the CN. Figure 7.2 shows the basic UTRAN architecture with different logical interfaces connecting various elements of the network. In the UTRAN, Node-B is connected to the RNC using the I_{ub} interface. The I_{ub} interface allows the RNC and the Node-B to negotiate radio resource usage to support different connections according to the QoS requirements of services supported by these connections. The I_{ub} link supports three types of information transfer: (1) various user data, (2) signaling related to user data transmission, and (3) logical operations and maintenance data of Node Bs. The I_{ur} interface is used to connect two RNCs located in different radio network subsystems to exchange data and signaling information. The I_{ur} interface provides capabilities to support the mobility of UEs between RNSs. The I_{ur} interface could connect the RNCs over a direct physical connection or through a virtual connection. In the UTRAN, RNCs are connected to the core network using the I_u interface. All traffic to and from external networks passes through the I_u interface, which supports three distinct service domains. These domains are packet switched (PS), circuit switched (CS), and broadcast (BC). According to the 3GPP standard, at any time there shall not be more than one I_u-PS interface toward the PS domain except under certain

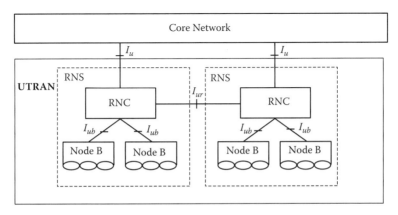

Figure 7.2 UTRAN architecture showing the different elements of the core network. (3GPP, TS25.401, R8, UTRAN Overall Description, v.8.2.0, Fig. 4, 16, 2008. With permission from ETSI.)

Table 7.1 Summary of Radio Resource Management Functionalities of the Different Elements of UTRAN

Node-B	SRNC	DRNC
Packet scheduling Dynamic resource allocation QoS provisioning Load control mechanisms Uplink power control	QoS parameter mapping Handover control Uplink outer-loop Power control (for HSUPA connections)	Admission control Initial uplink power setting Radio resource allocation Code distribution Load control

special network configurations. Similarly, each RNC shall not have more than one I_u-CS interface toward the core network except in circumstances where inter- or intra-system handovers or the SNRS relocation operation is supported.

An RNC in an RNS may take on multiple networking roles. The RNC could be a *serving* RNC (SRNC) connecting UEs to the CN via the UTRAN using different I_u interfaces. The SNRC also performs L2 (Layer 2) processing and all basic resource management functionalities, such as radio access bearer (RAB) mapping, handover decision, and outer-loop power control [5]. The RNC can also act as a *controller* RNC (CRNC), which controls its own Node-B in an RNS, and is responsible for load and congestion control on the I_{ub} link. The CRNC also executes the admission control and code allocations of new connections in its own cell. Another RNC configuration known as the *drift* RNC (DRNC) is responsible for performing micro-diversity combining and splitting, as well as routing data between I_{ub} and I_{ur} interfaces without performing any L2 processing of user plane data. For HSDPA and HSUPA networks, the radio resource management functionalities are split between the Node-B and the RNC. The radio resource management functionalities of different elements of the UTRAN are summarized in Table 7.1.

7.2.1 UTRAN Functionalities

UTRAN functionalities have been developed based on the following principles. In the UTRAN, the CRNC will own all the radio resources of a cell that can be borrowed by a serving RNC to handle all UE connections in the cell. The SNRC is also responsible for dynamic power control of dedicated channels that are admitted by the corresponding CNRC. For HSUPA connections, fast power control function is coordinated by the Node-B. The resource scheduling task within the UTRAN is performed by various RNCs. Dedicated channels are scheduled by the SRNC, whereas the common channels are scheduled by the CRNC. However, the packet scheduling task on the U_u interface remains with the Node-B. Admission control is performed by the DRNC, which allocates new radio access bearers or new radio links to HSPA connections. The admission control algorithms should try to avoid overload situations and make a decision based on interference conditions and resource management criteria. On the UTRAN side, the admission control technique is applied on all I_u and U_u interfaces. The UTRAN also introduces congestion control techniques to monitor, detect, and handle situations when a system may be reaching an overload circumstance or may have exceeded the overload condition with existing traffic load. Both the admission and congestion control algorithms could be applied on different interfaces based on local traffic and radio transmission conditions to optimize the performance of various interfaces. It is also possible to use joint admission and congestion control algorithms on multiple interfaces to optimize the overall performance of the UTRAN. An end-to-end HSPA connection can be seen as a multi-hop connection involving a number of interfaces (U_u, I_{ub}, I_{ur}, and I_u) where a congestion control algorithm could be applied locally or jointly between different interfaces to meet the QoS requirements of the HSPA connections. Admission and congestion control algorithms are not specified by 3GPP standards, and it is left to the vendors to develop efficient algorithms within the UTRAN framework.

The UTRAN architecture for the HSPA network has been enhanced compared to the Release 99 standard by incorporating transport channels and associated functionalities to support these enhanced packet-based services. The HS-DSCH (High-Speed Downlink Shared Channel) was introduced to support the HSDPA traffic and the E-DCH (Enhanced Dedicated Channel) was introduced to support the HSUPA traffic. The HSDPA standard introduces several data rate adaptation and control mechanisms to support the peak data rates and high spectral efficiency, as well as higher QoS for asymmetric data traffic. These functionalities are distributed within UTRAN elements and include RNCs, Node-B, and various interfaces. These functionalities are described in following subsections.

7.2.2 UTRAN Interface Functionalities

7.2.2.1 I_u Interface

The I_u interface was designed to support the interconnection of RNCs to the CN access points offering services in PS, CS, and BC domains. These interfaces offer independence between protocol layers as well as independence between control and user planes [5]. The I_u interface was developed as an open and multivendor interface to allow the separate evolution of

O&M (Operations and Management) functionalities. The main capabilities of the interface are that it

- Offers procedures to establish, maintain, and release radio access bearers
- Offers SRNS (serving RNS) relocation, intra- and inter-system handovers
- Supports cell broadcast services
- Offers separation of each UE connection on the protocol level for user-specific signaling management
- Offers location-based services using geographical area identifier or global coordinates with uncertainty parameters
- Offers simultaneous access to multiple core network domains for each UE
- Transfers signaling between the UE and the CN
- Offers simultaneous access to multiple CN domains for a single UE
- Offers mechanism for resource reservation for packet data streams
- Has procedures to support multicasting services using MBMS (Multicast Broadcast Multimedia Services) bearer services

7.2.2.2 I_{ur} Interface

The I_{ur} interface supports the interconnection of RNCs developed by different manufacturers. It also supports the continuation of services between RNSs via the I_u interface [6]. The main capabilities of the I_{ur} interface are

- The interface provides capability to support mobility of UEs between different radio network subsystems. The mobility is offered by supporting handover, radio resource management, MBMS connection management, and synchronization between different RNSs.
- The interface supports different types of data streams such as DCH, RACH (Random Access Channel), HS-DSCH, E-DCH transport frames carrying user data and control information between the SNRC and Node-B, and the DRNC.
- Within the UTRAN, the HSDPA traffic is carried by the HS-DSCH frame. A UE may have multiple HS-DSCH data streams on the I_{ur} interface. The I_{ur} interface provides a means for transporting MAC-d (dedicated MAC) PDUs (Protocol Data Units). In addition, the interface provides a means to the SRNC for queue reporting and a means for the DRNC to allocate capacity to the SRNC.
- The HSUPA traffic within the UTRAN is carried by E-DCH data frames. The I_{ur} interface provides the means for transporting I_{ub}/I_{ur} E-DCH frames carrying user data between Node-B and the SRNC via the DRNC.

A UE may have multiple E-DCH data streams supporting different applications and traffic sources. In addition to the above functionalities, the interface provides the following additional options:

- A means for the Node-B to indicate the number of HARQ transmissions to the SRNC
- A means to indicate to the SRNC for the purpose of reordering connection frame number and subframe numbers that are added by the Node-B.

7.2.2.3 I_{ub} Interface

The interface provides an interconnection between the Node-B and the RNC by separating radio network and transport network functionalities to facilitate the introduction of future transmission technologies [7]. This interface has been developed based on several key principles, as follows:

1. The functional division between the RNC and the Node-B.
2. Neither the physical structure nor any internal protocols of the Node-B shall be visible over the I_{ub} link, thus helping the introduction of new technologies in the future.

The I_{ub} interface capabilities are similar to the I_{ur} interface supporting various data streams. The interface provides the means for transporting HSDPA traffic using HS-DSCH frames between the RNC and the Node-B. Similarly, the interface supports HSUPA traffic between the Node-B and the RNC using the E-DCH frames. Each HS-DSCH data stream on the I_{ub} link is carried on a single transport bearer. For each HS-DSCH data stream, a transport bearer must be established over the I_{ub} interface. The I_{ub} interface also supports features similar to the I_{ur} interface to support HARQ transmission features.

7.2.3 Radio Resource Management

Radio resource management is concerned with the allocation and maintenance of communication resources. In a UTRAN network, resources must be shared between circuit-switched and packet-switched connections for resource optimization. To efficiently allocate radio resources, the UE and UTRAN network entities will survey the radio environment by measuring its own and surrounding cells' SNR (signal-to-noise ratio), BER (bit error rate), power profiles, interference, synchronization status, etc. Based on these measurements, various transmission resources are allocated by the RNCs and the Node-B in the UTRAN. Some of the key radio resource management functionalities of a UTRAN are summarized below.

7.2.3.1 Combining/Splitting

This function controls information streams to receive or transmit the same information through multiple physical channels transmitted or received from a single mobile terminal. Multiple physical channels may receive information from different cells particularly in case of a soft handover. Depending on the physical network configuration the combining/splitting function could be carried out at the SRNC, DRNC, or Node-B level.

7.2.3.2 Connection Setup and Release

This function is responsible for the control of end-to-end connection setup and release. The function also maintains the end-to-end connection by using appropriate algorithms to support call QoS and handover executions. This function is located both in the UE and in the RNC.

7.2.4 Radio Bearer Allocation

This function allocates appropriate physical radio channels to users by matching the QoS of a radio bearer with the connection setup request. During the connection setup, users will specify the QoS of the required connection. The function is located in the CRNC and the SRNC. The I_u interface provides extensive RAB (radio access bearer) management functionalities that control connections between UEs and the CN. Depending on subscription, service, and requested connection QoS, a particular RAB will be selected from a pool of RABs. The CN controls connection establishment, modification, or release of an RAB by initiating appropriate protocol functionalities. These are CN-initiated and UTRAN-executed functionalities. A RAB release request could be generated by the UTRAN when it fails to keep the RAB established for a UE. The RAB characteristics mapping function is used to map appropriate radio access bearers to the U_u interface and the I_u interface transport bearer. When the PS domain traffic is carried by the UTRAN, the RNC will perform the mapping between the radio access bearers and the IP-layer QoS functions.

An HSPA network requires support of priorities to accommodate different types of multimedia traffic. The allocation of a priority level of an RAB is determined by the CN based on subscription information, QoS information, etc. According to the above information, the CN requests the UTRAN to establish or modify RAB parameters with an indicated priority level, preemption capability, and the queuing preferences. Queuing and resource preemption can be carried out by the UTRAN according to the preference set up during the call establishment procedure. The radio resource admission control is carried out by the UTRAN when it receives a request to modify or to establish a radio access bearer from the CN. Based on the request, the UTRAN analyzes the load situation, and then the admission control module either accepts or rejects the request. If a call request is queued, then the request is handled by the RAB queuing, preemption, and priority functions.

The admission control on the I_{ub} link is based on uplink interference and downlink power information. The admission control module is located in the CRNC. The Node-B reports uplink interference measurements and downlink power information over the I_{ub} interface. The CRNC controls the reporting procedures, including the reporting frequency. On the I_{ub} link traffic on high-speed channels (HS-DSCH and E-DCH) is managed by the Node-B.

7.2.5 Radio Protocol Functions

This function provides user data and signaling transfer capabilities across the UMTS radio interface by adapting the QoS of various services according to radio transmission conditions. The QoS of different services will depend on the selection and allocation of radio bearers. The UMTS radio protocol includes the following main functionalities:

- Multiplexing of different services using the allocated connection to a UE
- Multiplexing of UEs on various radio bearers
- Segmentation and assembly of information blocks
- Delivering acknowledged and unacknowledged services according to the allocated radio access bearer

7.3 UTRAN Protocol Structure

UTRAN functionalities are layered into the radio network layer (RNL) and the transport network layer (TNL). All UTRAN logical nodes and interfaces between them are defined as a part of the RNL. All UTRAN-related functions are only visible in the RNL. For all UTRAN interfaces, their related transport protocol and functionalities are specified in [1, 8]. The TNL provides services for user plane transport, signaling transport, and specific O&M transport. The TNL represents standard transport technology without any UTRAN-specific changes. The general UTRAN protocol architecture is shown in Figure 7.3. The protocol is structured into three vertical planes implementing all TNL and RNL functionalities. The protocol model supports three standard vertical planes: control, user, and signaling. The control plane is mainly responsible for the UMTS control signaling, which includes implementation of the application protocol of different logical interfaces, and the signaling bearer for transporting the application protocol messages.

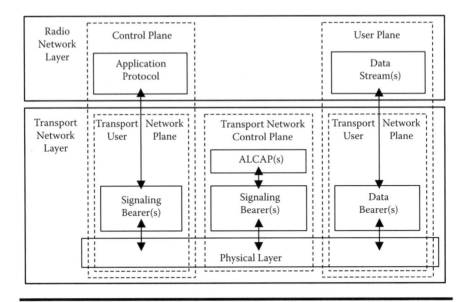

Figure 7.3 General protocol architecture for UTRAN. (3GPP, TS25.321, R8, Medium Access Control (MAC) Protocol Specification, v.8.4.0, Fig. 10, 39, 2008. With permission from ETSI.)

One of the main tasks of the application protocol is setting up the bearers of the I_u interface and then subsequently the radio link for the I_{ur} and I_{ub} interfaces. In the three-plane structure, the bearer parameters in the application protocol are not directly tied to the user plane technology. The user plane is responsible for transporting user information, which includes coded voice or voice calls and various types of data streams. The transport network control plane is responsible for control signaling of the transport layer. This layer does not handle any RNL signaling. However, this layer does include the signaling protocol used to set up data bearers and the signaling bearer needed for the ALCAP (Access Link Control Application Part). The ALCAP may not be needed for all types of data bearers. In absence of any ALCAP signaling transaction, transport network control is not required. This situation arises when either preconfigured data bearers or IP-based UTRAN or CN nodes are used. When the transport network control plane is used to set up transport bearers for data traffic in the user plane the following approach is used. First, the control plane establish signaling transactions which is triggered by the set-up of data bearers by the ALCAP protocol specific to the user plane technology. The transport network user plane allocates data bearers in the user plane and signaling bearers for the application protocol. Data bearers in the transport network user plane are directly controlled by the transport network control plane for real-time services.

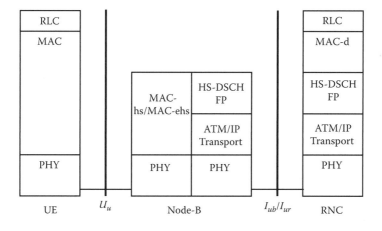

Figure 7.4 HSDPA protocol architecture.

7.3.1 HSDPA Protocol Structure

The UTRAN uses two new transport channels (HS-DSCH and E-DCH) compared to release 99 of 3GPP protocols to support HSDPA and HSUPA traffic flows. The HSDPA user plane protocol architecture is shown in Figure 7.4. The figure shows the distributed nature of the protocol. The Node-B is responsible for scheduling, radio resource allocation, and QoS provisioning and load control mechanisms. The RNC is responsible for QoS parameter mapping, handover control, and maintaining the link to the CN. The HSDPA physical layer at the U_u interface has been enhanced by incorporating the fast link adaptation procedure by using the adaptive modulation and coding (AMC), and HARQ (Hybrid Automatic Repeat Request) techniques in the Node-B and in the UE. To support the advanced physical layer capabilities, two specialized MAC entities have been introduced, known as MAC-hs (Medium Access Control High Speed) and MAC-ehs (Medium Access Control Enhanced High Speed). The MAC-hs/ehs provide enhanced functionalities such as fast scheduling to facilitate the efficient usage of radio resources to adapt to channel conditions and the network load [8]. To implement HSDPA features, three new channels have been introduced in the physical layer. These channels are HS-DSCH, HS-SCCH (High Speed Shared Common Channel), and the HS-DPCCH (High Speed Dedicated Physical Control Channel). The HS-DSCH carries data in the downlink direction, with the peak rate reaching up to 28.8 Mbps. The HS-SCCH is used to support signaling and the physical layer retransmission procedures. The HS-DPCCH carries various control signals, including the ARQ acknowledgment and the downlink quality feedback information. The HS-DSCH is dynamic in nature, wherein resource sharing can be achieved after every 2-ms transmission time interval (TTI) utilizing packet schedulers. The physical layer is also responsible for selecting the HS-DSCH parameters such as the transmission data rate based on the terminal capability, number of code allocation, modulation type, and HARQ parameters.

The distributed HSDPA architecture splits the MAC layer between the RNC and the Node-B. MAC PDUs, generated by the RNC and called MAC-d PDUs, are aggregated and sent to the Node-B over the I_{ub} interface using HS-DSCH frames. The delivery of PDUs over the I_{ub} interface is managed by a flow control protocol that can act independently for each cmCHPI (Common Transport Channel Priority) used for different UEs. The RLC (radio link control) layer at the RNC is responsible for segmentation and retransmission of user and control data on the downlink. The RNC receives data from the CN, which is appropriately segmented using the RLC protocol. The RLC layer is involved in the data link layer retransmission process when the physical layer retransmission known as HARQ fails or exceeds its retransmission limits. The MAC-d layer is retained in the RNC to handle transport channel switching. The MAC-hs layer is located in the Node-B and is responsible for scheduling, priority handling, HARQ, and selection of an appropriate transport format and resources. The HS-DSCH is the transport channel that carries the user data. The HS-DSCH FP (Frame Protocol) handles the data transport between the RNC and the Node-B. For an HSDPA connection, data received from the CN is encapsulated in a MAC-d PDU and then transmitted over the I_{ub} link using the HS-DSCH frame. The HS-DSCH FP entity adds header information to form a HS-DSCH FP PDU that is transported to Node-B over a transport bearer. The MAC-hs or MAC-ehs entity in Node-B transfers MAC-hs/ehs PDU to the peer MAC entity in the UE over the U_u interface. The MAC/L2 protocol at the RNC and at the Node-B could use different types of transport protocols, such as ATM (Asynchronous Transfer Mode) or UDP/IP (User Datagram Protocol/Internet Protocol). The I_{ub} link requires a flow control mechanism to ensure that Node-B buffers are appropriately utilized and high throughput is obtained on the downlink. The flow control mechanism will be supported by MAC layers at both ends of the I_{ub} link.

7.3.2 HSUPA Protocol Structure

The enhanced uplink was introduced to significantly enhance the capacity of the uplink as well as to improve the QoS (throughput and delay) of uplink traffic [9, 10]. To improve the performance of the uplink it is necessary to enhance the dedicated channels by offering priorities to streaming, interactive, and background services. Appropriate QoS mechanisms must be developed to support packet services such as streaming, interactive, and background data transmission. The enhanced uplink was developed

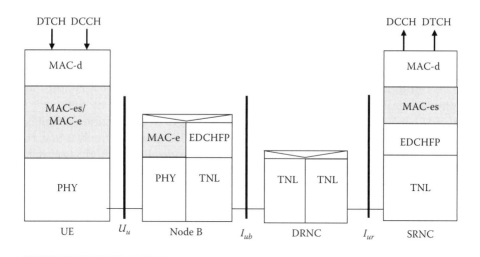

Figure 7.5   HSUPA protocol stack.

with the view that enhanced terminals can coexist with legacy terminals (specified by Release 99, Release 4, and Release 5). The enhanced uplink was developed to enhance the overall system performance by operating alongside the HSDPA standard. At the same time, it is assumed that the enhanced uplink should operate without any dependency on the deployment of HSDPA features. To support the HSUPA features, the UTRAN protocol introduces new MAC entities: (1) new entities MAC-es/e and MAC-is/i are added in the UE below the MAC-d layer; and (2) new entities MAC-e/MAC-i in the UE handle HARQ retransmissions, scheduling and multiplexing, and the TFC selection tasks. In the Node-B, these entities support HARQ retransmission, scheduling, and demultiplexing operations. In the SRNC, the new entity MAC-es is used to provide in-sequence delivery of PDUs and a data combining function when data is received from multiple Node-Bs to support a soft handover. A reordering function within the MAC-es is used to provide in-sequence delivery of data blocks to the RLC layer. In the CRNC, MAC-e/i entities provide functionalities to support in-sequence delivery of PDUs, disassembly and reassembly of PDUs, and the collision detection function. Figure 7.5 shows the HSUPA protocol architecture, highlighting the new entities in the MAC layer. HSUPA connections use the E-DCH transport channel in the UTRAN to carry user data. The E-DCH MAC entity in the UE transfers the MAC-e/i PDU to its peer MAC entities in the RNC. The E-DCH FP entity adds header information to form the E-DCH FP PDU, which is transported to the RNC over a transport bearer. The PDU generation process is controlled by the interworking function (IWF) at Node-B, which receives data from the UE. The E-DCH channel scheduling is performed by the MAC-e/i entity in the Node-B, and reordering is performed by the

MAC-es/is entity in the RNC. The Node-B performs two types of resource scheduling known as the absolute grant (AG) and the relative grant (RG) [11]. The UE sends a scheduling request to Node-B for permission to transmit; in response to this request, the Node-B sends a grant specifying the data rate of the approved connection. The scheduler uses the E-DCH absolute grant channel (E-AGCH) for transmitting primary or secondary grant (PAG or SAG) information, which caters to large changes in connection data rates. The RG information is sent on the E-DCH relative grant channel (E-RGCH) to send relative grants, which reflect minor changes in the connection data rate. The relative grant allows the Node-B scheduler to incrementally adjust the services grant of the UE under its control. The RG values can take one of three different values: UP, DOWN, or HOLD. The new MAC entity MAC-e/i also performs the task of resource scheduling. This option provides the serving node (Node-B/RNC) a better view of the resource requirements of UE and the amount of resources it can actually use. When any scheduling information is transmitted, its contents must be updated in new transmissions with the buffer status after the application of E-TFC (E-DCH transport format combination).

7.3.3 UTRAN MAC Protocol

This section briefly discusses the basics of the UTRAN MAC protocol for the HSDPA and HSUPA standards from a resource management point of view. In the UMTS/HSPA network, the MAC protocol is split between a UE and different entities of the UTRAN. In the MAC layer, the logical channels are mapped on transport channels. The MAC layer is also responsible for selecting the appropriate transport format (TF) for each channel, and this depends on the instantaneous data rate requirements of logical channels. The MAC layer also performs several other important tasks, including priority handling and scheduling of data flow, multiplexing and demultiplexing of PDUs, traffic measurements, and control as specified by the RRC (radio resource controller). Figure 7.6 shows the MAC architecture of the HS-DSCH channel on the UTRAN side. The logical channels are shown on top of the MAC module, whereas the transport channels are shown below the MAC module. In this section the different logical channel characteristics are not discussed; detailed descriptions of these channels can be found in [9, 11]. Newly introduced MAC-hs/MAC-ehs entities handle HS-DSCH channel-specific functions. All data received on HS-DSCH channels are mapped onto MAC-hs/MAC-ehs entities. The MAC-hs/MAC-ehs is configured by the RRC using the MAC control service access point (SAP). The RRC sets the transport format combinations for the HS-DSCH channels. The associated downlink signaling carries information to support the operation of HS-DSCH channels, while the uplink signaling carries feedback information

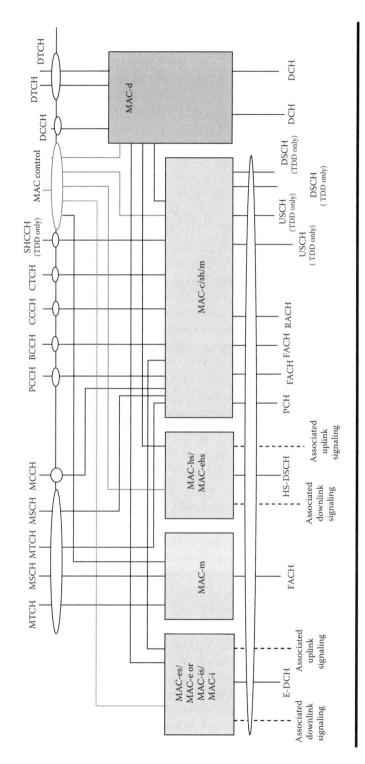

Figure 7.6   MAC protocol architecture of the HS-DSCH channel. (3GPP, TS25.321, R8, Medium Access Control (MAC) Protocol Specification, v.8.4.0, Fig. 4.2.3.1, 16, 2008. With permission from ETSI.)

such ACK/NACK and CQI (channel quality indicator). On the UTRAN side, one MAC-hs entity is used for each cell to support HS-DSCH data transmission. The MAC-hs is configured by upper layers and is responsible for handling data transmission on the HS-DSCH channel. This entity is also responsible for the management of HSDPA physical resources. The MAC-hs handles the priority of MAC-d PDUs and performs the following two resource management functions:

- *Flow control.* This is a companion flow control function of the MAC-c/sh/m entity. These functions jointly offer a controlled flow of data between the MAC-hs located in the Node-B and the Mac-d located in the RNC in a dynamic manner by taking into account the transmission capabilities of the air interface. This flow control mechanism essentially regulates data flow between the RNC and the Node-B. This flow control algorithm also affects the data flow on the U_u logical interface. The flow control function needs to be developed to reduce the Layer 2 signaling latency, as well as to reduce missing and retransmitted data that could be caused by HS-DSCH channel congestion. The flow control mechanism is important for maintaining the QoS of uplink and downlink data transmission.
- *Scheduling/priority control.* This function manages HS-DSCH resources between HARQ entities and data flow according to the connection priority. Feedback signaling based on ACK/NACK determines whether transmission or retransmission needs to be initiated.

The MAC-e entity is used to support the HSUPA operation. There is one MAC-e entity in the Node-B for each UE and one E-DCH scheduler function in the Node-B. When the MAC-e is configured by the RRC layer, the MAC-e and the E-DCH scheduler jointly handle the HSUPA functions in the Node-B; they jointly perform the following functions:

- *E-DCH scheduling.* This function manages resources of UEs. The scheduler grants resources to UEs based on their requests. A grant could be absolute or relative as described in the earlier section.
- *E-DCH control.* This entity is responsible for the reception and transmission of scheduling requests.
- *Demultiplexing:* This function is used to demultiplex MAC-e PDUs at the RNC. After demultiplexing the MAC-es PDUs are forwarded to the associated MAC-d flows.
- *HARQ.* A HARQ entity with the MAC-e is capable of supporting multiple stop and wait HARQ processes. Each process is responsible for generating ACKs and NACKs indicating delivery status of each E-DCH PDU.

7.4 E-UTRAN Architecture

The Evolved UTRAN (E-UTRAN) framework was developed to support the evolution of 3GPP radio access technology toward a high data rate, lower latency, packet-optimized radio access networking technology [4, 12]. The E-UTRAN standard is developing to support future evolved air interfaces that will support wider transmission bandwidths up to 20 MHz instead of the current 5 MHz, new transmission schemes, and advanced multi-antenna configurations using MIMO (multiple-input, multiple-output) technology. The main targets for the evolution of the radio interface and radio access network architecture include the following:

- Significantly increased peak data rate on both the up- and downlinks. Peak data rates of 100 Mbps on the downlink and 50 Mbps on the uplink is expected in the initial E-UTRAN standard. The plan is that the peak data rate will scale with the bandwidth. It is also expected to offer increased "cell edge bit rate" with the current deployment of Node-Bs.
- Possibility of UTRAN latency (UE to RNC/RNC to UE) of 10 ms or less.
- Support of scalable bandwidth 5, 10, 20, and possibly 15 MHz.
- Support of interworking with 3G systems and other non-3GPP-specified systems such as IEEE802.11 WLAN and IEEE802.16/WiMAX.
- Efficient support of various types of standard and advanced services in the PS domain. New services such as VoIP are contemplated in the near future.
- Support of lower power consumption of user terminals by introducing discontinuous uplink transmission [3, 13].

The E-UTRAN architecture is similar to the UTRAN, with a reduced number of network elements. The overall architecture of E-UTRAN is shown in Figure 7.7. The eNB (E-UTRAN Node-B) replaces Node-B and is connected

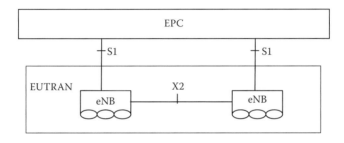

Figure 7.7   Evolved UTRAN architecture (3GPP, TS36.401, R7, Evolved Universal Terrestrial Radio Access Network (E-UTRAN); Architecture Description, v.8.5.0, Fig. 6.1-1, 10, 2009. With permission from ETSI.)

via the S1 interface to the evolved packet core (EPC) network. Two logical interfaces, S1 and X2, are based on the IP protocols to efficiently support packet-switched traffic. The S1 logical interface replaces the two-hop I_{ub}/I_u links of the UTRAN. This architecture also eliminates the need for the RNC. An eNB can support FDD (frequency division duplex), TDD (time division duplex), or dual-mode operation. The E-UTRAN architecture is based on an LTE flat network structure where eNB directly communicates with the enhanced core network without the need for the RNC. The eNB will host the following functions:

- *Radio resource management:* radio bearer control, radio admission control, connection mobility control, dynamic resource allocation/ scheduling.
- *Mobility management entity:* distribution of paging messages to the eNBs.
- *User plane entity:* IP header compression and encryption of user data streams, termination of U-plane (user plane) packets for paging tasks, switching of U-plane to support UE mobility.

7.5 UTRAN Design: I_{ub} Link Resource Management Algorithms

This section presents UTRAN/I_{ub} specific link resource management algorithms. The I_{ub} link allocation and its resource management algorithms located in the Node B and the RNC. The resource management algorithms significantly influence the traffic QoS on the U_u interface. Also, the I_{ub} link design influences UTRAN resource requirements and the capital cost of system development [14, 15]. In this section, we review the I_{ub} flow control techniques and analyze the effect of the flow control algorithm on service QoS and UTRAN resource requirements. The UTRAN architecture discussed in the previous section shows that proper dimensioning of the HSDPA RAN is becoming important due to the increasing data rate of the HSDPA air interface and the introduction of more dynamic/bursty traffic sources. The I_{ub} link is an expensive RAN resource, particularly from the operator's point of view [14]. Some early work by Toskala et al. [16] has shown that with increasing radio link utilization, there is a considerable reduction in the I_{ub} link utilization. Their work showed that at least 20% over-dimensioning of the I_{ub} link is necessary to achieve 95% of the maximum available HS-DPA air interface capacity. With the increasing HSDPA data rate, this 20% over-dimensioning factor will significantly increase the operational cost of the UTRAN. The over-dimensioning factor could vary with different traffic types and user QoS requirements. To avoid the over dimensioning of the

I_{ub} link, we introduce a joint resource management algorithm that allocates the I_{ub} link capacity based on the air interface throughput of each HSDPA connection.

As discussed, the HSDPA architecture uses a distributed RAN architecture that requires a flow control algorithm between the RNC and the Node B. Node-B controls the traffic flow on the I_{ub} link. The RNC sends the buffer space request to the Node-B using the HS-DSCH Capacity Request frame protocol. The Node-B initially allocates a transmission capacity to the RNC for each UE connection using the HS-DSCH Capacity Allocation frame protocol, as shown in the flow diagram in Figure 7.8 [17]. Node-B sets up the initial credit window size and the MAC-PDU size for the RNC. The RNC can send a request to alter the allocated credit size and the MAC-PDU size. The main objectives of the I_{ub} flow control algorithms include the following:

■ Minimize the end-to-end latency between the RNC and the UE.
■ Increase the downlink air interface and the I_{ub} link throughput.
■ Support different types of packet schedulers at the Node-B and maintain increased air interface and RAN efficiencies.
■ Minimize the I_{ub} interface signaling traffic.
■ Avoid the Node-B buffer overflow or buffer starvation.

Recently, a number of researchers have studied the impact of the I_{ub} link resource allocation techniques on HSDPA air interface performance. Work by Legg [18] developed a flow control mechanism known as **K x *max bits*** to maximize throughput and packet delay. The parameter **K** is an integer used to control the data flow on the I_{ub} link. The value of **K** depends on the I_{ub} link latency. Simulation results show that the packet delay and signaling load on the I_{ub} link can be reduced by adjusting the value of **K.** The results also show that the flow control mechanism could reduce packet loss on the I_{ub} link. Necker and Weber [19] studied a flow control algorithm to analyze HSDPA system performance. The authors discuss the effect of control dead times on the link that arise due to signaling delays. The dead time refers to the time when no resource allocation update is made due to high signaling delay. Because of the dead time, the flow control algorithm may not be able to react to data rate fluctuations on the air interface. Simulation studies showed the influence of resource allocation update delay and control dead time on IP packet delays. They also studied the effect of control dead time on the Node-B buffer performance and resource allocation [19]. Bajzik et al. [20] developed a congestion control mechanism on the I_{ub} link. Using the back-pressure algorithm, a control signal is sent from the Node-B to the MAC-d layer at the RNC to adjust the I_{ub} link allowable transmission rate. The allowable transmission rate is adjusted based on a threshold-based the Node-B queue management technique. Results show that the back-pressure mechanism can significantly reduce I_{ub} link packet delays and

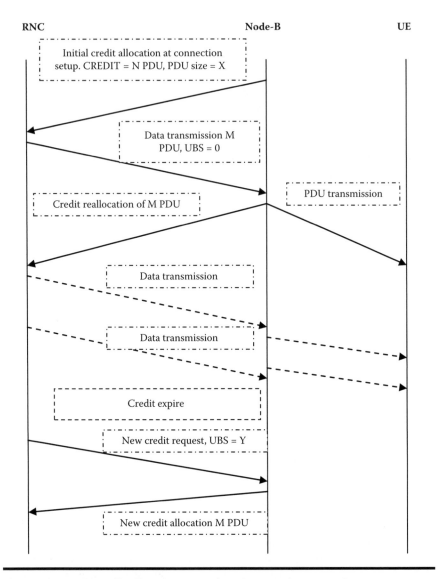

Figure 7.8   RNC credit allocation procedure for HSDPA connections.

retransmission rate. Also, the back-pressure algorithm moderately increases the air interface throughput.

7.5.1 Joint I_{ub} Link Flow Control Model

In this section we present a joint I_{ub} link flow control algorithm. The performance of the algorithm is initially analyzed by using an analytical model. Further performance of the flow control is analyzed using a simulation

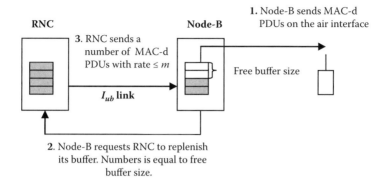

Figure 7.9   Flow control technique employed between Node-B and the RNC. (X. Yan, Khan, J.Y., and B. Jones. An Adaptive Radio Access Network Resource Management Technique for a HSDPA Network. 1–5, IEEE, *Proceedings of PIMRC.* 2007. With permission from IEEE.)

model. In the analytical model for each HSDPA downlink connection, we allocate a fixed space in the Node-B buffer, as shown in Figure 7.9. We assume that n represents the buffer size for each HSDPA connection in the Node-B main buffer, where n is represented in number of MAC-d PDUs that could be sent over the air interface at the peak transmission rate in every TTI under the best transmission channel conditions. The Node B's main buffer will accommodate all HSDPA connections by allocating buffer space to each connection based on their Qos requirements. We also assume that the I_{ub} link bandwidth is m (virtual circuit capacity) for each user connection, where m is the number of MAC-d PDUs sent in each TTI over the I_{ub} link. Another assumption used in this model is that the I_{ub} link creates a separate virtual circuit for each downlink user. After each downlink transmission event on the air interface, Node-B immediately requests the RNC to send more MAC-d PDUs to replenish its buffer. The request size R_s (in number of PDUs) is equal to the free buffer size. If the requested update size is larger than the I_{ub} link virtual circuit bandwidth m (i.e., $R_s > m$), then only m MAC-PDUs are delivered to Node-B because attempting to deliver a larger value than the allocated virtual circuit capacity would lead to either I_{ub} link congestion or a greater delay in refilling the buffer due to the reallocation of I_{ub} link resources. The value of R_s can vary between 1 and n. To avoid low I_{ub} link utilization, we maintain the relationship that $m < n$. Using the above flow control algorithm, the Node-B buffer is used to adjust the HSDPA user throughput on the air interface based on the radio link condition. It is possible that a peak air interface transmission rate may not be sustainable due to the I_{ub} link and the air interface transmission rate mismatch.

Due to the time-variant nature of radio channels, the transmission data rate of the air interface could be highly variable. We assume that the transmission rate (expressed in number of MAC-d PDUs per TTI) of an HSDPA connection is distributed between 0 and a maximum number n. The value of n depends on the UE category. Let r represent the air interface rate that the radio channel condition can support in terms of PDUs. Let R_i represent the probability of air interface transmission rate, where i represents the number of PDUs transmitted at that rate. After the Node-B transmits PDUs on the air interface, it is assumed that the Node-B immediately requests the RNC to send more packets to replenish its buffer before the next transmission opportunity, as shown in Figure 7.9. Let l_n be MAC-d PDU numbers in the Node-B buffer measured after the buffer is replenished and waiting for the next transmission opportunity over the air interface in nth TTI. The buffer length of a user could vary between m and n, that is, $m \leq l_n \leq n$. We can form a Markov chain using the values of l_n. Consider the stationary state where all $\{l_n\}$ have the same probability distribution. Let l denote the random variable in MAC-d PDU numbers in the Node-B buffer.

$$\text{Let } \pi_i = P\{l = m + i\}, \quad i = 0, ..., n - m \tag{7.1}$$

The transition matrix P is shown in the Equation (7.2).

$$\begin{pmatrix} \sum_{i=m}^{n} R_i & R_{m-1} \ldots R_{2m-n+1} & \sum_{i=0}^{2m-n} R_i \\ \sum_{i=m+1}^{n} R_i & \ddots & \sum_{i=0}^{2m-n+1} R_i \\ \vdots & & \vdots \\ R_n & R_{n-1} \cdots R_{n-m-1} & \sum_{i=0}^{m} R_i \end{pmatrix} \quad (m \leq n \leq 2m) \tag{7.2}$$

In the stationary state, we can write $\pi = \pi P$, so

$$\pi_0 = \pi_0 \sum_{i=m}^{n} R_i + \pi_1 \sum_{i=m+1}^{n} R_i + ... + \pi_{n-m} R_n$$

$$\pi_1 = \pi_0 R_{m-1} + \pi_1 R_m + ... + \pi_{n-m} R_{n-1}$$

$$\vdots$$

$$\pi_{n-m-1} = \pi_0 R_{2m-n+1} + \pi_1 R_{2m-n} + ... + \pi_{n-m} R_{n-m-1} \tag{7.3}$$

$$\pi_{n-m} = \pi_0 \sum_{i=0}^{2m-n} R_i + \pi_1 \sum_{i=0}^{2m-n+1} R_i + ... + \pi_{n-m} \sum_{i=0}^{m} R_i$$

$$\pi_0 + \pi_1 + \cdots + \pi_{n-m} = 1$$

We now present a brief analysis to calculate the HSDPA air interface resource utilization. We define the HSDPA air interface resource utilization η as the ratio of the actual number of MAC PDUs (which is limited by the number of PDUs in the Node-B buffer) that can be transmitted via the air interface to the maximum number of MAC PDUs that can be supported by the air interface in current radio transmission conditions in each TTI, as shown in Equation (7.4). The HSDPA air interface utilization takes into account the radio channel condition, buffer length, and the I_{ub} link effective bandwidth. For example, on one occasion, if 8 PDUs of an HSDPA connection are waiting in the Node-B buffer but the radio link condition permits transmission of 10 PDUs, then we obtain a η value of 0.8 (8/10). In that case, the downlink has the opportunity to transmit 10 PDUs but the buffer has only 8 PDUs, causing under-utilization of the link.

$$\eta = \frac{Actual\_MAC\_PDU\_Tx}{Max\_MAC\_PDU\_air} \qquad (7.4)$$

From a statistical point of view, the air interface capacity of a cell will be approximately normally distributed according to the Central Limit Theorem. Because the distribution of air interface rate is discrete, to simplify our analysis, we approximate the HSDPA air interface transmission data rate using the following binomial distribution as shown in Equation (7.5):

$$R_i = \binom{n}{i} p^i (1 - p)^{n-i} \quad 0 < i \le n; \quad 0 < p \le 1 \qquad (7.5)$$

$$mean = n.p \qquad (7.6)$$

$$HSDPA\_throughput\_at\_RLC\_layer = \frac{n \times p \times RLC\_packet\_size}{TTI} \qquad (7.7)$$

where R_i represents the probability of supported air interface transmission rate, i represents the number of PDUs transmitted at that rate, and n is equal to the maximum number of MAC-d PDUs that can be sent in every TTI. The value of n will depend on the UE category. The value of p depends on cell parameters. We assume a very good and controlled coverage in a small cell, and low user mobility could lead to a higher value of p, which represents a higher air interface transmission rate for users. The value of p also depends on the modulation and coding rate of an HSDPA user. The best radio channel condition is represented by the value of $p = 1$.

Figure 7.10 shows the probability distribution of air interface transmission rate per TTI for $p = 0.8$, and $n = 10$ for a category 12 UE. The value of $p = 0.8$ represents an above-average transmission channel condition. The plot shows that for $p = 0.8$, there are high probabilities that a UE will transmit 7 to 9 PDUs per TTI more often than other values. However, there is about a 10% probability that the link will transmit either 6 or 10 PDUs. The peak of the plot will shift toward the left for lower values of p. A similar probability distribution of transmission data rate was reported by an

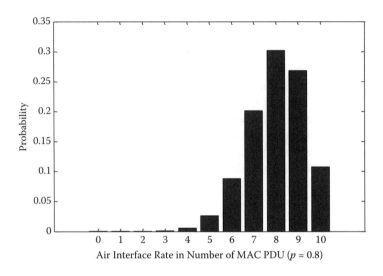

Figure 7.10 Probability distribution of air interface represented in number of MAC-d PDUs for $p = 0.8$ and $n = 10$. (X. Yan, Khan, J.Y., and B. Jones. An Adaptive Radio Access Network Resource Management Technique for a HSDPA Network. 1–5, IEEE, *Proceedings of PIMRC*. 2007. With permission from IEEE.)

Ericsson research group who measured HSDPA network performances in different transmission conditions [21]. The measurement used a test bed of 40 HSDPA sites with category 12 terminals. A measurement was made at the MAC-hs layer with a 10% BLER (block error rate) and a maximum bit rate of 1.5 Mbps. Measurements show that the packet transmission rate varies similarly to our simulation probability distribution with a high probability of transmission between data rates of 0.7 and 1.2 Mbps.

Using the above analytical model described, we can calculate the value of η for a category 12 UE for different air interface conditions and I_{ub} link capacities. Analysis results are shown in Figure 7.11, where the value of η is plotted against the I_{ub} link capacity for different air interface throughput and channel conditions. The figure shows that value of η increases with the I_{ub} link data rate for an RR (Round Robin) packet scheduler. The figure shows that if the I_{ub} link virtual circuit is allocated the same average capacity as the air interface connection capacity, then we achieve $\eta = 0.97$, that is, 97% utilization. If the virtual circuit capacity is lowered, then the value of η decreases. Recall the result of Figure 7.11: for a certain channel condition, the number of MAC PDUs transmitted could vary and, hence, momentary buffer starvation or link under-utilization is possible. The results show that if an HSDPA connection's air interface average throughput is known, and the I_{ub} link bandwidth is allocated accordingly, then the HSDPA resource utilization factor could reach as high as 97%. In a realistic HSDPA network, the air interface throughput will vary depending on the transmission channel

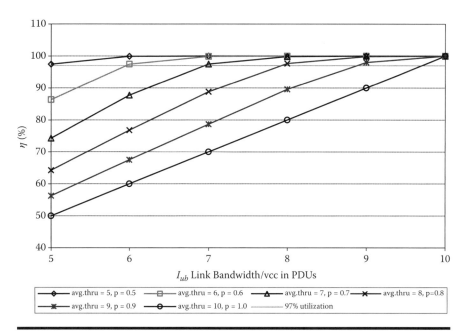

Figure 7.11   HSDPA resource utilization factor dependency on the I_{ub} link bandwidth allocation. (X. Yan, Khan, J.Y., and B. Jones. An Adaptive Radio Access Network Resource Management Technique for a HSDPA Network. 1–5, IEEE, *Proceedings of PIMRC.* **2007. With permission from IEEE.)**

condition. Hence, a fixed I_{ub} bandwidth allocation will either reduce the air interface utilization or it will reduce the I_{ub} link utilization. For example, if the air interface throughput varies between 6 and 8 PDUs, then the I_{ub} capacity should be matched to obtain higher values of η.

7.6 Air Interface and I_{ub} Parameter Interdependencies

This section examines the relationship between the HSDPA air interface and the I_{ub} link parameters. The results are used to examine the interdependencies of two interface parameters. To examine this interaction we develop a simulation model in addition to the analytical model described in the previous section.

7.6.1 Simulation Model

A simulation model was developed to further study the RAN performance for different air interface conditions. The simulator consists of an I_{ub} interface and a physical link, a traffic generator, an air interface link, and the proposed flow controller as shown in the Figure 7.12. We assume that the core network (CN) can deliver enough packets to the RNC so that a sufficient

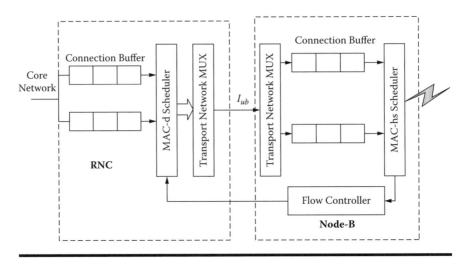

Figure 7.12 Simulation model block diagram.

number of packets are waiting in the RNC buffer during the active period of a data burst. The flow controller is implemented according to the
description given in the "UTRAN Protocol Structure" section. The Node-B
allocates each HSDPA connection a transmission bandwidth according to
the radio channel condition that is simulated by the binomial distribution as
shown in Equation (7.5). The I_{ub} link is simulated using the AAL2/ATM and
UDP/IP transport protocols. The model separately simulates three packet
scheduling algorithms: RR, Max C/I (Carrier-to-Interface Ratio), and PF (Proportional Fair). Key simulation parameters are listed in Table 7.2.

7.6.2 Performance Analysis

We gathered several simulation and analytical results to study the RAN and
air interface parameter interactions. Figure 7.13 shows the relationship between the air interface transmission condition p and the HSDPA efficiency
η for category 5/6 and category 12 UEs using an RR packet scheduler with
15 UEs. The figure shows that the analytical and simulation results are in
close agreement. In this simulation, each VC (virtual circuit) capacity on the
I_{ub} link was matched with the average throughput of the respective connections on the downlink. Figure 7.13 shows that the value of η decreases
as the transmission capacity of the air interface increases, as represented by
the value of p. The figure also shows that the category 5/6 UE connections
perform better than category 12 terminals because of the higher I_{ub} link
data rate. Table 7.2 shows that the category 5/6 terminals were allocated
with 18 PDUs for each Virtual Circuit (VC). Category 12 terminals receive
12 PDUs/TTI for each VC. As mentioned earlier, a connection's buffer in
the Node-B is updated by the RNC after every transmission opportunity

Table 7.2   Key Simulation Parameters

Simulation Parameter	Value/Features
UE category	5, 6, and 12
Number of UEs	15
I_{ub} link data rate	2.048 Mbps (Cat 5/6), 4.096 Mbps (Cat 12)
I_{ub} link transport protocol	AAL2/ATM and UDP/IP
Maximum buffer allocation for each connection	21 PDU/connection for Cat 5/6 UE, 10 PDU/connection for Cat 12 UE
Virtual circuit capacity	18 PDU/VC for Cat 5/6 UE, 8 PDU/VC for Cat 12 UE
Scheduling algorithm	RR, Max-C/I, PF
Air interface model	Binomial distribution
Traffic model	Web browsing, Pareto distributed based on the UMTS traffic model
PDU size	320 bits

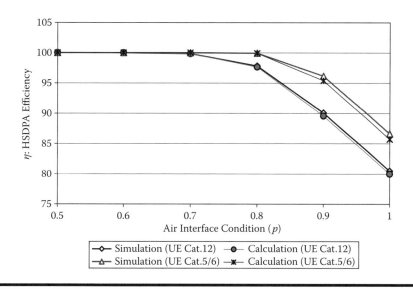

Figure 7.13   HSDPA efficiency factor variation for different air interface conditions. (X. Yan, J. Y. Khan and B. Jones, Study of Interdependency Between the HSDPA Air Interface and the Radio Access Network, *Proc. of the IEEE 18th PIMRC*, 3–7 September 2007. With permission from IEEE.)

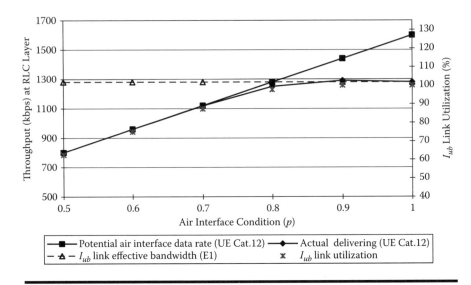

Figure 7.14 HSDPA throughput measured at the RLC layer for different air interface conditions. (X. Yan, J. Y. Khan and B. Jones, Study of Interdependency Between the HSDPA Air Interface and the Radio Access Network, *Proc. of the IEEE 18th PIMRC*, 3–7 September 2007. With permission from IEEE.)

by the same number of PDUs that were transmitted in the immediate past transmission opportunities. Hence, it is possible that if the air interface transmission rate peaks under a better channel condition, then the I_{ub} link may not be able to replenish the Node-B buffer with the required number of PDUs, thus resulting in a lower value of η.

Next we observed the relationship between the transmission channel condition and HSDPA throughput measured at the RLC layer. Figure 7.14 shows that it is theoretically possible that the combined air interface throughput can linearly increase with an increasing value of p. However, the I_{ub} link latency arises because of the fixed VC capacity allocation and use of fixed-size ATM cell transmissions on the I_{ub} link. The result shows that the actual throughput of the air interface could reach up to 1.3 Mbps for a 2.048-Mbps link used for the category 12 terminals. Similar results were obtained when a 4.096-Mbps (2xE1) link was used. Results also show that the I_{ub} link utilization remains low for the lower value of p due to low throughput of the air interface.

Furthermore, we examined the effect of queue occupancy of each HSDPA connection for different air interface transmission conditions. Figure 7.15 shows the buffer occupancy value for different values of p. The graph shows that for $p = 0.7$, the buffer occupancy level is 10 PDU for 80% of the time where the occupancy level is distributed between the minimum size of 8 PDUs to 10 PDUs for $p = 0.8$. For a better channel condition with $p = 0.9$,

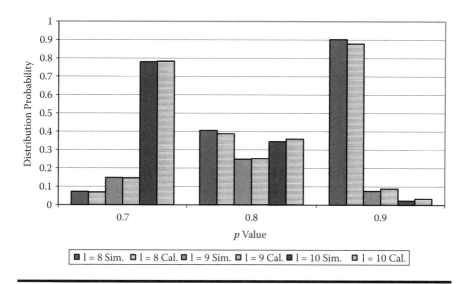

Figure 7.15 **Connection buffer length distribution at Node-B for different air interface conditions. (X. Yan, J. Y. Khan and B. Jones, Study of Interdependency Between the HSDPA Air Interface and the Radio Access Network,** *Proc. of the IEEE 18th PIMRC,* **3–7 September 2007. With permission from IEEE.)**

the buffer occupancy level drops down to 8 PDU due to the higher drain rate by the air interface which is higher than the buffer replenishment rate by the I_{ub} link. Note that the 8 PDU is the minimum buffer size maintained by each connection.

Next we examined the impact of Node-B scheduling algorithms on the HSDPA throughput. We used a total of 10 category-12 users. In this simulation the value of p is controlled by a uniform distribution where the value could vary between 0.5 and 1.0. Figure 7.16 shows the cumulative distribution function (CDF) of potential air interface throughput and the actual throughput using three packet schedulers. The Max-C/I fully exploits the available capacity, thus offering the highest potential throughput to some users. The RR scheduler offers a lower potential throughput than Max-C/I but a fairer distribution of resources. The PF scheduler provides a trade-off between the fairness and the capacity. In this case, a modest gain is achieved over the RR scheduler. Throughputs of all these schedulers are affected by the I_{ub} link resource allocations. It is possible to reduce the latency of the I_{ub} link by over-allocating virtual circuit capacities but the approach will reduce the link utilization and increase the transmission rate requirements and the cost of a RAN. Figure 7.17 shows the maximum achievable HSDPA efficiency factor for three different packet schedulers. From the RAN resource allocation point of view, we can see that the RR scheduler can operate at a reasonably higher capacity level because of the cyclic air interface transmission schedule, which allows the RAN to supply

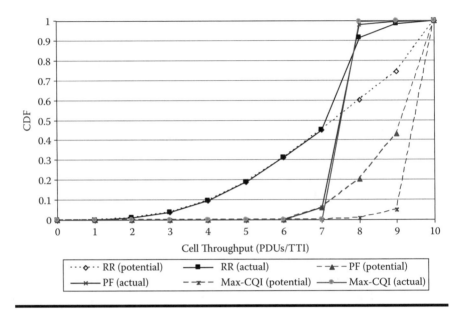

Figure 7.16 HSDPA cell throughput for every TTI using different packet schedulers. **(X. Yan, J. Y. Khan and B. Jones, Study of Interdependency Between the HSDPA Air Interface and the Radio Access Network,** *Proc. of the IEEE 18th PIMRC***, 3–7 September 2007. With permission from IEEE.)**

a sufficient number of PDUs within the required time. The importance of an adaptive RAN resource allocation technique becomes more important with increasing HSDPA air interface data rates. When the HSDPA network migrates from the current transmission rate of 14.4 Mbps to 40 Mbps in a few years' time, then the network will potentially support 4 to 5 times more users on the downlink. In addition, it is safe to assume that more and more

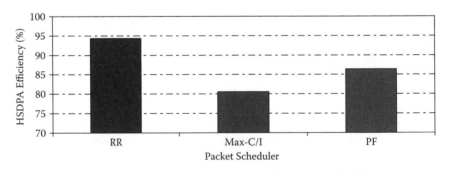

Figure 7.17 HSDPA efficiency factor for different packet scheduling techniques. **(X. Yan, J. Y. Khan and B. Jones, Study of Interdependency Between the HSDPA Air Interface and the Radio Access Network,** *Proc. of the IEEE 18th PIMRC***, 3–7 September 2007. With permission from IEEE.)**

bursty applications such as video games will be supported on wireless networks in the future, which will increase the ratio of average to peak traffic levels on the RAN. In that case, if an I_{ub} link is over-dimensioned, then it may be possible that I_{ub} link utilization will significantly decrease at the same time that the cost of the RAN will increase significantly. Instead of over-dimensioning a RAN, a more prudent approach will be to use adaptive resource allocation techniques on the I_{ub} link.

In the following section we propose an air interface state information-based I_{ub} link resource allocation technique.

7.6.3 Adaptive I_{ub} Link Management Algorithm

In this section we propose a simple but effective adaptive I_{ub} link bandwidth allocation technique based on the air interface average throughput. The adaptive resource management algorithm will be located at Node-B, which will measure every HSDPA connection throughput and, using the measured information, will allocate the data rate of the corresponding VCs of the I_{ub} link by sending the resource allocation packet from the Node-B to the RNC. The feedback-based estimation can offer better utilization of the I_{ub} link as well as improve the HSDPA efficiency factor.

7.6.3.1 Estimating HSDPA Air Interface Average Throughput

We estimate each HSDPA connection average throughput using a simple averaging equation. Equation (7.8) shows the averaging process. The connection throughput is measured for every burst transmission, and the running average of the air interface throughput is used to allocate on the I_{ub} link VC bandwidth for that particular connection. The average throughput value is calculated separately for every data burst. When a connection goes to the idle mode, its average throughput value is reset to the default value of *m*.

$$\overline{Throughput} = \frac{1}{n}\sum_{i=1}^{n} x_i \tag{7.8}$$

where x_i represents the throughput in number of MAC PDUs for the ith transport opportunity of a connection, and n represents the transport opportunity numbers.

When $X_1, X_2, \ldots, X_i$ – is binomially distributed (n,p) as in Equation (7.9),

$$var\left(\overline{Throughput}\right) = var\left(\frac{1}{N}\sum_{i=1}^{N} X_i\right)$$
$$= \left(\frac{1}{N}\right)^2 \sum_{i=1}^{N} var\left(X_i\right) = \left(\frac{1}{N}\right)^2 Nnp(1-p) = \frac{np(1-p)}{N} \tag{7.9}$$

7.6.3.2 Adaptive I_{ub} Bandwidth Allocation

The scheduler in the Node-B selects an active HSDPA connection to transmit packets on the air interface in every TTI. After each transmission, the Node-B calculates the I_{ub} connection bandwidth $\overline{m}$ using throughput value of Equation (7.8). Node-B also calculates the value of R_s by measuring the buffer length. The Node-B then sends a control message by incorporating the value of the I_{ub} link's allocated capacity and the request size R_s for the next transmission. Using the proposed algorithm it is possible to match the air interface throughput rate and the I_{ub} link throughput. The required aggregate I_{ub} link bandwidth is calculated using Equation (7.10):

$$Bandwidth = \sum_{i=1}^{N} s_i \times \overline{m}_i \qquad (7.10)$$

where s_i represents the HSDPA connection i repetition rate that depends on the scheduling technique used by the Node-B. N represents the total number of VCs mapped on the I_{ub} link

Next we compare the HSDPA air interface resource utilization figure for different I_{ub} link bandwidth allocation techniques. In this simulation we used a total of 15 category-12 UEs. We grouped these 15 UEs into three groups, with each group having 5 UEs. One group was assumed to work in good channel conditions, which support an average of 9 MAC-PDUs per TTI. The second group of 5 UEs was assumed to operate in medium chan- nel conditions with an average of 7 MAC-PDUs per TTI. The third group consisted of 5 UEs operating in poor channel conditions with an average throughput of 5 MAC-PDUs per TTI. In this simulation we used a Round Robin (RR) scheduling technique at the Node-B. In this simulation, first we allocated each VC equal to its downlink peak data rate in terms of number of PDUs. For category-12 UEs, the peak data rate was 10 MAC-d PDU/TTI. After obtaining the simulation results for the fixed virtual capacity, we em- ployed the adaptive virtual circuit capacity allocation algorithm to analyse the performance of both interfaces. Results presented in Figure 7.18 show that the adaptive algorithm reduces the HSDPA air interface utilization only by about 3% but the algorithm reduces the effective I_{ub} link capacity re- quirements significantly. Figure 7.19 shows that the I_{ub} physical link band- width requirements and the effective bandwidth utilization for the adaptive allocation and peak rate allocation techniques using the AAL2/ATM and UDP/IP transport protocols. Figure 7.19 also shows that to support the same amount of air interface traffic, the adaptive allocation technique requires about 0.8 Mbps less I_{ub} link data rate for the AAL2/ATM transport protocol and about 0.7 Mbps less data rate for the UDP/IP protocol. Also, if we compare the link utilization figure, then we see that the adaptive algorithm increases the link utilization by 27%. The proposed algorithm increases

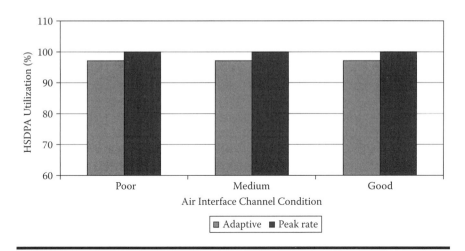

Figure 7.18 Effect of I_{ub} link capacity allocation techniques on the HSDPA utilization factor. (X. Yan, Khan, J.Y., and B. Jones. An Adaptive Radio Access Network Resource Management Technique for a HSDPA Network. 1–5, IEEE, *Proceedings of PIMRC*. 2007. With permission from IEEE.)

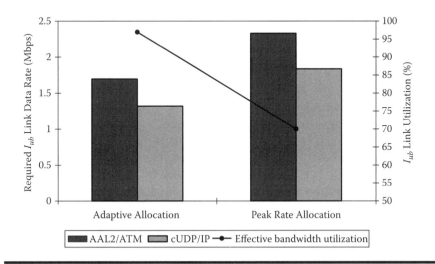

Figure 7.19 I_{ub} link transmission data rate requirements for different link capacity allocation techniques. (X. Yan, Khan, J.Y., and B. Jones. An Adaptive Radio Access Network Resource Management Technique for a HSDPA Network. 1–5, IEEE, *Proceedings of PIMRC*. 2007. With permission from IEEE.)

the link utilization, at the same time reducing the I_{ub} link required transmission rate. Simulation results show that the effect of a variable channel condition can be effectively minimized by implementing appropriate resource allocation techniques presented in this chapter. For a high data rate HSDPA network, the proposed algorithm becomes more appropriate. The proposed algorithm can significantly reduce the operating cost and bandwidth requirements of a RAN.

The proposed joint resource management algorithm can also be extended to HSUPA connections. In case of HSUPA connections, the Node-B to RNC transmission will be supported by the I_{ub} link. In this case, the I_{ub} link must be properly dimensioned so that Node-B buffer overflow can be avoided; the overflow could be caused either by the fluctuating radio transmission conditions on the air interface or by I_{ub} link congestions. Node-B will be able to monitor uplink traffic as well as the uplink transmission condition by measuring the SINR (signal-to-interference-noise ratio) value of each connection. Based on the traffic information and transmission conditions, the Node-B can allocate the appropriate buffer and the I_{ub} link transmission resources. A recent work by Li et al. [22] investigated the backhaul dimensioning for the HSUPA traffic. They studied an ATM-based I_{ub} link to support various HSUPA traffic and found that the dimensioning factor of the I_{ub} link was directly related to the burstiness of the uplink traffic. The dimensioning factor is defined as the ratio of the allocated I_{ub} link bandwidth to the total HSUPA traffic demands of all UEs in all cells. The work considered the soft handover scenario where Node-B could accept HSUPA traffic from several cells. The authors also concluded that increasing the QoS of HSUPA traffic demands a higher dimensioning factor. Simulation results showed that the dimensioning factor for low to high QoS requirements could vary from 2 to 10. From the operator's point of view, such a dimension factor will be quite expensive; hence, the backhaul network needs to be optimized using adaptive resource allocation algorithms as described for the HSDPA network in this section. The work of Li et al. also showed that congestions on the downlink could affect the throughput and QoS on the uplink. Hence, it is very important that for an HSPA network, joint optimization of resource allocation techniques is necessary to maintain high network throughput and QoS and to reduce the operational cost of a network.

7.7 Summary

This chapter introduced the UTRAN and the E-UTRAN architectures for the HSPA standard. Various key functionalities of the UTRAN are presented. The chapter also discussed the E-UTRAN architecture, which will introduce a significant change in the radio access network design. A gradual increase in packet-switched traffic volume in the HSPA network will demand more flexible IP-based radio access network design. Results presented in this chapter showed that the introduction of IP-based radio access networks will reduce the operational cost of the RAN as well as improve the network resource utilization. Discussions in this chapter showed that it is very important to jointly optimize the resource allocation algorithms in the radio access network to support high data rate connections and QoS for packet-based services on the HSPA air interface. The joint resource management algorithm introduces a new network design paradigm that can be further extended to optimize the UTRAN/E-UTRAN resources.

References

[1] 3GPP, TS25.401, R8, UTRAN Overall Description, v.8.2.0, December 2008.

[2] P. Lescuyer and T. Lucidarme, *Evolved Packet System (EPS): The LTE and SAE Evolution of 3G/UMTS*, Chapter 2, John Wiley & Sons, 2008.

[3] H. Holma, A. Toskala, K. Ranta-aho, and J. Pirskanen, High speed packet access evolution in 3GPP release 7, *IEEE Commun. Mag.*, 45(12): 29–35, December 2007.

[4] 3GPP, TR25.913, R8, Requirements of Evolved UTRA (E-UTRA) and Evolved UTRAN (E-UTRAN), v.8.0.0, December 2008.

[5] 3GPP, TS25.410, R8, UTRAN I_u Interface: General Aspects and Principles, v.8.0.0, December 2008.

[6] 3GPP, TS25.420, R8, UTRAN I_{ur} Interface: General Aspects and Principles, v.8.1.0, December 2008.

[7] 3GPP, TS25.430, R8, UTRAN I_{ub} Interface: General Aspects and Principles, v.8.0.0, December 2008.

[8] 3GPP, TS25.913, R8, High Speed Downlink Packet Access (HSDPA): Overall Description, v.8.4.0, December 2008.

[9] H. Holma and A. Toskala (Eds.), *WCDMA for UMTS- HSPA Evolution and LTE*, 4th edition, John Wiley & Sons, 2007.

[10] 3GPP, TS 25.319, R8, Enhanced Uplink Overall Description, v.8.0, March 2009.

[11] 3GPP, TS25.321, R8, Medium Access Control (MAC) Protocol Specification, v.8.4.0, December 2008.

[12] 3GPP, TS36.401, R7, Evolved Universal Terrestrial Radio Access Network (E-UTRAN); Architecture Description, v.8.5.0, March 2009.

[13] 3GPP, TR25.903, R7, Continuous Connectivity for Packet Data Users, v.7.0.0, March 2007.

[14] Whitepaper, Cost Optimized Transport Evolution—Making the Right Choice, Nokia Corporation, 2006.

[15] X. Yan, J.Y. Khan, and B. Jones, Study of interdependency between the HSDPA air interface and the radio access network, *Proc. of the IEEE 18th PIMRC*, 3–7 September 2007.

[16] H. Holma and A. Toskala (Eds.), *HSDPA/HSUPA for UMTS: High Speed Radio Access for Mobile Communications*, John Wiley & Sons, 2006.

[17] 3GPP, TS25.435, R7, UTRAN I_{ub} Interface: User Plane Protocols for Common Transport Channel Data Streams, v.7.8.0, March 2008.

[18] P.J. Legg, Optimised I_{ub} flow control for UMTS HSDPA, *Proc. of the IEEE VTC 2005,* 4: 2389–2393, 2005.

[19] N.C. Necker and A. Weber, Impact of Iub Flow Control on HSDPA System Performance, *Proc. IEEE PIMRC 2005,* September 2005, 3: 1703–1707, 2005.

[20] L. Bajzik, L. Korossy, K. Veijalainen, and C. Vulkan, Cross layer backpressure to improve HSDPA Performance, *Proc. IEEE PIMRC 2006,* September 2006, pp. 1–5.

[21] J. Derksen, R. Jansen, M. Mayala, and E. Westerberg, HSDPA performance and evolution, *Ericsson Rev.,* 3: 117–120, 2006.

[22] X. Li, Y. Zaki, T. Weerawardane, A. Timm-Giel, and C. Goerg, HSUPA backhaul bandwidth dimensioning, *Proc. IEEE 19th PIMRC,* 15–18 September 2008.

Chapter 8

Automated Optimization in HSPA Radio Network Planning

Iana Siomina, Di Yuan, and Fredrik Gunnarsson

Contents

8.1 Overview

Automated optimization of radio network configurations can be considered a cost-effective means to improve mobile broadband performance, provided that accurate spatial network information is available. This chapter addresses automated optimization of antenna configurations in evolved 3G networks featuring High-Speed Packet Access (HSPA). Detailed radio link and network models are described, together with performance metrics in terms of link models relating radio link quality to data rates. The network planning task is formulated as an optimization problem maximizing HSDPA throughput with HSUPA performance as a soft constraint, while taking into account resource sharing among common channels, HSPA, and Release 99 traffic. A search algorithm is developed to solve the resulting optimization problem effectively and time efficiently. Performance analysis for a realistic large-scale planning scenario is provided to demonstrate the benefit of the automated optimization approach and the impact of optimizing different antenna configuration parameters, including antenna tilt and azimuth. For example, cell-edge data rates are improved by more than 50% by adopting automated optimization compared to the baseline configuration, and the largest performance gain is achieved by means of electrical antenna tilting.

8.2 Introduction

The increasing interest in mobile broadband services implies requirements on efficient data transport through radio networks. In part, this is handled by radio resource management mechanisms that dynamically adjust the network operations to cater for the current user needs. However, such mechanisms cannot fully compensate for poor network planning, which may substantially limit the potential radio network performance. An investment in careful radio network planning and network-wise configuration optimization is therefore well justified when it enables potentially greater revenues and improved user service experiences in the operational network.

8.2.1 Background

Radio networks of the third generation are continuously evolved by the 3rd Partnership Project (3GPP) specification work. A 3G network can be divided into a radio access network (RAN) and a core network (CN). The core network handles billing, higher layer services, data transport to other networks, etc. The radio access network consists of one or several radio network subsystems (RNSs), each managed by a radio network controller (RNC). The actual radio links to the mobiles (or user equipment, UE) are maintained by base stations (or Node Bs).

In the first releases, both conversational and data services were mainly provided via dedicated channels (DCHs). A DCH is managed on a higher level by RNC, while Node B handles physical link operations, such as power control, to maintain an acceptable link quality. In addition, Node B transmits common channels (CCHs) to support other physical channels and handle connection initiations.

To enable faster and more flexible data transmissions to mobiles, HSDPA (High-Speed Downlink Packet Access) was introduced. Data for HSDPA users is carried over a shared channel, also called the high-speed downlink shared channel (HS-DSCH), and Node B decides which user to target with the channel (scheduling), and at what data rate (link adaptation). These decisions are based on feedback from UEs reporting the perceived radio channel quality.

Means to enable higher uplink data rates were also introduced as Enhanced Uplink (EUL) or High-Speed Uplink Packet Access (HSUPA). Similar to HSDPA, Node Bs are more involved in radio resource management. The combination of HSDPA and EUL/HSUPA is commonly referred to as High-Speed Packet Access (HSPA).

8.2.2 HSPA Network Planning and Dimensioning

A common tool in the planning and dimensioning process is a cell planning tool, typically featuring radio signal propagation prediction functionality. The accuracy of such predictions depends on the level of detail in the input information (terrain, land usage, buildings, antenna models, propagation models, expected services, traffic load, etc.), and also to what extent the models are adjusted to match the region considered in the planning procedure. If detailed information can be obtained, using optimization methods in the network planning phase becomes reasonable and beneficial.

In traditional network optimization with dedicated services, the focus is typically on the number of served users that are satisfied. Considering data transport over shared channels, such as HS-DSCH, different users will have different data rates. This calls for a modified set of key performance indicators. In particular, the focus shifts to data rate distributions, often with a special attention paid to performance-wise worst user locations.

8.2.3 Chapter Scope

The scope of the chapter is to introduce relevant models for HSPA planning and dimensioning, both with respect to common channels, dedicated channels, as well as HSPA. In particular, performance metrics designed to address HSDPA performance take into account radio link quality as well as the scheduling strategy in Node B. Furthermore, various optimization algorithm considerations are discussed, including complexity and suitable search algorithms. A simulated annealing approach is described to optimize the antenna orientation parameters, and this approach is used to optimize a radio network in a case study.

8.3 Modeling and Optimization

Planning cellular networks deals with locating and configuring radio network elements to meet specified performance targets. Obtaining satisfactory results has to rely on some optimization process that approaches the planning problem systemically. This is a challenging engineering task, particularly for large-scale planning scenarios having many network elements, each with a large set of configuration parameters. In this context, an automated optimization engine, designed to tackle the complexity of large-scale planning with very little need for manual intervention or tuning, is highly desirable. The optimization engine alone, however, is not sufficient for accomplishing the planning task. Two additional key elements—system modeling and performance assessment—are equally important. In this chapter we practice a planning process formed by these three elements, as shown in Figure 8.1, to HSPA radio network planning.

Prior to system modeling, the scope of planning decisions is defined. Typically, the network planning stage deals with decisions that have significant mid-term and long-term impacts on performance, whereas decisions to respond to system dynamics, such as tuning handoff parameters and adjusting scheduling policies, lie in the domain of network operation. Major planning decisions include the location of base stations; antenna configuration in terms of height, tilt, and azimuth, coverage pattern; and for some cellular systems, channel assignment. Many of these decisions have

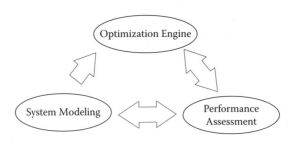

Figure 8.1   Key elements of an optimization process.

a multi-cell effect, even being applied in a single cell and, once taken, are costly to revise afterward; hence, the optimization process used in this stage is of vital significance.

System modeling translates the planning problem into a mathematical form based on expertise. The model identifies and quantifies how the performance indicators will respond to the design and configuration parameters considered in the planning (e.g., the effect of antenna tilting on signal-to-interference-plus-noise ratio (SINR)) and thereby data throughput. Typically and particularly for large-scale planning, the optimization process targets network-wise performance indicators. To this end, the goal of system modeling is to capture the essence of system behavior, rather than deriving an exact picture with full details, which, if doable at all, will result in planning problems that are too complex to be dealt with.

The optimization engine has the task of delivering a design solution. In this chapter the solution consists of tilt and azimuth configurations of cell antennas. The core of the engine is one or several optimization algorithms that are designed to produce high-quality solutions using a reasonable amount of computing effort. What is considered a reasonable amount varies from one planning case to another. For relatively easy planning problems, applying an off-the-shelf solver to a solution-oriented mathematical programming model [5] is a promising approach. For large-scale cellular network planning, however, stochastic search algorithms [4] are often more appropriate.

Performance assessment evaluates the optimization engine by numerical tests on planning scenarios with real-life or close-to-real-life data. In addition to the numerical values of the performance indicators reported by the optimization process, engineering expertise can be used to make a judgment of the planning solution and its performance from the practical point of view. Network simulation, if applicable to the planning scenario, is yet another useful tool for performance assessment. In addition to evaluating the optimization engine, performance assessment aims at analyzing the system model. The system model is revised accordingly if performance assessment identifies additional elements or constraints that are of significance to the overall performance. In the chapter, this aspect of performance assessment is illustrated by a comparative study of the impact of uplink consideration in HSPA planning.

8.3.1 The System Model

8.3.1.1 Preliminaries and Overview

For HSDPA, data throughput is the primary performance consideration. Several factors influence the downlink user data throughput: the SINR, the number of channelization codes, and resource sharing decisions, in particular scheduling. Among them, SINR is the key performance aspect from a network planning viewpoint. The SINR of HS-DSCH is a result of radio propagation condition, interference, noise, and transmission power. Although SINR is of primary concern in implementing both 3GPP Release 99 (R99) and HSDPA services, there is a significant difference in system modeling, namely that the closed-loop power control on the R99 DCH does not apply to HS-DSCH. Thus, HSDPA users will enjoy higher data throughput by more power at downlink.

The system model in this chapter accounts for networks with coexisting R99 and HSDPA. For this type of planning scenario, the output power of Node B is shared between HS-DSCH, DCH, and CCH, including common pilot channel (CPICH) defining network coverage and cell ranges. A common power-sharing strategy is the fill-up approach [11], which allocates all the power left from supporting CCH and DCH to HS-DSCH. By this approach, HSDPA power is dynamic, yielding better resource utilization in comparison to a constant amount of HS-DSCH power. As a result of the fill-up approach, the cells constantly operate at full power at the downlink in the system model.

The two planning parameters in the optimization framework, tilt and azimuth, refer to the angle of the main beam of the antenna below the horizontal plane (downtilt), and the horizontal angle between the north and the antenna's main lobe direction, respectively. For tilting, mechanical tilting, electrical tilting, as well as their combinations are considered. With mechanical tilting, the physical angle of the brackets used to mount the antenna is adjusted, while with electrical tilting, the antenna pattern is adjusted by varying the weights of antenna elements. As no physical adjustment is required, electrical tilting can be controlled remotely and is therefore a more cost-effective way of reducing interference.

Denote by $\mathcal{C} = \{1, \ldots, C\}$ the set of cells, where C is the number of cells. The total transmission power available in cell $i \in \mathcal{C}$ is denoted by P_i^{TOT}. To simplify the discussion, we consider one directional antenna in every cell. In the subsequent text, the two terms (antenna and cell) are often used interchangeably. The configuration of antenna i is a tuple $\langle t_i^m, t_i^e, a_i \rangle$, where t_i^m, t_i^e, and a_i are the mechanical tilt, electrical tilt, and azimuth angles, respectively. Each of the three parameters has a discrete set of candidate values. For antenna i, all possible combinations of the three parameter values form a set of candidate configurations $\mathcal{K}_i = \{1, \ldots, K_i\}$. The network-wise antenna configuration is a vector $\mathbf{k} = (k_1, \ldots, k_C)$, where $k_i \in \mathcal{K}_i$ is the configuration of antenna i.

The service area is modeled by a regular grid of pixels $\mathcal{J} = \{1, \ldots, J\}$, where J is the number of pixels. A pixel $j \in \mathcal{J}$ is a small square area within which radio propagation is considered uniform in the network planning context. Higher pixel resolution yields more accurate planning, at the

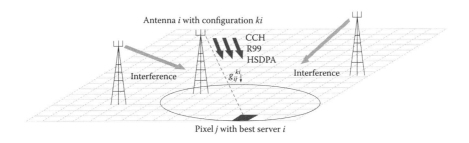

Figure 8.2   An illustration of some elements of the planning framework.

price of increasing the size of the planning problem. We use $g_{ij\downarrow}^{k_i}$ and $g_{ij\uparrow}^{k_i}$ to denote the total average power gain between antenna i under configuration k_i and pixel j at downlink and uplink, respectively. The gain values are defined by environment-dependent signal propagation conditions, hardware characteristics (e.g., maximum antenna gain, directional antenna losses, cable loss, body loss, etc.), and depend also on configuration (e.g., antenna tilt and azimuth). The number of candidate antenna configurations and pixel resolution together set the amount of data required for planning. In Figure 8.2, we provide an illustration of some concepts and elements that have been introduced.

For a user demand map (i.e., distribution of active users of each service type), let $P_i^{CCH}(\mathbf{k})$, $P_i^{DCH}(\mathbf{k})$, and $P_i^{HS}(\mathbf{k})$ denote the power consumed by CCH, DCH, and HS-DSCH in cell i, respectively, under network antenna configuration $\mathbf{k}$. We have

$$P_i^{TOT} = P_i^{CCH}(\mathbf{k}) + P_i^{DCH}(\mathbf{k}) + P_i^{HS}(\mathbf{k}), \qquad i \in \mathcal{C} \qquad (8.1)$$

Figure 8.3 gives an overview of the components in the optimization framework. Given an antenna configuration vector, we compute the power needed on CCH for coverage and the best server pattern, that is, the serving antenna of each pixel. The best server pattern and the R99 user demand map determine the downlink power consumed by R99. This, together with the power on CCH, give the HS-DSCH power of each cell by Equation (8.1). The achievable HSDPA data throughput over the service area can then be computed from the HS-DSCH power, the antenna configuration, and a given HSDPA demand map. In the following sections we detail the relationship between the system components in Figure 8.3.

As shown in Figure 8.3, the evaluation of HSDPA performance uses R99 and HSDPA demand maps as input. A demand map corresponds to a snapshot of the users served by the network over the service area. To achieve a robust solution in antenna configuration, the optimization can be

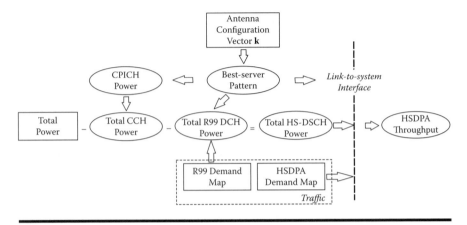

Figure 8.3 Downlink components in the optimization framework.

performed for a large number of demand maps following expected user distribution, or for an aggregated statistical demand map of each traffic type.

8.3.2 *Coverage and CCH Power Consideration*

Network coverage is provided by the presence of CPICH [2]. CPICH enables UE to perform channel estimation and cell selection/reselection. The requirement of CPICH presence can be modeled by $E_c/I_0 \geq \gamma^c$, where E_c/I_0 is the ratio between the received chip energy and the total received power spectral density at the UE's antenna connector [1], and γ^c is a threshold value. CPICH E_c/I_0 over the service area is a result of antenna configuration and CPICH power allocation. At present, a common practice is the uniform scheme, that is, to allocate the same amount of power to CPICH in all cells, which is therefore further assumed in the chapter. Nonuniform CPICH power has been studied, for example, in [18]. Let $P^{CPICH}(\mathbf{k})$ denote the CPICH power for configuration $\mathbf{k}$, and $v_\downarrow$ the total noise power in downlink. Under antenna configuration vector $\mathbf{k}$, antenna i covers pixel j with CPICH if the following inequality holds:

$$E_c/I_0 = \frac{P^{CPICH}(\mathbf{k})g_{ij\downarrow}^{k_i}}{\sum_{l\in\mathcal{C}} P_l^{TOT}g_{lj\downarrow}^{k_l} + v_\downarrow} \geq \gamma^c \tag{8.2}$$

Because CPICH is transmitted continuously, an excess power level of CPICH makes less power available to traffic channels and may also cause pilot pollution. Thus, the goal is to find the minimum level of $P^{CPICH}(\mathbf{k})$ to ensure the coverage of the service area. From Equation (8.2), coverage at j is secured only if $P^{CPICH}(\mathbf{k}) \geq \min_{i \in \mathcal{C}}(\sum_{l \in \mathcal{C}} P_l^{TOT} g_{lj\downarrow}^{k_l} + v_\downarrow)\gamma^c/g_{ij\downarrow}^{k_i}$. To cover all $j \in \mathcal{J}$, the required CPICH power is therefore [19]

$$P^{CPICH}(\mathbf{k}) = \max_{j \in \mathcal{J}} \min_{i \in \mathcal{C}} \frac{\left(\sum_{l \in \mathcal{C}} P_l^{TOT} g_{lj\downarrow}^{k_l} + v_\downarrow\right)\gamma^c}{g_{ij\downarrow}^{k_i}} \tag{8.3}$$

In addition to CPICH, power must be allocated to other common channels, for example, synchronization channel (SCH), paging indicator channel (PICH), and acquisition indicator channel (AICH). These common channels are typically powered in proportion to CPICH [14]. In the system model, the total power consumed by CCHs is estimated by scaling $P^{CPICH}(\mathbf{k})$ with a factor $\beta > 1$,

$$P_i^{CCH}(\mathbf{k}) = \beta\, P^{CPICH}(\mathbf{k}), \quad i \in \mathcal{C} \tag{8.4}$$

If a pixel $j \in \mathcal{J}$ is covered by the CPICH of multiple antennas, the one giving the highest E_c/I_0 is referred to as the best server. Equivalently speaking, the best server is the antenna having the largest gain value in j, because the numerator in Equation (8.3) is a constant for a given antenna configuration vector $\mathbf{k}$. Thus, there is a straightforward mapping between antenna configuration and the best server pattern. In the following, $\mathcal{J}_i(\mathbf{k})$ denotes the subset of pixels for which antenna i is the best server under configuration $\mathbf{k}$.

8.3.3 R99 Power Consideration

Denote by $\mathcal{S}$ the set of service types in R99. Each service $s \in \mathcal{S}$ is characterized by its SINR target γ^s and an activity factor v^s. The R99 demand map is specified by the number of admitted users in each pixel and of each service type, denoted by d_{js}^{DCH}, $j \in \mathcal{J}$, $s \in \mathcal{S}$. For each user of service type s, closed-loop power control strives to meet the SINR target γ^s. For antenna i and pixel j served by i (i.e., $j \in \mathcal{J}_i(\mathbf{k})$), denote by P_{ijs}^{DCH} the power consumed by supporting a user in j of service type s. Let α_j denote the orthogonality factor in j (with $\alpha_j = 1$ modeling full orthogonality, that is, no intra-cell interference). Under perfect power control, $P_{ijs}^{DCH}(\mathbf{k})$ can be obtained from the following equation:

$$\frac{P_{ijs}^{DCH}(\mathbf{k})g_{ij\downarrow}^{k_i}}{(1 - \alpha_j)\left(P_i^{TOT} - v^s P_{ijs}^{DCH}(\mathbf{k})\right) g_{ij\downarrow}^{k_i} + \displaystyle\sum_{l \in \mathcal{C}: l \neq i} P_l^{TOT} g_{lj\downarrow}^{k_l} + v_\downarrow} = \gamma^s. \tag{8.5}$$

In Equation (8.5), the inter-cell interference is calculated under total cell power at downlink, and lower activity factor gives higher intra-cell interference. Both are consequences of Equation (8.1). To provide service to the R99 users located in $\mathcal{J}_i(\mathbf{k})$, the DCH power of antenna i is

$$P_i^{DCH}(\mathbf{k}) = \sum_{j \in \mathcal{J}_i(\mathbf{k})} \sum_{s \in \mathcal{S}} d_{js}^{DCH} v^s P_{ijs}^{DCH}(\mathbf{k}), \quad i \in \mathcal{C} \qquad (8.6)$$

The power vector given by Equation (8.6) corresponds to the downlink load factor—one of the key performance indicators in a network based on WCDMA (wideband code division multiple access) [12]. For this reason, it is typically incorporated into the objective function in automated optimization for planning R99 networks [16]. For networks with both R99 and HSDPA services, it is reasonable to assume that, by R99 admission control, the R99 demand map will not lead to a DCH power consumption exceeding some specified percentage of the total cell power. Thus, some power will always be available to HSDPA data.

8.3.4 HSDPA Performance

From Equation (8.1), the power of HS-DSCH in cell i is $P_i^{HS}(\mathbf{k}) = P_i^{TOT} - P_i^{CCH}(\mathbf{k}) - P_i^{DCH}(\mathbf{k})$. Denote by $d_j^{HS}(j \in \mathcal{J})$ the number of active users in j in the HSDPA demand map. In addition to power and interference, throughput on HS-DSCH depends on several factors, such as the modulation scheme and the number of channelization codes. To model HSDPA performance, a suitable SINR metric is the narrowband SINR ratio after despreading the HS-PDSCH (High-speed Physical Down-Link Shared Channel) [7]. For antenna i and pixel $j \in \mathcal{J}_i(\mathbf{k})$, this SINR value reads:

$$SINR_{ij}^{HS}(\mathbf{k}) = SF^{HS} \times \frac{P_i^{HS}(\mathbf{k})}{P_i^{TOT}\left(1 - \alpha_j + \frac{I_{oc}}{I_{or}}\right)} \qquad (8.7)$$

In Equation (8.7), $I_{oc} = \sum_{l \in \mathcal{C}: l \neq i} P_l^{TOT} g_{lj\downarrow}^{k_i} + v_\downarrow$ is the received other-cell interference plus noise, and $I_{or} = P_i^{TOT} g_{ij\downarrow}^{k_i}$ is the received power from the serving cell i. The spreading factor SF^{HS} equals 16. Note that the definition of Equation (8.7) does not depend on the number of channelization codes or the modulation scheme. A function $\phi(SINR_{ij}^{HS})$ is then used to translate $SINR_{ij}^{HS}$ into single-user data throughput (bps), taking into account the effect of the number of codes, AMC (adaptive modulation and coding), and HARQ (Hybrid Automatic Repeat Request). The shape of function ϕ can be obtained empirically [7]. Merely for the sake of simplifying notation, consider the case where ϕ is of the same shape for all users (extension to user-specific mobility profile and number of codes is straightforward).

Then the average HSDPA single-user data throughput can be expressed as

$$\bar{\phi}(\mathbf{k}) = \frac{\sum_{i \in C} \sum_{j \in \mathcal{J}_i(\mathbf{k})} d_j^{HS} \cdot \phi(SINR_{ij}(\mathbf{k}))}{\sum_{j \in \mathcal{J}} d_j^{HS}} \tag{8.8}$$

It is worth noting that parameters d_j^{HS}, $j \in \mathcal{J}$ do not necessarily have to represent a specific HSDPA user demand map. Instead, these parameters can be used as weights to reflect the relative importance of different parts of the service area. For example, setting $d_j^{HS} = 1$, $j \in \mathcal{J}$, $\bar{\phi}(\mathbf{k})$ models the average single-user throughput over the entire service area. For planning scenarios where cell-edge throughput is of primary concern, zero weight can be assigned to pixels other than the ones forming cell edges.

In addition to single-user throughput, another interesting performance metric is the HSDPA cell-average throughput. In this case, the behavior of the Node B scheduler, which controls the time sharing among all simultaneously active users in a cell, should be accounted for. Two commonly used scheduling schemes are Round Robin (RR) and Proportional Fair (PF). Denote by τ_j the time share of a user in $j \in \mathcal{J}_i(\mathbf{k})$. Under RR scheduling, all users in $\mathcal{J}_i(\mathbf{k})$ have equal time share, whereas in PF scheduling the time allocated to a user in j is proportional to its channel condition, reflected by $\phi(SINR_{ij}^{HS})$. Hence, parameters τ_j for the two scheduling schemes, respectively, take the form

$$\tau_j = \frac{1}{\sum_{j \in \mathcal{J}_i(\mathbf{k})} d_j^{HS}} \quad \text{and} \quad \tau_j = \frac{\phi(SINR_{ij}^{HS}(\mathbf{k}))}{\sum_{j \in \mathcal{J}_i(\mathbf{k})} d_j^{HS} \cdot \phi(SINR_{ij}(\mathbf{k}))} \tag{8.9}$$

Let $\Phi_i(\mathbf{k})$ denote the average HSDPA throughput of cell i, and define cell-average throughput $\Phi(\mathbf{k}) = \frac{1}{C} \sum_{i \in C} \Phi_i(\mathbf{k})$. Given a scheduling strategy, $\Phi_i(\mathbf{k})$ can be computed by the following expression:

$$\Phi_i(\mathbf{k}) = \frac{\sum_{j \in \mathcal{J}_i(\mathbf{k})} d_j^{HS} \cdot \phi(SINR_{ij}(\mathbf{k})) \cdot \tau_j}{\sum_{j \in \mathcal{J}_i(\mathbf{k})} d_j^{HS}} \tag{8.10}$$

8.3.5 Uplink Consideration

Given antenna configuration k_i of cell i, the uplink SINR requirement for user j being provided service s in cell i can be formulated as follows:

$$\frac{p_j g_{ji\uparrow}^{k_i}}{\sum_{j' \in \mathcal{J}: j' \neq j} \omega_{j'} p_{j'} g_{j'i\uparrow}^{k_i} + v_\uparrow} \geq \gamma^s \tag{8.11}$$

where p_j is the transmission power of a user in pixel j, $g_{ji\uparrow}^{k_i}$ is the total power gain between a user transmitter in pixel j and the cell i receiver, $\omega_{j'}$ is the activity factor in pixel j', and $v_\uparrow$ is the total noise power received at Node B antenna i. To simplify the formulation, we assume the same noise power at all Node Bs and the same SINR threshold as in the downlink. Note that the activity factor $\omega_{j'}$ takes into account the service s activity, the number of users in pixel j', and the single-user activity in the pixel. Assuming the maximum UE power p^{max} and a known noise-rise factor $r_i(\mathbf{k})$ in cell i, the following condition necessary to provide service s in pixel j can be derived from Equation (8.11):

$$g_{ji\uparrow}^{k_i} \geq \frac{\gamma^s}{1 + \omega_j \gamma^s} \cdot \frac{r_i(\mathbf{k}) v_\uparrow}{p^{max}} \tag{8.12}$$

where $r_i(\mathbf{k}) = \frac{1}{v_\uparrow} \cdot \left(\sum_{j' \in \mathcal{J}} \omega_{j'} p_{j'} g_{j'i\uparrow}^{k_i} + v_\uparrow \right)$. In detailed planning, noise-rise factors for a given antenna configuration vector $\mathbf{k}$ can be found by solving the following system of C linear equations (see, e.g., [16]):

$$\left(1 - \sum_{j \in \mathcal{J}_i(\mathbf{k})} \frac{w_j \gamma^s}{1 + w_j \gamma^s} \right) r_i(\mathbf{k}) - \sum_{l \in \mathcal{C}: l \neq i} \left(\sum_{j \in \mathcal{J}_i(\mathbf{k})} \frac{g_{ji\uparrow}^{k_i} w_j \gamma^s}{g_{jl\uparrow}^{k_i} (1 + w_j \gamma^s)} \right)$$

$$\times\ r_l(\mathbf{k}) = 1, \quad i \in \mathcal{C} \tag{8.13}$$

In some cases the network can be dimensioned and configured for a given maximum noise-rise factor r^{max}, that is, assuming $r_i(\mathbf{k}) = r^{max}$, $i \in \mathcal{C}$. The approach can be used for higher-level planning, in situations with high traffic variation or uncertainties, and can also be adopted for HSUPA assuming that data is always available for transmission.

In radio network planning and optimization, there is often a target coverage degree ξ. With the presented uplink model, the uplink coverage degree $\xi(\mathbf{k})$ for antenna configuration $\mathbf{k}$ is the fraction of the total service area where the coverage condition of Equation (8.12) is satisfied for all services for the found noise-rise factors. The antenna configuration is said to provide the desired UL coverage when the fraction is at least ξ, otherwise an UL coverage loss of $\xi - \xi(\mathbf{k})$ occurs. In practice, the coverage requirement can be service specific, which is a straightforward extension to the presented model.

8.4 Optimization Algorithm

8.4.1 Optimization Task

To tackle a planning problem, the optimization engine needs as input a system model that captures the system behavior depending on the decision variables, and an optimization task specified as a set of objectives and constraints. The system model in the previous sections links antenna configuration $\mathbf{k}$ (a vector of decision variables) to HSDPA performance metrics $\bar{\phi}(\mathbf{k})$ and $\Phi(\mathbf{k})$. The former does not depend on the assumption of the scheduling policy, and for this reason it will be used as the objective function to illustrate the optimization process. The optimization task is hence to solve the combinatorial optimization problem of *finding an antenna configuration that maximizes* $\bar{\phi}(\mathbf{k})$:

$$\max_{k \in \mathcal{K}_1 \times \cdots \times \mathcal{K}_C} \bar{\phi}(\mathbf{k}) \tag{8.14}$$

The desired UL coverage is introduced as a soft constraint with which the optimization task can be reformulated as *finding an antenna configuration that maximizes* $\bar{\phi}(\mathbf{k})$ *while minimizing the UL coverage loss* $\xi - \xi(\mathbf{k})$. The new objective function is $F(\mathbf{k}) = \bar{\phi}(\mathbf{k}) \left(1 - \rho \cdot \max\{\xi - \xi(\mathbf{k}), 0\}\right)$, where the introduced weighted penalty makes configurations with a coverage loss of at most $1 - \xi$ preferred, and those with a higher loss feasible but left for a later judgment.

8.4.2 Complexity Consideration

From a computational complexity viewpoint, the problem in (8.14) is NP-hard and no more difficult than that with the coverage loss penalty in the objective function; a formal proof of this result can be obtained from a generalization of the 3-satisfiability problem, of which the NP-hardness is well known [8]. NP-hardness is a term used to denote a class of computationally difficult optimization problems. It implies that an algorithm guaranteeing global optimum will take, in the worst case, an amount of computing time growing exponentially in the problem size. In (8.14), the problem size is defined by the number of candidate configurations at each antenna and the number of pixels used to represent the service area. The hardness of the problem and the objective of dealing with large-scale planning scenarios justify the use of heuristic search algorithms, which, although do not guarantee optimality, are effective in finding high-quality solutions quickly.

8.4.3 Search Algorithms

There is a rich set of principles that have been developed for designing search algorithms, such as local search [4], tabu search [9], simulated annealing [13], greedy randomized adaptive search procedure (GRASP) [17], and genetic algorithms [10]. The general idea, shared by all the algorithms, is to perform an iterative search in the solution space. The objective function is used to drive the search toward better solutions. In every iteration, one solution (or a set of solutions) is kept, and a set of new solutions, generated by modifying parts of the current one, are evaluated. The scheme used for solution modification is hence a key algorithmic design aspect. The current solution is replaced by a new one, if the latter has a better performance value. For many search algorithms, this update operation is referred to as a move in the solution space. Taking solution-improving moves will eventually end up at a local optimum, that is, a solution for which the modification scheme is not able to improve, but it is not necessarily the global optimum. To deal with this issue, all algorithms, except local search, implement some mechanism to drive the search from local optima, by temporarily accepting moves leading to worse solutions. For tabu search, the mechanism is to allow only moves that tend to go away from local optima. The other algorithms incorporate randomization to achieve a similar effect.

To design a search algorithm for (8.14), the most natural way of solution modification is to adjust the tilt and/or azimuth angles of one or several antennas. Thus, a general recipe of algorithmic design has the following key components:

1. Configuration adjustment of one or several antennas
2. Performance evaluation by the system model
3. Solution update

In the rest of this section, an implementation of the recipe is demonstrated through a simulated annealing algorithm.

8.4.4 Simulated Annealing

A simulated annealing algorithm generates and evaluates one solution per iteration. It adopts a probabilistic move to deal with local optima. Suppose **k** is the current antenna configuration, and **k'** is a new configuration from adjustment of **k**. If the objective function improves, that is, $F(\mathbf{k'}) > F(\mathbf{k})$, the algorithm moves to **k'**. Otherwise, the move is made with a probability. The standard function for determining the probability is $e^{-\delta/T}$. Here, δ is the amount of worsening in solution performance, that is, $F(\mathbf{k}) - F(\mathbf{k'})$ or the relative difference $\left(F(\mathbf{k}) - F(\mathbf{k'})\right)/F(\mathbf{k})$. The denominator T, called temperature, is a parameter used by the algorithm to control the willingness of accepting the worsening. By the probability function, it is more likely to

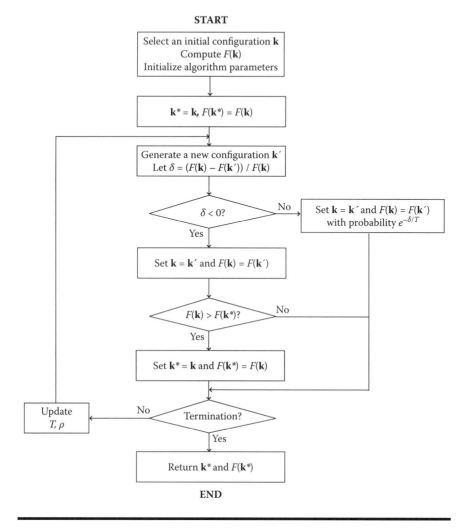

Figure 8.4   A flowchart of the optimization algorithm.

move to a worse solution if $F(\mathbf{k}) - F(\mathbf{k}')$ is small or T is large. The algorithm stops after a maximum allowed number of iterations, of which the value is set to make the computation time-affordable. Parameter T should be high enough at the algorithm start to enable moving away from local optima. During the search, the value of T and consequently the probability of accepting non-improving moves decrease gradually. Typically, T approaches zero when the number of performed iterations gets close to the maximum allowed number. To control the importance of the soft UL coverage constraint, the weighting factor ρ in the objective function is increased after every new move with $\xi - \xi(\mathbf{k}') > 0$, and decreased otherwise. The flowchart of the algorithm is given in Figure 8.4.

8.4.5 Configuration Adjustment

The algorithm design is not complete without specifying how to generate a new configuration vector $\mathbf{k}'$ from $\mathbf{k}$. The most obvious choice is to randomly select an antenna i, and pick a configuration k_i', again randomly, from $\mathcal{K}_i \setminus \{k_i\}$. Vector $\mathbf{k}'$ then takes the form $(k_1, \ldots, k_{i-1}, k_i', k_{i+1}, \ldots, k_C)$. This type of purely random selection is, however, not effective; too often, it produces $\mathbf{k}'$ that performs worse than $\mathbf{k}$. Because the time required to evaluate a configuration vector grows by network size, for large-scale planning it becomes crucial that configuration adjustment will frequently yield solution improvement.

A refined strategy, in which configuration adjustment is restricted to a few cells that appear to be critical to HSDPA throughput under current configuration $\mathbf{k}$, turns out to be very effective in exploring the solution space. The underlying idea is to consider one of the pixels having lowest SINR and hence poor throughput, and a small subset of cells having largest impact on this SINR value. Let j be the pixel under consideration. The set of critical cells is composed by the best server and the cells generating heavy interference, that is, cells having highest gain values in j (cf. the scenario illustrated by Figure 8.2). Each critical cell receives a value representing its importance, set in proportion to the gain. The algorithm selects a cell from the critical set, with a probability distribution following the cells' relative importance values. Once a cell is selected, one of its alternative candidate configurations is chosen randomly. The SINR at j is then reevaluated. Because this evaluation is made for j only, it demands very little computation. The algorithm makes a number of such minor iterations, each of which is a reconfiguration trial of one critical cell for j. The output configuration is the one giving the highest SINR at j. This output configuration defines $\mathbf{k}'$ in the major algorithm iteration.

Although possible antenna adjustments are very restricted per iteration in the above procedure, diversity is not lost. The pixel used for defining critical cells varies by $\mathbf{k}$. As a result, the set of cells subject to configuration adjustment will move from one part of the network to another during the search.

8.5 A Case Study

As a case study for the presented optimization framework, we consider a test network originating from a planning scenario for the downtown area of Lisbon, provided by the MOMENTUM project [3]. The network parameters are summarized in Table 8.1. Given isotropic pathloss predictions [3] and antenna diagrams, the power gain parameters for each antenna configuration have been found by the model in [6]. An antenna configuration in

Table 8.1 Network Parameters

Parameter	Value
Number of sites/cells (C)/pixels (J)	60/164/52500
Pixel size	$20\,\text{m} \times 20\,\text{m}$
Total service area size	$4200\,\text{m} \times 5000\,\text{m}$
Total cell transmission power (P_i^{TOT}, all cells)	20 W
Total downlink noise power ($v_\downarrow$)	$8 \times 10^{-14}\,\text{W}$
Total uplink noise power ($v_\uparrow$)	$4 \times 10^{-14}\,\text{W}$
Downlink installation loss	3 dB
Node B antenna	Kathrein K742265, 2110 MHz, 18.5 dBi
UE antenna	omni, 0 dB
Maximum UE power (p^{max})	0.2 W
Total downlink CCH power/CPICH power (β)	1.8
CPICH E_c/I_0 target (γ^c)	0.01
SINR target for R99 speech (γ^{speech})	5.5 dB
UL target coverage degree (ξ)	0.95

this study is a combination of antenna electrical tilt, mechanical tilt, and azimuth. For each tilt type, the range of possible values is from $0°$ to $6°$. Because the studied network represents an urban environment with dense site locations (typically interference limited), the preference in downtilting is given to electrical tilting, that is, mechanical downtilt can be used only if electrical downtilt has reached its $6°$ maximum. Antenna azimuth can be adjusted in the range of $[-10°,+10°]$ with a $5°$ step. An installation loss of 3 dB is assumed in downlink, but not in uplink, for which we assume tower-mounted amplifiers. The UL coverage is evaluated over a statistical snapshot of the entire area; the snapshot mimics nonuniform user distribution while still accounting for the contribution of every pixel.

Five antenna configurations are considered for performance evaluation. In the reference configuration (REF), 2.0 W CPICH power, no tilt, and default azimuth are assumed. In the optimized configurations, the CPICH power is to be found together with network antenna configuration $\mathbf{k}^*$, where the optimization decision space is given by the range of either a single antenna parameter (electrical tilt in ET, mechanical tilt in MT, or azimuth

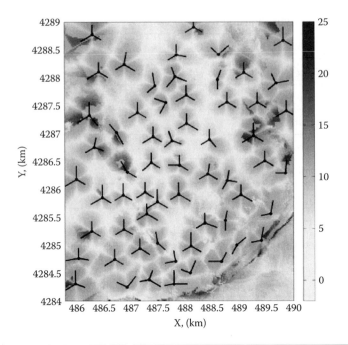

Figure 8.5 HS-DSCH SINR in REF configuration, (dB).

in AZ configuration) or all three parameters (MEA configuration). For each of the configuration scenarios, the optimization algorithm has been run for a statistical user distribution snapshot of R99 traffic for which we assume speech service with an activity factor $v^{speech} = 0.5$ and an SINR target of 5.5 dB. The orthogonality factor is nonuniform and has been provided together with the data. The assumed HSDPA service demand and user activity factors in UL are uniform, which is not a limitation of the presented framework, but allows us to obtain a generic solution independent of HSDPA traffic distribution and scheduling.

Node B locations and initial antenna configuration are depicted in Figure 8.5 where we also show HS-DSCH SINRs over the service area in REF configuration. The SINR distribution gives us a general picture of system performance without the need to consider any specific HS-user distribution or mapping function ϕ. Figure 8.6 and Figure 8.7 demonstrate the SINR distribution and HSDPA throughput, respectively, in MEA configuration, for which the largest improvement among all configurations has been achieved due to more configuration alternatives. For throughput mapping we used a function in [11] obtained by simulations for five channelization codes and the Pedestrain-A mobility model. We observe that optimizing antenna configurations allows us to achieve better cell isolation between

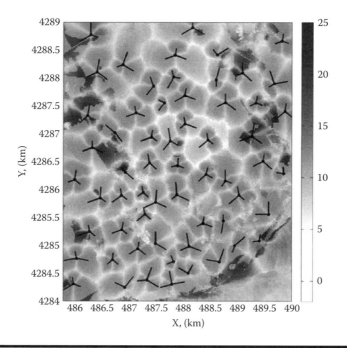

Figure 8.6   HS-DSCH SINR in MEA configuration, (dB).

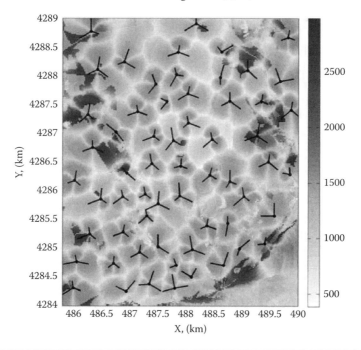

Figure 8.7   HSDPA throughput in MEA configuration, (kbps).

densely located sites and improve the received signal quality in areas with bad coverage. The difference is particularly significant in areas having low signal quality in REF antenna configuration, that is, areas where an increase in throughput gives more perceived service enhancement.

To evaluate HSDPA performance for a realistic HSDPA user distribution, we consider a snapshot with 2,380 active users nonuniformly distributed over the service area. Table 8.2 summarizes the power and coverage statistics for the five configurations. Table 8.3 summarizes the single-user and cell HSDPA throughput. In both tables, when relevant, we show the mean, and the 5th and 95th percentile values of the performance metrics. The last row in each table shows the maximum relative change (in %) obtained by optimization, that is, comparing MEA and REF configurations. A great reduction in uniform CPICH and thus CCH power can be observed in all scenarios. The improved interference also results in less power needed to support R99 traffic in the optimized configurations and therefore overall more power available for HSDPA traffic. The relative power budget improvement for HSDPA is, however, smaller than the relative gains in CCH and R99 power due to their smaller total weight in the cell power budget. The maximum (over all configurations) relative cell throughput improvement is approximately in the same range as the HSDPA power budget gain, although slightly higher for RR scheduler which is more sensitive to poor performance of cell-edge users. As expected, the largest difference between the two schedulers (8.5%) is in the fifth percentile, for which the single-user throughput improvement is also the largest. In a sense, the PF scheduler can be seen as a radio resource management means to improve cell-edge data rates compared to using RR. With the REF configuration, the fifth percentile data rate is increased by 7.4% when using PF compared to RR. In a better planned MEA, the fifth percentile data rate is increased by only 1.7%. Hence, the relative impact of more advanced radio resource management mechanisms is greater for worse planned networks. In other words, in well-planned networks, there may be less need for advanced radio resource mechanisms to provide good cell-edge data rates.

As already discussed, MEA configurations have the best performance among the five configurations. Among the configurations optimized for a single antenna parameter, the best performance is achieved with electrical tilting, which is the most effective way of reducing inter-cell interference, while AZ gives least performance gain because changing antenna azimuth just shifts interference, which may be inefficient in environments with densely located sites. Another observation is that ET configuration performance is very close to that of MEA; this is due to the effect of the soft UL coverage constraint, which prioritizes solutions with no excessive UL coverage loss. Note also that the target UL coverage has been achieved with all optimized configurations, but not in REF.

Table 8.2　Power and Coverage Statistics

Configuration	UL Coverage (%)	CCH Power (W)	R99 Power, (W)			HSDPA Power, (W)		
			5%	95%	Mean	5%	95%	Mean
REF	91.54	3.60	0.38	8.11	3.05	8.17	15.90	13.30
MT	95.73	1.80	0.31	5.94	2.38	12.16	17.82	15.78
ET	96.99	1.71	0.26	5.59	2.17	12.59	17.97	16.08
AZ	95.86	2.41	0.38	7.99	3.02	9.22	17.06	14.52
MEA	97.37	1.46	0.23	5.63	2.11	12.83	18.21	16.38
MEA vs. REF%		−59.44	−39.47	−30.58	−30.82	+57.04	+14.53	+23.16

Table 8.3 HSDPA Throughput Statistics

Configuration	$\tilde{\phi}(k^*)$, (kbps)			RR $\Phi_i(k^*)$, (kbps)			PF $\Phi_i(k^*)$, (kbps)		
	5%	95%	Mean	5%	95%	Mean	5%	95%	Mean
REF	458.38	1548.68	959.45	601.92	1320.69	932.94	646.69	1364.58	987.58
MT	676.92	1914.45	1175.51	866.87	1419.88	1146.25	903.63	1453.06	1193.56
ET	738.65	2033.55	1248.46	935.98	1448.16	1204.04	973.64	1496.81	1249.10
AZ	506.88	1616.08	1011.71	651.91	1366.55	984.71	687.55	1402.18	1038.77
MEA	750.59	2010.63	1248.85	965.55	1495.23	1215.33	982.28	1519.09	1261.18
MEA vs. REF%	+63.75	+29.83	+30.16	+60.41	+13.22	+30.27	+51.89	+11.32	+27.70

The optimized configurations have also been compared to that obtained by minimizing uniform CPICH power in [19]. Interestingly, the approaches demonstrate very similar single-user HSDPA performance, that is, optimizing antenna configuration using CPICH as the objective and defining bottleneck pixel as the one requiring the highest CPICH power [19] give a good solution for HSDPA throughput. CPICH minimization, however, results in a larger variation in cell sizes due to not taking into account the traffic amount in cells. Large cells have a negative impact when the cell HSDPA performance or user performance at a higher layer is evaluated, as these cells diminish the gain obtained from more HSDPA power available in cells. The solution obtained from HSDPA throughput optimization with the UL coverage constraint, on the other hand, has a higher CPICH power, but more balanced cell sizes.

Finally, we investigate a trade-off between the gain from configuration optimization and the amount of efforts needed to configure the network in terms of the number of cells that require reconfiguration. For each optimized configuration, we collect the information during the optimization search in which uniform HSDPA service demand is assumed. The results are shown in Figure 8.8. Statistics for improving moves only have been

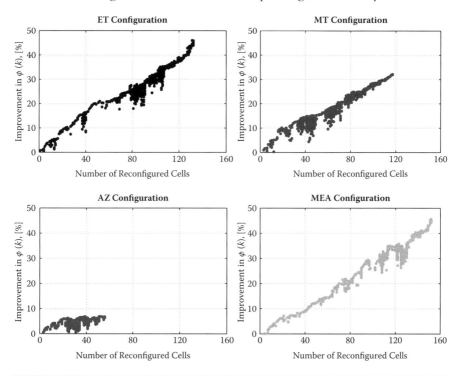

Figure 8.8   Single-user HSDPA throughput improvement versus the number of reconfigured cells.

saved. Recall, however, that accepting worse solutions and moving from those is allowed with simulated annealing. Note also that the number of reconfigured cells does not necessarily increase with the number of search iterations, and points with the same number of reconfigured cells may or may not correspond to the same set of cells. In addition to earlier noted higher gain with electrical tilting, we make the following observations. The Pareto 20/80 rule, by which 20% of efforts gives 80% of the result, seems to not apply to network reconfiguration in our case study where the gain increases almost linearly with the number of cells involved, although a concave shape in the left part of the curves can still be observed. It becomes evident that network optimization is a complex task and simple cell-by-cell tuning is not likely to lead to the desired result. Also, once configured, replanning a network can be a very costly procedure for an operator, which motivates for using advanced radio network planning tools with large-scale automated optimization capabilities already in earlier network planning stages.

8.6 Conclusions

An optimization framework for automated HSPA cell planning was presented in this chapter, including a detailed system model and an optimization algorithm that is particularly efficient for large network optimization. Moreover, the modeling approach is well-suited for planning networks in heterogeneous environments because the optimization framework does not make any restriction on cell structure or radio propagation characteristics. Our numerical experiments for a realistic case study demonstrate that the presented optimization approach can provide a significant gain in HSPA performance, especially at the cell edge. More efficient radio resource utilization, network performance gain, higher service quality, and improved user satisfaction are the other benefits of a well-planned radio network. These, combined with the complexity consideration of the planning task, the amount of work and reconfiguration cost for improving a poorly planned network, highly motivate for using automated radio network optimization.

References

[1] 3GPP, Requirements for Support of Radio Resource Management (FDD), 3GPP TS 25.133 V8.5.0, 2008.
[2] 3GPP, Physical Channels and Mapping of Transport Channels onto Physical Channels (FDD), 3GPP TS 25.211 V8.3.0, 2008.

[3] IST-2000-28088 Momentum Project, http://momentum.zib.de, 2005.

[4] E. Aarts and J. K. Lenstra (Eds.), *Local Search in Combinatorial Optimization*, Princeton University Press, 2003.

[5] A. Eisenblätter and H.-F. Geerdes, Wireless network design: Solution oriented modeling and mathematical optimization, *IEEE Wireless Commun. Mag.*, 13(6): 8–14, 2006.

[6] A. Eisenblätter, A. Fügenschuh, E.R. Fledderus, H.-F. Geerdes, B. Heideck, D. Junglas, T. Koch, T. Kürner, and A. Martin, Mathematical methods for automatic optimization of UMTS radio networks, Project Report D4.3, IST-2000-28088, Momentum, 2003.

[7] F. Frederiksen, H. Holma, T. Kolding, and K. Pedersen, HSDPA bit rates, capacity, and coverage, in H. Holma and T. Toskala (eds.), *HSDPA/HSUPA for UMTS: High Speed Access for Mobile Communication*, John Wiley & Sons, 2006.

[8] M.R. Garey and D.S. Johnson, *Computers and Intractability: A Guide to the Theory of NP-Completeness*, W.H. Freeman, 1979.

[9] F. Glover and M. Laguna, *Tabu Search*, Kluwer Academic Publishers, 1997.

[10] D.E. Goldberg, *Genetic Algorithms in Search, Optimization and Machine Learning*, Kluwer Academic Publishers, 1989.

[11] H. Holma, T. Kolding, K. Pedersen, and J. Wigard, Radio resource management, in H. Holma and T. Toskala (Eds.), *HSDPA/HSUPA for UMTS: High Speed Access for Mobile Communication*, John Wiley & Sons, 2006.

[12] H. Holma, K. Pedersen, and J. Reunanen, Radio resource management, in H. Holma and T. Toskala (Eds.), *WCDMA for UMTS: HSPA Evolution and LTE*, John Wiley & Sons, 2007.

[13] S. Kirkpatrick, C.D. Gelatt, and M.P. Vecchi, Optimization by simulated annealing, *Science*, 220(4598): 671–680, 1983.

[14] J. Laiho, A. Wacker, and T. Novasad (Eds.), *Radio Network Planning and Optimisation for UMTS*, John Wiley & Sons, 2002.

[15] Z. Michalewicz and D.B. Fogel, *How to Solve It: Modern Heuristics*, Springer-Verlag, 2004.

[16] M. Nawrocki, H. Aghvami, and M. Dohler (Eds.), *Understanding UMTS Radio Network Modelling, Planning and Automated Optimisation: Theory and Practice*, John Wiley & Sons, 2006.

[17] M.G.C. Resende and C.C. Ribeiro, Greedy randomized adaptive search procedures, in F. Glover and G. Kochenberger (Eds.), *Handbook of Metaheuristics*, pp. 219–249, Kluwer Academic Publishers, 2003.

[18] I. Siomina, Radio Network Planning and Resource Optimization: Mathematical Models and Algorithms for UMTS, WLANs, and ad hoc Networks, Ph.D. dissertation, Linköping University, LiU-Tryck, 2007.

[19] I. Siomina, P. Värbrand, and D. Yuan, Automated optimization of service coverage and base station antenna configuration in UMTS networks, *IEEE Wireless Commun. Mag.*, 13(16): 16–25, 2006.

Chapter 9

HSPA Transport Network Layer Congestion Control

Szilveszter Nádas, Sándor Rácz, and Pál L. Pályi

Contents

9.1 Introduction

The introduction of High-Speed Packet Access (HSPA) greatly improves the achievable bit rate but it presents new challenges to be solved in the Wideband Code Division Multiple Access (WCDMA) radio access network (RAN). For dedicated channels (DCHs), transport network bandwidth can be reserved by means of admission control. Bandwidth reservation is not efficient for HSDPA because of the higher peak rates and much higher variance of achieved bit rate; thus a new solution is needed to control congestion. In the Internet, such congestion is controlled by the end-user Transmission Control Protocol (TCP), but that is not possible in the access transport network because lost packets are retransmitted by WCDMA-specific lower layers.

This chapter describes the Iub/Iur transport network and congestion control solutions employed to efficiently utilize the transport network bottleneck. The main focus of the chapter is HSPA flow control, but it also describes other congestion avoidance mechanisms, namely buffering and quality-of-service differentiation, call admission control, and link dimensioning. It also provides insight into how the different congestion avoidance mechanisms cooperate. The architecture of flow control, the congestion detection mechanisms, and the possible congestion actions are described. Different flow control solutions are compared and a few case studies are presented. Finally, the HSPA Framing Protocol (FP), which provides the HSPA flow control framework, is described in detail and transport network overhead is evaluated.

9.1.1 Iub/Iur Transport Network Architecture

Figure 9.1 gives a schematic view of a WCDMA network, which consists of user equipment (UE), the WCDMA terrestrial radio access network (WCDMA RAN), and the core network. WCDMA RAN (WRAN) is also called Universal Mobile Telecommunications System (UMTS) Terrestrial Radio Access Network (UTRAN).

The WCDMA RAN handles all tasks that relate to radio access control, such as radio resource management and handover control. The core network, which is the backbone of WCDMA, connects the access network to external networks (e.g., PSTN, Internet). The user equipment (mobile

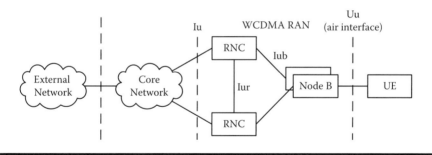

Figure 9.1   Schematic view of a WCDMA network.

terminal or mobile station) is connected to radio base stations (Node B) over the WCDMA air interface (Uu). During soft handover, one UE can communicate with several Node Bs simultaneously.

According to the WCDMA RAN specifications by the 3rd Generation Partnership Project (3GPP), all radio network functions and protocols are separated from the functions and protocols in the transport network layer (TNL). The WCDMA RAN transport network transmits data and control information between the radio network controller (RNC) and the Node Bs (Iub interface) or between RNCs (Iur interface). For a given UE there is always a single serving RNC (SRNC), which terminates the user and control plane protocols of that UE. In this chapter whenever we write RNC, we mean this SRNC. The transport network layer provides data and signaling bearers for the radio network application protocols between RAN nodes, and includes transport network control-plane functions for establishing and releasing such bearers when instructed to do so by the radio network layer.

The first release of the 3GPP specification was called Release 99. Figure 9.2 shows the Release 99 protocol stack at the Iub interface for transferring data streams on common transport channels (CCHs) and dedicated transport channels (DCHs) to the air interface.

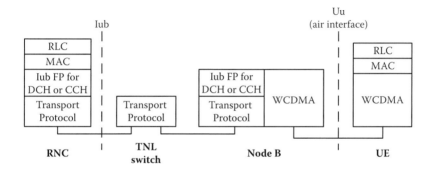

Figure 9.2   User plane protocol stack between RNC and Node B for Release 99.

The acknowledged retransmission mechanism of the radio link control (RLC) protocol ensures reliable transmission of loss-sensitive traffic over the air interface. RLC provides three types of service to the upper layer according to the standard [1]: transparent (no additional protocol information), unacknowledged (delivery not guaranteed), and acknowledged data transfer mode. The normal RLC mode for packet-type services is the acknowledged mode, so RLC works in this mode in the case of HSPA as well. The RLC protocol is used by both signaling radio bearers and radio bearers for packet-switched data services, but not by radio bearers for circuit-switched services. RLC Acknowledged Mode (AM) [1], which is a selective repeat automatic repeat request (SR-ARQ) protocol, provides transport service to upper layers between UE and the RNC. RLC AM does not include congestion control functionality because it assumes that RLC protocol data units (PDUs) are transmitted by the Medium Access Control–dedicated (MAC-d) layer according to the available capacity allocated on Uu and the transport network. The RLC status messages, which are being sent regularly based on preset events, trigger retransmission of all missing PDUs. The most dominant of these terms and events in the sending time interval. The receiver side detects missing RLC PDUs based on gaps in the sequence numbers. The receiver side RLC requests retransmission by sending back a status PDU, informing which PDUs within the receiving window have been acknowledged (ACK) or lost (negative ACK, NACK). Upon reception of a status message, the sender can slide its transmission window (Tx window) if one or more in-sequence frames are acknowledged, so that new PDUs can be sent. If there are NACKs in the status message, the sender retransmits the missing PDUs, giving them priority over new ones. Several unsuccessful retransmissions of the same PDU trigger an RLC reset and the whole RLC Tx window (maximum ~80 kBytes in the case of 42-Byte RLC PDUs) is discarded.

In Release 99, the MAC-d protocol forms sets of transport blocks in the air interface and schedules them according to the timing requirements of WCDMA. Each scheduled period, called a transmission time interval (TTI), is 10 ms in length or multiples thereof.

Release 99 WCDMA radio connections, or radio access bearers (RABs), have practical bit rate values between 8 and 384 kbps. The size of the MAC transport blocks and the length of the TTI are RAB specific.

For data transfer over the Iub interface, the MAC transport blocks are encapsulated into Iub frames according to the Iub user-plane (UP) protocol for CCH or DCH data streams. Each Iub UP data stream needs a separate transport network connection between the RNC and Node B. The transport network thus establishes one TNL connection for each data stream. In the TNL switch (optional) is used for building aggregating transport networks.

In response to the increased need for higher bit rate and more efficient transmission of packet data over cellular networks, WCDMA 3GPP Release 5 extended the WCDMA specification with High-Speed Downlink Packet

Access (HSDPA) [2]. The demand for uplink performance improvement was addressed by introducing Enhanced Dedicated Channel (E-DCH)—often referred as Enhanced Uplink (EUL) or High-Speed Uplink Packet Access (HSUPA)—in 3GPP Release 6 [3]. HSDPA and HSUPA together are called HSPA. The main architectural novelty of HSPA is that certain parts of the control of radio resources have been moved from RNC to Node Bs. In 3GPP Release 7, higher-order modulation and multiple-input, multiple-output (MIMO) are introduced for HSDPA to further improve the achievable bit rate [4].

For HSDPA, a new shared downlink transport channel, called High-Speed Downlink Shared Channel (HS-DSCH), is also introduced. This channel is dynamically shared among packet data users, primarily in the time domain; the TTI is 2 ms. The application of shared channel makes the use of available radio resources in WCDMA more efficient. HSDPA also supports new features that rely on the rapid adaptation of transmission parameters to instantaneous radio conditions. The main principles are fast link adaptation, fast Hybrid ARQ (HARQ) with soft-combining, and fast channel-dependent scheduling [5]. To support HSDPA with minimum impact on the existing radio interface protocol architecture, a new Medium Access Control (MAC) sublayer, MAC-hs, has been introduced for HS-DSCH transmission and is implemented in Node B and UE. In Release 7, higher-order modulation and MIMO are introduced. To support the higher bit rates achievable, layer-2 enhancements became necessary. Previously allowed RLC PDU sizes do not support these bit rates; therefore, RLC PDU size must be increased. However, increasing the fixed RLC PDU size would result in higher padding in RLC and over the air interface, and the increased fixed size would also result in smaller cell coverage, due to the larger minimum air interface transport format. Therefore, flexible RLC PDU sizes were introduced and an improved MAC layer, MAC-ehs, introduced segmentation and multiplexing [4].

For HSUPA, new MAC layers (MAC-e/-es) were introduced to support the new features, that is, fast Hybrid Automatic Repeat Request (HARQ) with soft combining, reduced TTI (2 ms), and fast scheduling. HSUPA was further improved with the possibility of higher-order modulation in Release 7 [6]. The Release 6 and 7 improvements allow layer-1 peak rates up to 5.7 Mbps and 11 Mbps in uplink, respectively.

Despite the fact that similar features have been introduced for HSDPA and HSUPA, there are several essential differences [3]. In the case of HSDPA, the HS-DSCH is shared in the time domain among all users; for HSUPA, the E-DCH is dedicated to a user. For HSDPA, the transmission power is kept more or less fixed and rate adaptation is used. However, this is not possible for HSUPA because the uplink is non-orthogonal; therefore, fast power control is needed for fast link adaption. Consequently, for HSDPA, the shared resources are the transmission power and the code space of the shared channel; but for HSUPA, the shared resource is the interference headroom. Soft handover is not supported by HSDPA, while for HSUPA, soft handover is used to decrease the interference from neighboring cells and to have macro diversity gain.

The main architectural novelty of HSPA is that the control of radio frame scheduling has been moved from RNC to Node Bs. While fixed capacity (e.g., 64 kbps) can be reserved for traditional DCH traffic in the access transport network, per-flow bandwidth reservation is not efficient for HSPA because air interface throughput is much higher and fluctuates more. If bandwidth reservation is not used, then congestion can happen in both the Iub transport network and the air interface. In the current architecture, TCP cannot efficiently resolve a congestion situation in the access network because RLC AM retransmissions between the RNC and the UE hide the congestion situations from TCP. Thus, a flow control function has been introduced to control the data transfer between the RNC and Node B. Originally, HSDPA flow control was designed to take only the transmission capabilities of the air interface into account and to limit the RLC round-trip time (RTT) and the number of RLC stalls [7]. However, in practice, the increased air interface capacity did not always come with similarly increased Iub transport network capacity. Network operators often upgrade the Node Bs first and delay the upgrade of the transport network until there is significant HSPA traffic. In some cases also, the cost of Iub transport links is still high; however, it decreases significantly with the introduction of new mobile backhaul technologies [8]. Thus it is a common scenario that the throughput is limited by the capacity available on the Iub transport network links and not by the capacity of the air interface. On these bottleneck transport network links, it is important to maintain high efficiency. For HSDPA, it has been identified in 3GPP that the flow control framework can also resolve these congestion situations if a transport network congestion detection functionality is available. The design of the HSUPA framing protocol already took this aspect into account. For this purpose, congestion detection-specific fields to the Iub HS-DSCH Data Frame [7,9–11] and a new Information Element for the Iub/Iur Framing Protocol E-DCH Data Frame were introduced [12]. The requirements and principles of HSDPA and HSUPA congestion control are summarized in [13].

TCP slow start normally increases the TCP congestion window (cwnd) size fast to its maximum when a new flow arrives it is kept at that value, because due to RLC retransmissions, TCP does not experience loss (except, for example, RLC reset). In case of transport network congestion, RLC keeps retransmitting lost PDUs until they are successfully received (or until RLC resets); however, these retransmissions even increase the congestion level. In this way, due to the aforementioned reasons, TCP is notified significantly later about the transport network congestion. As a summary, we need a system-specific congestion control (CC) because the RLC does not have congestion control functionality and TCP congestion control cannot operate

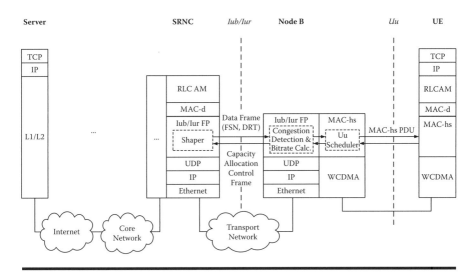

Figure 9.3   HSDPA protocol stack and flow control architecture.

efficiently above the RLC AM protocol. The locations of RLC AM and HARQ are depicted on Figure 9.5. Note that TCP is able to control congestion on the interfaces, which are outside the RLC retransmission. Such interfaces include the Iu interface in WCDMA and the S1 interface in the case of 3GPP Long-Term Evolution (LTE).

The nodes and protocol layers involved in the HSDPA and HSUPA flow control (FC) are depicted in Figure 9.3 and Figure 9.4, respectively [14]. The figures also show the location of the flow control-related functionalities in boxes with dashed lines.

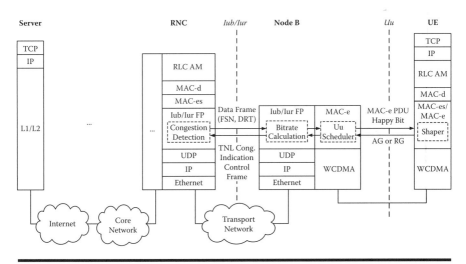

Figure 9.4   HSUPA protocol stack and flow control architecture.

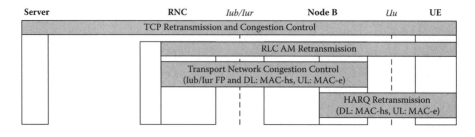

Figure 9.5 **Protocol layers performing congestion control and/or retransmission.**

The task of the HSDPA flow control is to regulate the transfer of MAC-d PDUs from the RNC to Node B. And the task of HSUPA flow control is to regulate the transfer of MAC-es PDUs on the Iub/Iur transport network toward the RNC. Several of these flows can share the same air interface (e.g., a cell) or transport network bottleneck (e.g., a link).

Note that for HSUPA, the regulation provided by FC is needed only when the Transport Network limits the performance. When the transport network is not limiting, the FC has no effect on the flows.

Figure 9.5 depicts the protocol layers that perform congestion control and/or retransmission. Over the air interface, fast Hybrid-ARQ is used to correct transmission errors. This results in a very small residual error probability on the air interface. Any transport network congestion is resolved by transport network congestion control and the retransmission of packets lost on transport network is done by RLC AM. TCP is used for congestion control and retransmission between the server and the UE.

9.1.2 Requirements, Different Traffic Types

The WCDMA RAN transport layer services must meet stringent quality-of-service (QoS) requirements. The most important measure of QoS performance on the Iub and Iur interfaces is maximum packet delay [15].

When HSPA is carrying moderate-speed QoS-sensitive traffic, QoS can be guaranteed by transport network bandwidth reservation by means of transport network admission control; flow control is not used. For best-effort (BE) traffic, bandwidth reservation is not efficient and flow control is used instead. HSPA carries BE traffic but HSPA Iub requires better than BE transport network connection because part of the control of radio resources is in RNC. When QoS-sensitive and BE traffic coexist in a system, the QoS-sensitive traffic is usually prioritized over BE traffic, and the capacity not used by the QoS-sensitive traffic can be utilized by the BE traffic.

9.1.3 Transport Network Solutions

With the "Iub cloud" logic [13], the traffic injected by the sending node (e.g., the RNC in the case of HSDPA) is less tightly controlled; that is, it is likely to inject too much traffic into the network, thus yielding a congestion situation. This approach should allow for statistical multiplexing in some scenarios without complex configuration of TNL topology in the RNC. However, this requires congestion detection in the receiving node (e.g., Node B in case of HSDPA) for congestion control. This is assumed to be the mainstream solution in this chapter because it supports a general transport network.

Using "Iub pipe" logic [13], the sending node enforces the traffic limit injected on the Iub interface, so it is able to instantaneously detect any congestion situation. The advantage of this approach is that there is no need to use congestion detection mechanisms in the receiving node because congestion detection is instantaneous—the only place it can occur is at the "pipe" entry. The drawback of "pipe" logic is that in some scenarios it may require complex configuration of TNL topologies in the sending node in order to leverage statistical multiplexing.

AAL2/ATM (ATM Adaptation Layer 2/Asynchronous Transfer Mode) or UDP/IP (User Datagram Protocol/Internet Protocol) is used as a transport protocol; in Figures 9.3 and 9.4, the UDP/IP/Ethernet solution is depicted as an example.

The initial WCDMA RAN specifications specified that the transport network layer must be based on ATM and AAL2 technology. However, Release 5 of the 3GPP specification also includes the option of an IP-based transport network.

For the AAL2/ATM transport solution, the data frames (DFs) are segmented into AAL2 common part sublayer (CPS) PDUs. These CPS PDUs are then fit into one or two ATM cells. There is no early packet discard for AAL2 queues; consequently, the end of the data frames can be lost, while the beginning of the data frame is still using the transport network capacity in vain because they will be discarded in the Node B anyway. We call these frames as *destroyed frames*. This behavior can be disadvantageous in the case of system overload. A detailed description of AAL2/ATM can be found in [16]. In the case of UDP/IP, the DFs to be transmitted can be larger than the maximum transfer unit (MTU) of the system, especially in case of high throughput. In this case, IP fragmentation is needed and DFs might be destroyed, as in case of AAL2/ATM. In most cases, however, the size of the DF is smaller than the MTU, and then a DF is either completely lost or transmitted. In the case of UDP/IP, the most commonly used is the layer-2 protocol Ethernet, but other layer-2 protocols are possible (e.g., MLPPP, Multi-Link Point-to-Point Protocol).

9.1.3.1 Typical Capacities and Architecture

For both HSDPA [17] and HSUPA, the Iub and Iur transport network links could be a bottleneck in the RAN. The transport network links are expected to be bottleneck e.g. in the case of E1, T1 transmissions[1] and ADSL transmission. For higher notes, e.g., E3 transmission[1] or 100 Mbps Ethernet access, the TN bit rate is higher than the typical achievable air interface throughput; however, in case of very good air interface conditions, TN can still be a bottleneck. In most networks, it is expected that in a significant percentage of the cases, the throughput is limited by the Iub/Iur transport network, especially in the initial deployment phase. As HSPA traffic is increasing in the network, most operators will expand their transport network to further enhance user experience.

9.2 Congestion Avoidance Mechanisms

Congestion occurs whenever the available link capacity is exceeded by the sum of the current flow rates. Congestion situations should be resolved in some way because they result in data loss. There are different means of controlling congestions. Congestion avoidance mechanisms can be classified based on the layer (e.g., data link routing or transport layer congestion control) where the mechanism operates, or on the duration of congestion. Note that solutions good for short-term congestion are not always good in the case of long-term overload, and vice versa. In most cases, a combination of mechanisms is used because overloads of various durations can occur in all networks [18].

Figure 9.6 depicts the congestion avoidance mechanisms proposed for use in the WCDMA RAN transport network. For very short data bursts, the most appropriate solution is to apply adequate buffers and QoS differentiation. This provides small delay for high-priority traffic, while buffering of low-priority, less delay-sensitive traffic increases utilization. For traffic with no fixed bandwidth requirement, flow control can be used to adapt the bit rate to the actually available resources. For guaranteed bit rate (GBR) services call admission control (CAC) can be used to check whether or not there is enough free capacity in the system. With adequate link dimensioning, the desired grade-of-service (GoS) can be guaranteed for a given traffic model.

Not all of these solutions are necessary in all cases; the optimal solution depends on several factors (e.g., on the cost of the transport network links and on the functionality available in the different equipments). For example, if the transport network links are very cheap, then they can be

---

[1] The bit rate available for ATM cells is 1920 kbps in the case of E1, 1536 kbps in the case of T1, and 33,920 kbps in the case of E3.

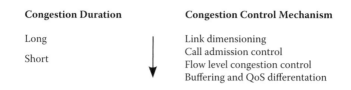

Congestion Duration	Congestion Control Mechanism
Long	Link dimensioning
	Call admission control
Short	Flow level congestion control
	Buffering and QoS differentation

Figure 9.6   Congestion duration versus congestion avoidance mechanism.

dimensioned according to the peak air interface capacity, and in this case probably no flow control or CAC is needed. However, in cases when the transport network links are costly, cooperating congestion control solutions on the different timescales can be used to optimally utilize the transport network links.

9.2.1 Buffering and QoS Differentiation

Buffering and QoS differentiation can improve packet-level performance. For example, we can give priority to the delay-sensitive packets or we can use larger buffers for best-effort packets. Both methods can improve system performance.

If we have more traffic classes with different QoS/GoS requirements, one option is to have completely separate (dedicated) resources for each traffic class. In this solution there is no interaction among traffic classes. As another option, we can share the resource among the traffic classes, but in this case we have to use proper scheduling and a buffering method that takes into account the QoS/GoS requirement of the traffic classes. In general, the resource-sharing method outperforms the application of separate resources and requires smaller link capacity.

Popular packet scheduling methods include [19]

- *Round Robin. This algorithm serves the queues in a cyclic order, ignoring the channel quality conditions. This method is outstanding in terms of simplicity and fairness of resource-sharing among queues.*
- *Priority. Always the queue with the higher priority is scheduled.*
- *Weighted Fair Queuing (WFQ). To each queue, a weight (w_i) is assigned. Queues are scheduled according their weights, so that user i achieves an average data rate of $\frac{Cw_i}{\sum_{j=1}^{N} w_j}$, where C denotes the link capacity.*
- *Rate shaping. Each queue has a given maximum transmission rate. If the incoming bit rate to the given queue is higher than the specified maximum, then the excess packets are dropped. This method is usually used in combination with other methods (e.g., priority scheduling).*

Popular buffering methods include

■ Finite buffer with FIFO or LIFO dropping
■ Random early detection dropping to inform TCP about congestion

CAC and the link dimensioning method must take into account applied buffering and QoS differentiation. The CAC method is developed for a given buffering and QoS differentiation method. Link dimensioning must take into account how the resources are shared among different flows and how well the flows with different QoS requirements are differentiated.

9.2.2 Flow-Level Congestion Control

The task of flow-level congestion control is to adapt the bit rate used by already ongoing flows to the available capacity. The used rate of GBR flows can be adapted to the available capacity, for example, in the case of adaptive multirate (AMR) voice codec or adaptive video streams. For Packet Switched (PS) traffic, the bit rate of the flow can be adapted more smoothly. TCP congestion control is a well-known example. In WCDMA RAN transport network HSPA, flow control is used to adapt the PS bit rate to the actually available capacity. HSPA flow control is detailed in "HSPA Flow Control" later in this chapter.

9.2.3 Call Admission Control

The task of Call Admission Control (CAC) is to ensure the desired QoS for traffic classes which have strict QoS requirements, usually GBR traffic. Whenever a new flow arrives, the CAC can decide to admit or reject it. A flow setup is rejected if the QoS requirements of the flows cannot be met, assuming that the arriving flow increases the traffic load. An already-established connection may also be released if it has lower priority than the arriving one in order to admit the new connection. The CAC can be based on calculation and/or measurements.

CAC algorithms based on analytical methods can vary in complexity. If the contribution of the GBR traffic to the total traffic load is high and detailed information about the flows is available, then the QoS of the flows can be evaluated using analytical calculations; and based on that, CAC decisions can be made. Such an algorithm is described in [20]. Otherwise, a simple admission control can be done, for example, reserving the peak or average bit rate of a GBR flow from the total link capacity. While a complex algorithm can allow high transport network utilization in a GBR-dominated case, a simple algorithm is often sufficient to reserve capacity because the traffic load of BE classes is high.

A CAC algorithm can also be based on measurement. In this case, the QoS is monitored and when the desired QoS is degraded, new flows are rejected and/or existing flows are released.

CAC can be end-to-end when it is executed only at the endpoints of the TNL flows. It can also be link-by-link when the TNL supports it, for example, in the case of AAL2/ATM-based transport network [16].

9.2.4 Link Dimensioning

Depending on traffic mix and the cost of the transport network capacity, different link dimensioning methods can be used. If transport network capacity is cheap, then *reasonable peak allocation* can be done. In this case, the total transport network capacity is configured to be high, such that the achievable air interface capacity is rarely higher than the dimensioning link capacity. On the other hand, if transport network capacity is costly and a traffic model is available, then *traffic model-based dimensioning* can be used. In case of higher-level aggregation, when the traffic of several Node Bs is aggregated, *overprovisioning* [21] can be used. In this case, the traffic of the transport network is continuously monitored, and the transport network links are upgraded as soon as the utilization reaches a limit.

Traffic model-based dimensioning can vary based on the properties of the dimensioned traffic. Traffic model-based dimensioning usually requires a target GoS. For a GBR service, such a GoS parameter can be, for example, the call rejection probability by the CAC; while for non-GBR services, for example, the experienced average throughput can be used. For GBR flows, the Kaufman-Roberts formula [22,23] can be used to determine the required capacity; while for PS traffic, an elastic calculation [24] can be used.

9.2.5 Congestion Avoidance Example

This section shows through an example how different congestion avoidance mechanisms cooperate.

We assign different traffic to different QoS levels, Figure 9.7 shows the used *buffering* and *QoS differentiation*. QoS level 0 and QoS level 1 have strict priority, whereas QoS level 2 and QoS level 3 share the remaining capacity based on the WFQ discipline. The vertical arrow in the figure indicates strict priority among the queues, QoS level 0 having the highest priority and the WFQ scheduler of QoS level 2 and QoS level 3 having the lowest priority.

Network synchronization and high-priority signaling belong to QoS level 0. Because of its highest priority, a short (e.g., 10 ms) buffer is sufficient for QoS level 0. As its volume is low compared to aggregated traffic demand, no *CAC* or *flow-level congestion control (FC)* is needed.

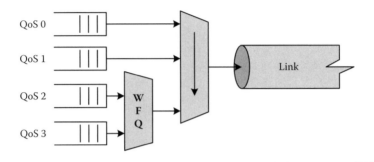

Figure 9.7   Example for buffering and QoS differentiation.

QoS level 1 is used for GBR services. It also has a short buffer, but is slightly longer (e.g., 20 ms). A simple CAC is used that reserves the peak rate of each flow end-to-end. The bandwidth reserved but not used can be reused by lower-priority traffic.

QoS level 2 is used for packet-switched interactive services, and QoS level 3 for packet-switched background services. To avoid starvation of background services caused by interactive ones. PS services can tolerate higher delay, therefore a larger buffer (e.g., 100 ms) can be used. This allows higher utilization of the bottleneck capacity. The peak rate of PS connections is usually high so no CAC is used for them. Instead, HSPA flow control is used to gracefully degrade the capacity used by PS flows.

For QoS level 0, the peak of network synchronization and high-priority signaling traffic can be reserved. However, the remaining capacity from this level can be used by lower-level traffic. For QoS level 1, the Kaufman-Roberts formula can be used to determine the required capacity. For PS traffic, elastic calculation can be used to determine the required capacity. The WFQ weights can be set by taking into account traffic volumes on the different QoS levels and also the fact that response time for interactive services is important. This can be done by, for example, multiplying the interactive traffic volume by a factor greater than 1.

9.3 HSPA Flow Control

Originally, HSDPA FC was intended to control radio scheduler queues in Node B (priority queue, PQ). On the one hand, these queues should be kept short enough to ensure that retransmitted data reaches its destination in a short time. On the other hand, these queues should be long enough to maximize air interface efficiency. In the case of HSDPA, the MAC-hs PQ in Node B can be long when the Uu interface is the bottleneck; in the case of HSUPA, there is no queuing in Node B, and so this is not an issue.

Later, the transport network became a potential bottleneck in the system because the increased air interface (Uu) capacity did not always come with

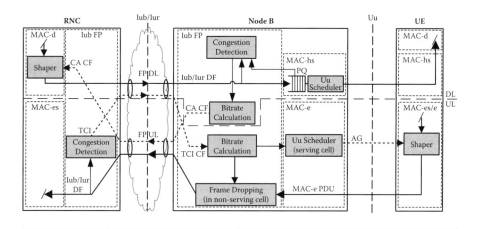

Figure 9.8 HSPA flow control architecture.

similarly increased transport network capacity in practice. Therefore, in addition to flow control, congestion control is also needed to avoid buffer overflow in the transport network. Thus, the algorithm used is often a combined flow and congestion control solution, which is often abbreviated as HSDPA/HSUPA flow control. In the transport network, it is the task of the flow control to provide high utilization, small data frame delay and loss. Flow control will share the TNL resources in a fair way among flows of the same priority.

Figure 9.8 depicts the HSPA flow control-related protocols and how the functionality is divided among the different protocol layers and nodes. The HSDPA part is shown in the upper part of Figure 9.8, whereas HSUPA is shown in the bottom part. In Table 9.1 the main differences between the mainstream (standardized) HSDPA and HSUPA flow controls are shown.

In the case of HSDPA, the Shaper in the RNC shapes the arriving MAC-d PDUs according to the signaled maximum flow bit rate. The Iub Framing

Table 9.1 Location of Different HSPA Flow Control Functionalities

	HSDPA FC	*HSUPA FC*
Congestion detection	Node B	RNC
Congestion indication	(to RNC)	To Node B in TCI CF
Bit rate calculation	Node B	Node B
Bit rate signaling	To RNC in CA CF	To UE in AG
Shaping	in RNC (MAC-d)	In UE (MAC-e)

Protocol (FP) puts them into Iub DFs and sends them to Node B. Each of these frames contains frame sequence number (FSN) and some of them contain the delay reference time (DRT). The Congestion Detection part in Node B aims at finding out the congestion level of the transport network, based on these pieces of information and on the length of the PQs. If transport network congestion is detected, the bit rate calculation part will be informed. The Bitrate Calculation part in Node B calculates the allowed maximum bit rate at which the RNC can send data to Node B to resolve the congestion situation. If the calculated bit rate of the flow changes, then the Shaper will be informed about the new bit rate through the capacity allocation (CA) control frame (CF).

In the case of HSUPA, the Shaper part is in the UE and Congestion Detection is in the RNC. The Shaper in the UE shapes the outgoing MAC-e PDUs according to the signaled maximum flow bit rate, and sends them to Node B. The MAC-e layer in Node B demultiplexes MAC-es PDUs and sends them to the Iub FP layer. The Iub FP in Node B puts arriving MAC-es PDUs to Iub DFs and forwards them to the RNC where Congestion Detection is performed based on DF fields similar to those of HSDPA. The Bitrate Calculation part is notified about congestion in the TNL Congestion Indication Control Frame (TCI CF). The Uu scheduler sends scheduling grants to the UE. There are two types of scheduling grants: absolute grant (AG) and relative grant (RG). AG is used to signal the maximum allowed bit rate by the transport network at which the UE may send data to Node B taking into account congestion information from the Congestion Detection part. The RG can modify this rate up/down in the serving cell, or only down in the non-serving cell. We assume that RG is not used for flow control purposes.

In the next section, "Congestion Detection and Network Monitoring," we discuss how *congestion detection and network monitoring* is performed. In the subsequent section, we discuss congestion action. ("Per-Flow HSPA Flow Control Solutions" and "Aggregated HSPA Flow Control Solutions") HSPA flow control algorithms can be implemented in different ways. Later sections describe *per-flow* and *aggregated* flow control. At the end, three case studies show examples for HSPA flow control.

9.3.1 Congestion Detection and Network Monitoring

Transport network congestion detection can be based on fields in Iub DFs or transport network-specific information. Flow control-related DFs and CFs are standardized in [7,12].

HS-DSCH Iub DFs contain the user data and transmit information about the amount of user data waiting in the RNC, called the user buffer size (UBS). They also contain information for congestion detection, the FSN, and the optional delay reference time (DRT).

The Iub/Iur Framing Protocol E-DCH DF contains the user data, the FSN, the connection frame number (CFN), and the subframe number. The CFN and subframe number are used for reordering, but can also be used to calculate the DRT. The FSN and DRT can be used for transport network congestion detection.

The DRT field, which can be found in the data frame payload, contains the value of a 16-bit reference counter in the sending node when the data frame is sent. The DRT counter is increased every 1 ms. The DRT can be used for dynamic delay measurements. In the case of HSDPA, the RNC does not send DRT in each frame. There is a flag to specify whether the frame contains DRT. In the case of HSUPA, each frame contains DRT information.

Apart from congestion detection based on DF fields, transport protocol-specific congestion detection techniques are also possible.

Transport network congestion detection for HSPA should be performed whenever a DF arrives at the receiving node. The following methods can be used:

- *Destroyed Frame Detection (DFD)*. Due to the fact that ATM cells (not the data frames) could be lost in the ATM-based transport network, it is possible that only a part of a data frame is lost. When the segmented data frame is reassembled, it is possible to detect that a part of the data frame was lost.
- *FSN gap detection*. The 4-bit FSN in the data frame can be used to detect missing data frames (which were fully lost).
- *Dynamic Delay Detection (DDD)*. In the case of HSDPA, the DRT field in the data frame contains the value of a reference counter in the RNC when the data frame was sent. The DRT is compared to a similar reference counter in Node B when the data frame is received. In the case of HSUPA, the Node B DRT is compared to a similar reference counter in RNC when the data frame is received. The difference between the two counters increases when the transport network buffer is built up. Congestion is detected when this difference increases too much compared to the minimum difference.
- *Other transport network protocol-specific methods*. For example, Explicit Congestion Notification (ECN) field [25] is used in IP header for lossless TCP congestion control but can also be used for HSPA transport network congestion control.
- *Direct TNL information*. For example, TNL buffer length can be used directly, but this makes congestion detection dependent on the transport network and requires communication between the TNL and the congestion control.
- *Data frame rate measurement*. This method can be used to determine the severity of the congestion and to get information about the state of the transport network.

Non-congestion-related data frame loss in the transport network can result in false congestion detection; therefore, it should be avoided. If it is not possible to keep it very small, then one might consider adding filtering for the loss-based congestion detection, depending on the used TNL technology and typical loss patterns. However, this is not needed in the case of the common TNL technologies used.

Dynamic Delay can be differentiated by its severity by defining different dynamic delay limits. These DDD limits must be configured by taking into account the frame delay variation caused by higher-priority traffic. The limits must be set higher than the non-congestion-related frame delay variation; otherwise, congestion will be detected even when there is no congestion in the transport network, and these false congestion detections will result in performance degradation. In the case of very small transport network buffers or very high frame delay variation caused by other traffic, it is recommended not to use DDD at all.

Different congestion events can be classified and different actions can be taken. For example, dynamic delay exceeding a small limit can be classified as *soft congestion* and all other congestion events (DFD, FSN, and DDD for large limit) can be classified as *hard congestion.*

HSDPA and HSUPA congestion detection are based on similar principles but in the case of HSUPA, congestion detection and bit rate calculation are separated. HSUPA congestion detection takes place in the RNC, whereas the bit rate calculation takes place in Node B. Therefore, congestion indication is not used in the case of HSDPA. In the case of HSUPA, the detected congestion and its severity are reported to Node B by a TCI CF, if no TCI CF was sent for a given time. The TCI contains a congestion status field, which can indicate no congestion, congestion due to delay build-up, or congestion due to frame loss.

9.3.2 Congestion Action

Congestion action refers to all the actions needed to resolve congestion, that is, bit rate calculation, notification, and shaping. The bit rate calculation algorithm itself is not standardized; each vendor can implement its own solution. Shaping and notification are standardized.

While the purpose of HSDPA flow control [17] and HSUPA flow control is similar, there are significant differences. First, for HSUPA, only the transport network bottleneck must be regulated, while for HSDPA there are also Uu scheduler queues in Node B to be regulated (called MAC-hs PQs [17]). This means that HSDPA FC must deal with Uu and transport network bottleneck; but in case of HSUPA FC, the Uu bottleneck is completely handled by the Uu scheduler. Second, HSUPA can be in soft handover, while HSDPA cannot. For HSUPA, this means that for the same radio bearer, there can be several (one serving and zero or more non-serving) flows to be controlled. Third, in the case of HSDPA, the FC has a more direct impact on flow rates than Iub. For HSDPA, the explicit rate should be given, whereas for HSUPA it can be given but other ways are also possible.

For HSDPA, based on UBS, Uu-related radio scheduler queue information and the output of congestion detection the FC algorithm in Node B decide how many MAC-d PDUs can be transmitted from the RNC for a given flow. This is reported to the RNC using the HS-DSCH CA CF. The CA CF defines maximum MAC-d PDU *length*, HS-DSCH *credits*, and HS-DSCH *interval* and HS-DSCH *repetition period* values. The MAC-d shaping in RNC ensures that within a given interval, not more PDUs are sent than specified in the *credits* field. Repetition period defines how many times the interval and the credits are repeated. A newly received CA CF overrides the old one. Setting the repetition period unlimited allows one to define an allowed shaping rate; e.g., max. PDU length 42 octets, credits 4, and interval 10 ms settings define a shaping rate of 134.4 kbps. In the case of flexible RLC, instead of checking the number of PDUs sent within an interval, the number of octets is limited per interval to avoid bit rate loss due to smaller RLC PDUs (see details in "HSPA Framing Protocol").

The CA CF can be used in a rate- or credit-based manner for HSDPA. The main advantage of rate-based usage is that it generates moderate CA load. High CA load should be avoided because it requires high processing power from the nodes. There are practical problems with using CA in a credit-based manner. It requires more intense signaling; moreover, CAs can be lost. Therefore, we do not know by when all the credits will have been used. A window-based solution would solve this problem but an acknowledgment window is not supported by the standard. Therefore, we discuss rate-based CA in this chapter.

The HSUPA air interface scheduler (Uu scheduler) operates by sending scheduling grants to UE and receiving scheduling requests from UE [3]. Only the scheduling framework is standardized, the scheduling algorithm itself is not. There are two types of scheduling grants: AG and RG. AGs can be sent only by the serving cell and transmitted over the E-DCH Absolute Grant Channel (E-AGCH), which is a shared resource among all users of the cell. The AG defines how many bits can be transmitted every TTI and thus a maximum limit of the data rate. The AG is valid until a new scheduling grant is received. The RG can modify this rate up/down in the serving cell, or only down in the non-serving cell. The UE indicates, by a flag called Happy Bit, whether or not it would benefit from a higher rate grant.

In the case of HSUPA, until the first rate reduce request is received, the Uu scheduler behavior is not affected by FC. Based on air interface conditions, hardware resources, and Happy Bit information, the Uu scheduler decides the granted bit rate represented by the AG. Upon receiving a rate reduce request, the scheduler decreases the granted bit rate.

HSUPA FC is designed to provide fair throughput sharing among the flows sharing the same transport network bottleneck when the transport network is limiting the throughput. The behavior of flows is regulated by the Uu scheduler until a transport network congestion is detected. The reason for this is that as long as the transport network is not a bottleneck, it is the task of the Uu scheduler to utilize the air interface as much as possible and to provide fairness among the flows. The Uu scheduler increases the granted bit rate with reasonable speed to avoid large interference peaks. This also ensures that sudden overload of the transport network is avoided.

9.3.3 Per-Flow HSPA Flow Control Solutions

For per-flow flow control, congestion detection and bit rate calculation are done independently for all flows. This approach is flexible for different transport network topologies, solutions, and configurations, that is, regardless of how the transport network bottlenecks are distributed. Consquently, per-flow solutions need less maintenance with future updates and changes of the WRAN system.

Thus, a per-flow solution supports different transport network bottlenecks for flows of the same Node B and TNL QoS differentiation among the flows. Some flows of a Node B might experience different bottlenecks; for example, when some flows are transmitted over Iur links, others are not; or when some flows can be routed on different paths on the Iub interface.

Per-flow solutions may partly rely on some aggregated information, as in [26]. In that reference, the flow control itself works per flow, but the initial rate for each flow is determined based on Node B-level aggregated information.

To achieve fair bandwidth sharing among flows sharing the same bottleneck, Additive Increase/Multiplicative Decrease (AIMD) can be used. Other congestion control styles such as Additive Increase/Additive Decrease (AIAD), Multiplicative Increase/Multiplicative Decrease (MIMD), and Multiplicative Increase Additive Decrease (MIAD) have also been studied in the literature. However, assuming synchronous congestion signals and static bandwidth, AIMD proved the only fair and stable choice among them [27]. In [27], it is shown that AIMD guarantees convergence to fairness; all flows converge to an equal share of resources in steady state, where no flows join or leave.

9.3.4 Aggregated HSPA Flow Control Solutions

Aggregated flow control solutions aim to share the available bandwidth among the flows based mainly on aggregated network information. Aggregation may be performed on multiple levels, for example, per cell or per Node B. Per-flow parts may also be present in such an algorithm but the dominant part is performed based on aggregated information. Aggregated solutions have an overall picture of the system; that is why they can react faster to changes in some cases and provide a fairer distribution of resources.

An aggregated solution requires detailed information about the transport network bottleneck(s) and QoS solution; it also should support aggregated transport network connections, where flows of several Node Bs can experience bottleneck. While such a solution is not impossible, its complexity is too high compared to the achievable gains. The aggregated solution requires continuous maintenance when the transport network topology or the transport network QoS solution used changes.

It is possible to have an aggregated solution that does not have full information about the transport network. In this case, however, many of the above advantages disappear.

9.3.5 Case Study: Aggregated HSDPA Flow Control Solution

In [17] an aggregated HSDPA flow control algorithm is proposed. Apart from controlling the MAC-hs PQ, it also solves the congestion situation on the transport network. The algorithm is a rate-based FC solution, which includes a per-flow part, and per-cell and per-Node B-level aggregated information. The per-flow part is responsible for the fast reaction to a congestion situation in the transport network or long PQ delays. The cell-level aggregation estimates the frequency of scheduling a PQ in the given cell. This is used for estimating the air interface bit rate of the PQ. The Node B (or transport network) level aggregation approximates the available transport network capacity for HSDPA and distributes it among the flows. The notations introduced in this section reflect the different aggregation levels and the types of variables. Variables without superscripts denote per-flow variables, while superscript *cell* denotes cell level and superscript *Node B* means Node B-level aggregation.

In the case of large transport network buffers, DFD and FSN cannot provide efficient congestion detection because protocol problems occur before the transport network buffer becomes full. DDD provides congestion detection in this case, and it also decreases transport network delay and loss in the case of small transport network buffers. The shaping rate is calculated separately for every active flow, once per 100 ms. The 100-ms value is a compromise between fast reaction, low CA frequency, and low calculation complexity. The time of this calculation, every 100 ms, is called the tick time, which is not synchronized for the different flows. The FC distributes the estimated transport network capacity among the flows proportional to their estimated available Uu rate. Thus, a newly arriving flow gets its share of the transport network capacity, while the rate of ongoing flows decreases.

The short-term transport network congestions are taken into account by a coefficient. The calculated rate is translated to CA format and sent to the RNC. The estimated transport network link capacity ($BW_{estTN}^{Node\ B}$) and the sum of estimated Uu capacities (BW_{estUu}^{NodeB}) are provided by Node B-level aggregation. The estimated Uu rate (BW_{estUu}) is provided by the cell-level aggregation. The FC calculates the shaping rate as follows:

$$BW_{calc} = \min\left(\frac{BW_{estUu}}{BW_{estUu}^{NodeB}} \cdot BW_{estTN}^{NodeB}, BW_{estUu}^{NodeB}\right) \cdot Q_{Iub}, \qquad (9.1)$$

where Q_{Iub} is a coefficient that reacts fast on a detected congestion. Q_{Iub} is set to 0.5 if transport network congestion has been detected since the last tick; otherwise, it is increased by 0.05 at each tick. The maximum value for Q_{Iub} is 1.

9.3.5.1 Node B-Level Aggregation

The purpose of Node B-level aggregation is to estimate the available transport network capacity. The value of BW_{estTN}^{NodeB} is updated once per second. It is based on the number of transport network congestions detected by the per-flow ticks of the different flows during the last second, called $N_{IubCong}^{NodeB}$. As there is one tick in every 100 ms per flow, one flow can contribute, at most, ten flags during the evaluation period.

BW_{estTN}^{NodeB} is limited by a preconfigured minimum and maximum rate (called BW_{minHS}^{NodeB} and BW_{maxHS}^{NodeB}, respectively), which must be configured according to the transport network configuration. If $N_{IubCong}^{NodeB}$ is greater than or equal to a predefined threshold (e.g., 5) or the number of active flows (N_{flow}^{NodeB}), then the $BW_{estTN}^{Node\ B}$ is decreased by 2% of BW_{maxHS}^{NodeB}. If the $BW_{estTN}^{Node\ B}$ was not decreased in the last 10 seconds, then it is increased by 1% of R_{maxHS}^{NodeB} every second. The constants are determined by considering how frequently and to what extent the transport network capacity typically changes.

BW_{estUu}^{NodeB} equals the sum of BW_{estUu} for all flows in Node B. This variable is used for distribution of R_{estTN}^{NodeB} among the flows.

9.3.5.2 Uu Rate Estimation

To predict the Uu rate of a flow (BW_{estUu}), the possible peak Uu rate of the flow (BW_{peakUu}) is divided by the average number of competing flows (N_{flow}^{cell}). For long PQs, the estimated values are further reduced. The BW_{peakUu} is reported by the radio scheduler.

$$BW_{estUu} = Q_{t_{PQ}} \cdot \frac{BW_{peakUu}}{N_{flow}^{cell}} \qquad (9.2)$$

where $Q_{t_{PQ}}$ is a coefficient calculated from the estimated time to serve all PDUs in the PQ (t_{PQ}). The time t_{PQ} is calculated as follows:

$$t_{PQ} = \frac{b_{PQ}}{\frac{BW_{peakUu}}{N_{flow}^{cell}}} \qquad (9.3)$$

where b_{PQ} is the length of the PQ (in bits). The value of $Q_{t_{PQ}}$ is set to keep the delay at an appropriate level. Its value is 1 if t_{PQ} is smaller than 50 ms; it is 0 if t_{PQ} is larger than 150 ms and decreases linearly if t_{PQ} is between 50 ms and 150 ms.

9.3.6 Case Study: Per-Flow HSUPA Flow Control Solution

In [28,29], a rate-based per-flow HSUPA transport network congestion control solution is presented.

When transport network congestion is detected, FC dominates the behavior. During this time, flows are regulated according to an algorithm, which conforms with the AIMD property. Multiplication with a coefficient provides the multiplicative decrease, and a constant increase in rate after reduction provides the additive increase property. The AIMD property is met only for the serving cell behavior. However, a MAC-e PDU is normally received in the serving cell with a higher probability; thus end-user fairness is dominated by serving-cell behavior.

9.3.6.1 Transport Network Congestion Detection

The transport network congestion detection part of the algorithm is performed whenever a data frame arrives at the RNC. Two different congestion detection methods are used: FSN gap detection and DDD.

9.3.6.2 Bit Rate Calculation

Whenever a TCI is received by Node B, it triggers a *congestion action* by the Flow Control entity. Depending on the severity of the congestion, a reduce request with a certain coefficient is issued. Different coefficients are applied in the case of soft and hard congestions. A different coefficient can also be used for the first TCI received for a flow. The motivation for this is that when there is no TNL congestion at all, the Uu scheduler increases the granted bit rate with a higher speed. Consequently, these UEs can potentially overload the transport network more so than UEs already limited by the effect of flow control.

Depending on whether it is a serving-cell flow or a non-serving-cell flow, the rate reduce request is issued to the Uu scheduler or to the Frame dropping functionality.

9.3.6.3 Congestion Action in the Serving Cell

Until the first rate reduce request is received, the Uu scheduler behavior is not affected by Flow Control at all. Based on air interface conditions, hardware resources, and Happy Bit information, the Uu scheduler determines the granted bit rate represented by the AG.

Upon receiving a rate reduce request, the scheduler decreases the granted bit rate by sending a new AG. Additionally, when a rate reduce request is issued for a flow, the scheduler is not increasing the absolute grant of that flow with more than a predefined rate (e.g., 20–200 kbps/s). The value is determined based on typical transport network bit rate and the typical number of parallel flows.

The Uu scheduler maintains an allowed bit rate according to the above algorithm, and the bit rate represented by the sent AG must be lower than this allowed bit rate. Note that there are only a certain number of different possible AG values to send. Consequently, the reduction in allowed bit rate and the reduction in AG may be different, according to the granularity of the possible AG values.

9.3.6.4 Congestion Action in the Non-Serving Cell

A TCI received in the non-serving cell will not trigger rate reduction by RG because a MAC-e PDU is received in the best cell (usually the serving cell) with a higher probability. Consequently, if we reduced the bit rate due to transport network limitations in the non-serving cell, we might reduce the bit rate of the end user unnecessarily. However, congestion action still needs to be taken; thus, a fraction of the received MAC-es PDUs are dropped. If these PDUs are not received in the serving cell, then RLC AM still retransmits these missing PDUs.

A *forwarding coefficient* determines the probability that a received MAC-e PDU is forwarded. It is 100% at initialization, and each received reduce request decreases it. It is gradually increased to 100% afterward. Note that this behavior does not conform to AIMD, but a MAC-e PDU is normally received in the serving cell with a higher probability; thus, end-user fairness is dominated by serving-cell behavior.

9.3.7 Other Flow Control Solutions

In [30], the authors introduce *cross-layer backpressure* in the RNC, which allows good transport network utilization when the transport network bottleneck buffer is in the RNC. Using CA Credits calculation the algorithm not only takes into account the information provided by Node B, but also the state of the AAL2/ATM level queues in the RNC. Thus, they assume that the Iub transport network consists of one link, and the state of the AAL2 buffer at its input is known. In this way they can perform a pro-active, aggregated congestion control. The scope of the algorithm is to control the load of the low-priority AAL2 buffer so that the delay on the AAL2 layer is not exceeding the maximum allowed delay.

9.4 HSPA Framing Protocol

The Iub/Iur Framing Protocol defines the DFs and CFs to be used over the WCDMA RAN transport network to transmit user data and transport network layer control information.

For HSDPA, the DFs and CFs are defined in [7]. Two types of HS-DSCH Frame Protocols exist for the HS-DSCH data transfer procedure: (1) HS-DSCH Frame Protocol Type 1, including HS-DSCH Data Frame Type 1 (Figure 9.9) and HS-DSCH Capacity Allocation Type 1 Control Frame (Figure 9.10); and HS-DSCH Frame Protocol Type 2, including HS-DSCH Data Frame Type 2 (Figure 9.11) and HS-DSCH Capacity Allocation Type 2 Control Frame (Figure 9.12). Type 1 (often described without explicitly stating the type) is the original solution. FP Type 2 was introduced to support higher bit rates; it supports flexible RLC, higher PDU sizes, octet aligned PDUs, and multiplexing of several logical channels. It also decreases transport network overhead compared to Type 1. HS-DSCH Frame Protocol Type 2 is selected if the HS-DSCH MAC-d PDU Size Format Information Element (IE) in NBAP (Node B Application Part) is present and set to "Flexible MAC-d PDU Size."

HS-DSCH Capacity Allocation Control Frame (CA CF) is used by Node B to control the user data flow from the RNC. CA CF informs the RNC about *TN Congestion Status*, about the priority of the flow (*CmCH-Pi*), and about the *maximum PDU length* allowed. It defines the amount of PDUs (*Credits*) that can be sent by the RNC. The *Interval* indicates the time interval during which the Credits are granted. By setting the *Repetition Period* to a high value, *Credits* are regenerated periodically; the length of the period is

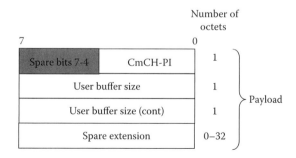

Figure 9.9 HS-DSCH data frame Type 1 structure. (*Source*: From 3GPP, TS 25.435 V7.1.0, UMTS UTRAN Iub Interface User Plane Protocols for Common Transport Channel Data Streams.)

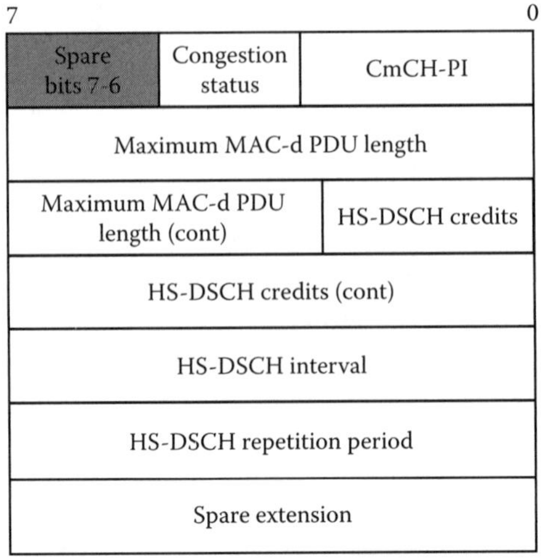

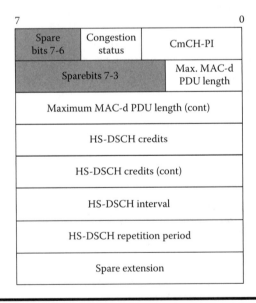

Figure 9.10   Capacity allocation control frame Type 1 payload structure. (*Source*: From 3GPP, TS 25.435 V7.1.0, UMTS UTRAN Iub Interface User Plane Protocols for Common Transport Channel Data Streams.)

Figure 9.11   HS-DSCH data frame Type 2 structure. (*Source*: From 3GPP, TS 25.435 V7.1.0, UMTS UTRAN Iub Interface User Plane Protocols for Common Transport Channel Data Streams.)

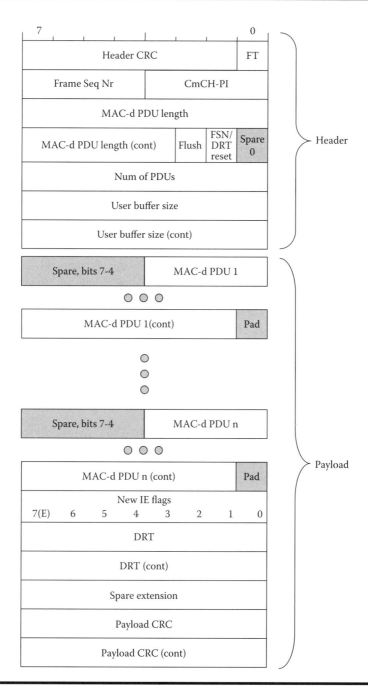

Figure 9.12 Capacity allocation control frame Type 2 payload structure. (*Source*: From 3GPP, TS 25.435 V7.1.0, UMTS UTRAN Iub Interface User Plane Protocols for Common Transport Channel Data Streams.)

defined by the *Interval*. The *Repetition Period* can be set to infinite. In this case, the CA CF represents a shaping bit rate because, in every Interval, the defined amount of PDUs can be sent until a new CA CF is received by the RNC. Special values can indicate zero or unlimited allocation.

A Type 2 CF allows longer PDU length (up to 1504 octets) than Type 1 because it has a longer *Maximum PDU length* field, and this length is defined in octets instead of bits. However, it requires octet-aligned PDUs, while Type 1 does not. A Type 2 CF supports flexible PDU sizes because the scheduling is defined in octets per interval instead of PDUs per interval. The CF fields are similar; but while for Type 1, the actual number of PDUs is restricted in each interval, for Type 2 the amount of octets is restricted. The number of octets is calculated in the shaper as the Maximum PDU length times the Credits. This, for example, allows sending two times *Credits* PDUs, if the actual PDU size is half of the maximum. This requires a more complex shaping algorithm, which was necessary to make the bit rate represented by the shaping independent of the actual PDU length in the case of flexible PDU size.

The HS-DSCH DF transfers PDUs between the Node B and RNC entities. The DF fields indicate the *FSN*, the priority of the flow *CmCH-Pi*, the *User Buffer Size* (UBS), and optionally the *Delay Reference Time (DRT)*. It contains the transmitted PDUs and information about their number and length. CRC provides error checking functionality for header and payload, and Spare Extension allows extending the frame in the future in a backward-compatible way.

The Type 1 DF allow only PDUs of the same size to be sent in the same DF. It defines PDU size in bits; and for octet-aligned PDUs, it has 8-bit *Spare* and *Pad* bits in total. The Type 2 data frame requires octet-aligned PDUs but allows PDUs of different size and logical channel to be sent in the same data frame. No extra bits per PDU are required, as in the case of Type 1. A Type 2 DF became necessary to allow higher PDU size and different-sized PDUs in the same data frame. By allowing different PDU sizes in a single data frame, the total overhead became smaller. (With flexible RLC, PDUs can be smaller than the maximum size to eliminate padding and the need for multiplexing parts of several SDUs to the same PDU.) The frame structure in Figure 9.11 is simplified; only the fields relevant for FDD and dedicated transmission are shown.

The HS-DSCH Capacity Request control frame (Figure 9.13) can be optionally sent by the RNC to Node B to indicate the *User Buffer Size* and request the sending of CA CF. As the data frame also includes this information, the use of this control frame is optional.

For E-DCH, the data and control frames are defined in [12].

The RNC uses the TNL Congestion Indication control frame (TCI CF, Figure 9.14) to signal detected transport network congestion to Node B. It informs Node B about *Congestion Status*, which can be no congestion,

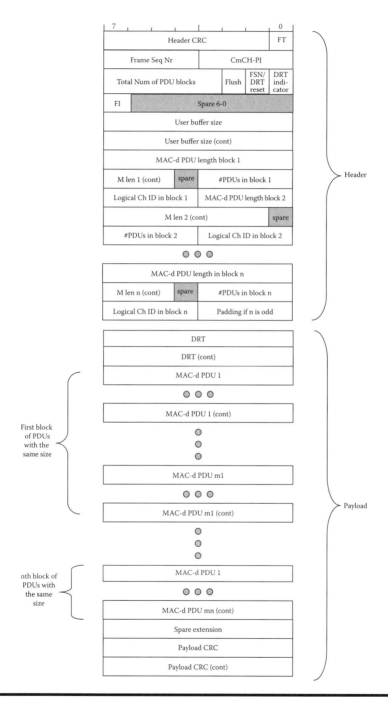

Figure 9.13 HS-DSCH Capacity Request payload structure. (Source: From 3GPP, TS 25.435 V7.1.0, UMTS UTRAN Iub Interface User Plane Protocols for Common Transport Channel Data Streams.)

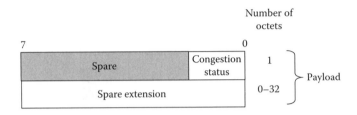

Figure 9.14 Structure of the TNL congestion indication (TCI) control frame. (*Source*: **From 3G PP, TS 25. 427 V 7.5.0, UTRAN Iub/Iur Interface user Plane Protocol for DCH Data Streams (Release 7). October 2007.)**

congestion due to loss or due to delay build-up. Upon receiving a TCI CF indicating congestion, Node B should reduce the bit rate of that user.

The E-DCH UL data frame (Figure 9.15) transfers MAC-es PDUs from the Node Bs to the RNC. Its fields indicate FSN Connection Frame Number (CFN), and Subframe Number. The CFN indicates the time when the data was received. Several subframes can be multiplexed to the same data frame, each having its separate header information. More than one subframe can be used in case of 2-ms TTI, when the data from up to five TTIs can be multiplexed in a single data frame. For each subframe, the Number of HARQ Retransmissions, the Subframe Number, the Data Description Indicator (DDI), and the Number of MAC-d PDUs (*N*) are indicated. The CFN and the last Subframe Number together can be used to measure dynamic delay; thus, no direct indication of *DRT* is needed. The data frame contains the MAC-es PDUs received in Node B. In the simplest case, the data frame contains a single subframe and a single MAC-es PDU. Cyclic Redundancy Check (CRC) provides error checking functionality for header and payload, and Spare Extension allows extending the frame in the future in a backward-compatible way.

9.5 TNL Overhead

We define the transport network overhead as the number of octets needed to be transmitted over the transport network, divided by the transmitted user-level IP octets within. It depends on the size of the data frame and the used transport network protocols. Apart from the transport protocol overhead, the overhead value also contains the Iub/Iur Framing Protocol and the RLC overhead. For E-DCH, the MAC-es overhead is also considered.

For the UDP/IP/Ethernet solution, the overhead depends much more on the data frame size, as in the case of AAL2/ATM. This is because for AAL2/ATM, most headers are on segmented PDUs, resulting in a fixed percentage, while for UDP/IP/Ethernet the headers are large, but apply to the data frame only once. If MLPPP is used as L2 for UDP/IP, then IP

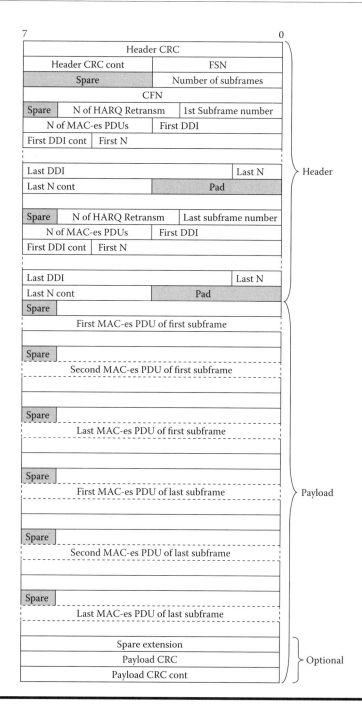

Figure 9.15   E-DCH UL data frame structure. (*Source*: From 3G PP, TS 25. 427 V 7.5.0, UTRAN Iub/Iur Interface user Plane Protocol for DCH Data Streams (Release 7). October 2007.)

Table 9.2   HSDPA Transport Network Overhead Examples

RLC PDU Size (octets)	DF Type	PDUs per DF	ATM OH	IP OH
42	1	10	1.32	1.27
82	1	5	1.28	1.24
302	2	2	1.23	1.14

header compression becomes possible and overhead in the case of small throughput is significantly reduced.

In the case of HSDPA, the transport network overhead primarily depends on the number of MAC-d PDUs carried in the HS-DSCH data frame. User data is segmented into RLC PDUs. For user data, the MAC-d layer does not add any headers so the RLC PDU is identical to the MAC-d PDU. RLC PDU size can take on one of the following fixed values: 42 or 82 octets, or it can be of flexible size.

Table 9.2 provides examples for HSDPA transport network overhead. The size of the data frame depends on the RLC PDU size and on the typical number of PDUs in a data frame. This is determined by the CA calculation and the CA shaping, and that is independent from air interface scheduling. For this example calculation, we assumed a given typical number of PDUs per data frame. Also, for the flexible PDU case, we assumed a maximum size of 302 octets and that all PDUs in the data frame are of the same

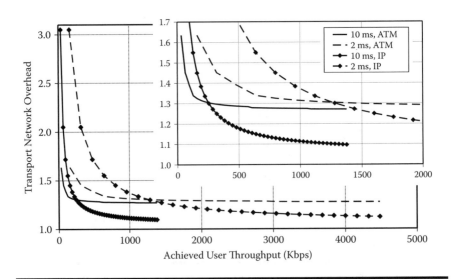

Figure 9.16   HSUPA transport network overhead as a function of the achieved user throughput.

maximum size. Because DRT use is not mandatory in all data frames, we assumed that the DRT is not present in the data frames.

In the case of HSUPA, the data frame size depends mainly on the achieved user throughput and on the TTI. This is because whenever a MAC-es frame is received, it is put into a data frame and transmitted over the transport network. For 2-ms TTI, the MAC-es PDUs from one or more TTIs may be bundled into one data frame before being transferred [12]. With bundling up to five PDUs, the Transport Network overhead in the case of 2-ms TTI can be decreased very close to the overhead in the case of 10-ms TTI. In the examples in this section, no bundling was assumed for 2-ms TTI. Figure 9.16 depicts the overhead in the case of AAL2/ATM and UDP/IP/Ethernet transport network for both 10-ms and 2-ms TTI. There is very large overhead in the case of small throughput and UDP/IP/Ethernet transport because the data frame size is very small for small bit rates. The aforementioned MLLPP with header compression can decrease this overhead if it becomes necessary.

References

[1] 3GPP, TS 25.322 V7.9.0, Radio Link Control (RLC) Protocol Specification (Release 7), 2009.

[2] 3GPP, TS 25.308 V7.9.0, UTRA High Speed Downlink Packet Access (HSDPA) Overall Description, 2008.

[3] S. Parkvall, J. Peisa, J. Torsner, M. Sågfors, and P. Malm. WCDMA enhanced uplink–principles and basic operation. In *VTC 2005—Spring,* pp. 1411–1415, 2005.

[4] J. Bergman, M. Ericson, D. Gerstenberger, B. Göransson, J. Peisa, and S. Wager. HSPA Evolution–Boosting the performance of mobile broadband access. *Ericsson Review,* 85(1): 32–37, 2008.

[5] S. Parkvall et al. WCDMA evolved–high-speed packet-data services. *Ericsson Review,* (2), 2003.

[6] 3GPP, TS 25.213 v7.6.0, Spreading and Modulation (FDD) (Release 7), September 2008.

[7] 3GPP, TS 25.435 V7.1.0, UMTS UTRAN Iub Interface User Plane Protocols for Common Transport Channel Data Streams, 2006.

[8] Ericsson White Paper. High-Speed Technologies for Mobile Backhaul. October 2008.

[9] 3GPP, TSG-RAN3 Meeting #43, Tdoc R3–041115, Introduction of an HS-DSCH data frame "frame number sequence" field. *Prague, Czech Republic,* August 2004.

[10] 3GPP, TSG-RAN3 Meeting #43, Tdoc R3–041118, Introduction of an HS-DSCH data frame "delay RNC reference time (DRT)" extension. *Prague, Czech Republic,* August 2004.

[11] 3GPP, TSG-RANWG3 Meeting #47, Tdoc R3–050763, Transport network congestion detection and control. *Athens, Greece,* May 2005.

[12] 3GPP, TS 25.427 v7.5.0, UTRAN Iub/Iur Interface User Plane Protocol for DCH Data Streams (Release 7), October 2007.

[13] 3GPP, TR 25.902 v7.1.0, Iub/Iur Congestion Control (Release 7), March 2007.

[14] 3GPP, TS 25.309 v6.6.0, FDD Enhanced Uplink; Overall Description; Stage 2 (Release 6), April 2006.

[15] 3GPP, TS 25.853 V4.0.0, Delay Budget within the Access Stratum, 2001.

[16] B. Karlander, S. Nádas, S. Rácz, and J. Renius. AAL2 switching in the WCDMA radio access network. *Ericsson Rev.,* 79(3): 114–123, 2002.

[17] S. Nádas, S. Rácz, Z. Nagy, and S. Molnár. Providing congestion control in the Iub transport network for HSDPA. In *Globecom 2007*, pp. 5293–5297, 2007.

[18] R. Jain. Congestion control and traffic management in ATM networks: Recent advances and a survey. *Comput. Netw. ISDN Syst.,* 28(13): 1723–1738, 1996.

[19] P.J. Ameigeiras. *Packet Scheduling and Quality of Service in HSDPA.* Ph.D. thesis, Aalborg University, 2004.

[20] S. Nádas, S. Rácz, S. Malomsoky, and S. Molnár. Connection admission control in all-IP UTRAN transport network. *Telecommun. Syst.,* 28: 9–29, January 2005.

[21] C. Fraleigh, F. Tobagi, and C. Diot. Provisioning IP backbone networks to support latency sensitive traffic. In *Proc. of IEEE Infocom,* pp. 375–385, 2003.

[22] J.S. Kaufman. Blocking in a shared resource environment. *IEEE Trans. Commun.,* 29(10): 1474–1481, 1981.

[23] J.W. Roberts. A service system with heterogeneous user requirements. In *Performance of Data Communications Systems and Their Applications.* North-Holland. 1981, pp. 423–431.

[24] S. Rácz, B.P. Gerö, and G. Fodor. Flow level performance analysis of a multiservice system supporting elastic and adaptive services. *Performance Eval.,* 49: 451–469, 2002.

[25] K. Ramakrishnan, S. Floyd, and D. Black. The Addition of Explicit Congestion Notification (ECN) to IP. RFC 3168 (Proposed Standard). September 2001. Available at http://www.icir.org/floyd/edn.html.

[26] P. Pályi, S. Rácz, and S. Nádas. Fairness—Optimal initial shaping rate for HSDPA transport network congestion control. In *11th IEEE Int. Conf. Commun. Syst. 2008,* pp. 1415–1421, Guangzhou, P.R. China, November 2008.

[27] D. Chiu and R. Jain. Analysis of the increase/decrease algorithms for congestion avoidance in computer networks. *Comput. Netw. and ISDN,* 17(3): 1–14, June 1989.

[28] S. Nádas, Z. Nagy, and S. Rácz. HSUPA transport network congestion control. In *4th IEEE Broadband Wireless Access Workshop,* December 2008.

[29] S. Nádas and S. Rácz. HSUPA transport network congestion control. *EURASIP Journal on Wireless Communications and Networking,* vol.2009, Article ID 924096, 10 pages, 2009. doi: 10.1155/2009/924096.

[30] L. Bajzik, L. Körössy, K. Veijalainen, and C. Vulkán. Cross-layer backpressure to improve HSDPA performance. In *2006 IEEE 17th Int. Symp. Personal, Indoor and Mobile Radio Communications,* pp. 1–5, Helsinki, September 2006.

Chapter 10

Performance Evaluation of HSDPA/HSUPA Systems

Dan Keun Sung and Junsu Kim

Contents

10.1 Introduction

The performance of HSDPA/HSUPA systems is of interest to research engineers to understand the characteristics of the systems and to further enhance performance. Link- and system-level simulations are pratical approaches to evaluate system performance. In this chapter the simulation methodology for the HSDPA/HSUPA systems is described and their performance results are presented.

10.2 Overall Procedure for Performance Evaluation

The overall system performance of HSDPA/HSUPA systems can be evaluated through two simulation steps: one is a *link-level simulation* and the other is a *system-level simulation*. Link-level simulation evaluates the performance of a single air-link between a single transmitter and a single receiver. The performance metrics of link-level simulation include bit error rate (BER), frame error rate (FER), and block error rate (BLER) for varying SINR (signal-to-noise ratio) values. The coding and decoding procedure of the physical layer and mobile radio channels are considered in link-level simulation. The main objective of link-level simulation is to evaluate the performance of a physical-layer coding chain. Therefore, the signal processing functionalities related to radio frequency (RF) blocks, A/D or D/A converting, and sampling functions can be assumed to be ideal.

Unlike link-level simulation, system-level simulation considers multiple transmitters and multiple receivers to analyze the various performance metrics, including throughput, delay, delay variation, etc. It also considers radio resource management schemes, including wireless scheduling, channel allocation, and rate/power control, which are the key functional blocks in determining overall system performance. The link-level simulation results, such as BER, FER, and BLER performance, are utilized in system-level simulation to model each air-link between a transmitter and a receiver. In terms of cell modeling, there is a Node B at the center of the cell and multiple UEs (user equipment) are distributed according to a random distribution. Therefore, a single transmitter (Node B) and multiple receivers (UEs) are implemented for the HSDPA system-level simulator, while multiple transmitters (UEs) and a single receiver (Node B) are considered in the HSUPA system-level simulator.

10.3 Link-Level Simulator

10.3.1 Common Function Blocks

The structure of a link-level simulator is based on a coding chain defined in the technical specifications of the HSDPA and HSUPA systems [1]. Because the objective of the link-level simulator is to evaluate the link performance of a traffic channel, the coding chains of HS-PDSCH for HSDPA and E-DPDCH for HSUPA are evaluated through link-level simulation. Figure 10.1 shows the overall procedure for HSDPA/HSUPA link-level simulators. Note that the function blocks marked "D" are implemented for HSDPA only. There is a transmitter at the Node B side and a receiver at the UE side in the HSDPA link-level simulator, and vice versa in the HSUPA link-level simulator.

To evaluate the performance of the coding chain, a random binary sequence is generated at the start of simulation. The performance metrics, including BER, FER, and BLER, are measured by comparing the input binary sequence with the decoded binary sequence after decoding. A detailed description for each block is omitted in this chapter. There is a counterpart block at the transmitter for each block at the receiver.

Link-level simulations are performed for various types of mobile radio channels, including AWGN, Rayleigh, ITU defined pedestrian A (PA), pedestrian B (PB), and vehicular A (VA) [2]. In particular the ITU-defined mobile radio channels represent the multi-path fading channels of which tapped-delay-line parameters are shown in Table 10.1. According to the assumption of link-level simulation, each single chip out of spreading sequence can be represented by a single real number because sampling is assumed to be perfect. Therefore, the relative delay of each path is quantized into an integer multiple of chip durations in the link-level simulation. For example, the base chip duration of WCDMA is 260 *ns* because the chip rate is 3.84 Mcps. Then the PA channel in Table 10.1 is modeled as a three-tap multi-path channel because the second and third taps are not resolvable. In other words, the relative delays of the PA channel are converted into an integer multiple of chip durations as follows: tap 1 → zero chip delay; tap 2 → one chip delay; tap 3 → one chip delay, tap 4 → 2 chips delay.

Figure 10.2 provides an example of link-level multi-path processing for the ITU-PA channel. Although four paths are defined in the PA channel, taps 1 and 2 have the same delay in terms of chip duration, as shown in the figure. The spread sequence for symbol *k* is delayed and distorted through the multi-path channel. Finally, the receiver obtains an aggregated sequence containing all the components of the different propagation paths. Multiple fingers at the RAKE receiver perform correlations to gather the signal energy of symbol *k*.

Figure 10.3 illustrates a RAKE receiver structure. The receiver can receive multiple distorted and delayed replicas of the transmitted radio frame due to the multi-path fading channel that causes inter-symbol interference (ISI). The *multi-path searcher and tracker* resolves each replica of the transmitted signal at the first step of the receiver, as shown in Figure 10.3. Using fingers, it is possible to resolve more paths. The distorted phase of each path is compensated by the *phase compensator* using the phase distortion

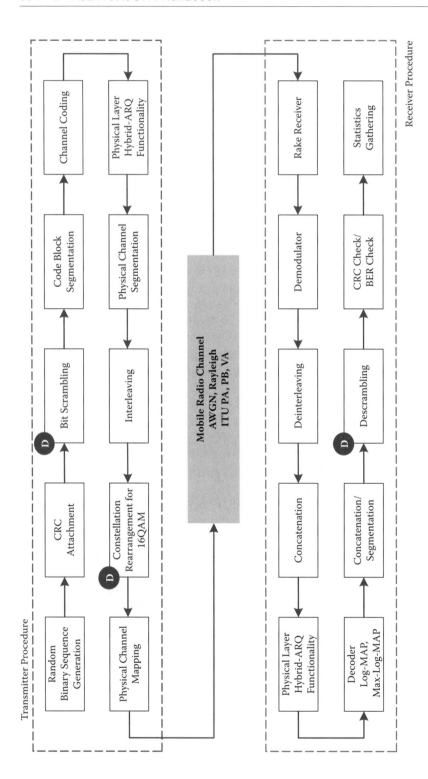

Figure 10.1   Procedure for HSDPA/HSUPA link-level simulation. D denotes HSDPA only.

Table 10.1   Parameters for ITU Mobile Radio Channels

	PA		PB		VA	
Tap	Relative Delay (ns)	Average Power (dB)	Relative Delay (ns)	Average Power (dB)	Relative Delay (ns)	Average Power (dB)
1	0	0.0	0	0.0	0	0.0
2	110	−9.7	200	−0.9	310	−1.0
3	190	−19.2	800	−4.9	710	−9.0
4	410	−22.8	1200	−8.0	1090	−10.0
5	–	–	2300	−7.8	1730	−15.0
6	–	–	3700	−23.9	2510	−20.0

information estimated by a *channel estimator*. Because the main objective of this link-level simulation is to evaluate the performance of the coding chain shown in Figure 10.1, we can assume a perfect channel estimation at the RAKE receiver. Finally, multiple *correlators* produce de-spread signals, and they are combined by a *multi-path combiner*. The combined signal is provided to the remainder of a decoding chain to obtain the decoded binary sequence.

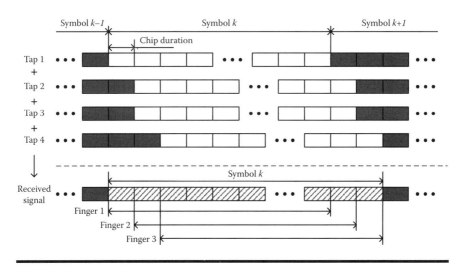

Figure 10.2   Example of multi-path processing with ITU-PA.

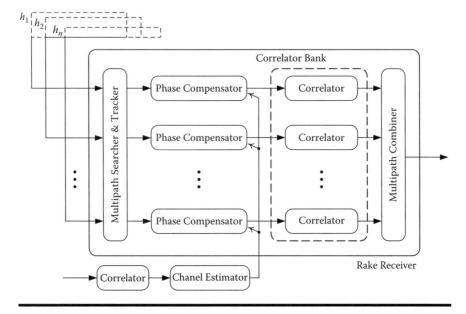

Figure 10.3 RAKE receiver structure.

10.3.2 HSDPA Link-Level Simulation

The HSDPA system supports QPSK and 16-QAM modulations, convolutional and turbo codes for channel coding, and multi-code transmission. There are various combinations of modulation and coding schemes and a number of multi-codes. Each combination is characterized as a channel quality indicator (CQI) [3]. The *CQI value* represents the *adaptive modulation and coding* (AMC) *level* generally used in wireless communications. Because a higher CQI value indicates a higher modulation order and weaker channel coding, it is possible to achieve a successful transmission with the higher CQI value only when the wireless channel status is sufficiently good. Therefore, the CQI value can be used as a good indicator representing the channel quality of each UE.

The HSDPA system defines a different CQI mapping table for a different category of UEs. The category is determined according to the capability of UE [4]. Categories 1 to 6 support up to 16-QAM and five multi-code transmission, Categories 7 and 8 support up to 16-QAM and ten multi-code transmission, Category 9 supports up to 16-QAM and 12 multi-code transmission, Category 10 supports up to 16-QAM and 15 multi-code transmission, and Categories 11 and 12 support QPSK only and up to five muti-code transmission. Among the twelve categories, Category 10 presents a maximum data rate of approximately 12.8 Mbps. Table 10.2 shows the representative CQI values for UE Category 10 [3]. The CQI values from 1 to 30 are defined

Table 10.2   Representative CQI Values for UE Category 10

CQI Value	Transport Block Size	Number of HS-PDSCH	Modulation Type
0	N/A	Out of range	
1	237	1	QPSK
. . .	. . .	. . .	. . .
7	650	2	QPSK
. . .	. . .	. . .	. . .
10	1262	3	QPSK
. . .	. . .	. . .	. . .
13	2279	4	QPSK
. . .	. . .	. . .	. . .
15	3319	5	QPSK
16	3565	5	16-QAM
. . .	. . .	. . .	. . .
23	9719	7	16-QAM
24	11418	8	16-QAM
25	14411	10	16-QAM
26	17237	12	16-QAM
27	21754	15	16-QAM
. . .	. . .	. . .	. . .
30	25558	15	16-QAM

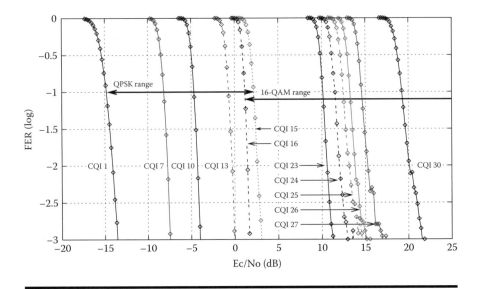

Figure 10.4 FER performance of the HSDPA system under a Rayleigh fading channel.

for data transmission, and CQI 0 indicates out-of-range status. For each CQI value, various parameters for the coding chain depicted in Figure 10.1 are determined. Therefore, it is possible to analyze the link performance for 30 CQI values.

Figure 10.4 depicts the FER performance of the HSDPA system for various CQI values under a Rayleigh fading channel. Higher values of CQI require more energy to achieve a certain level of FER. However, CQI 15 requires more energy than CQI 16 in a Rayleigh fading channel. CQI 15 is the maximum CQI for QPSK modulation, and CQI 16 is the minimum CQI for 16-QAM modulation. Therefore, CQIs 15 and 16 are in the overlapping area of the dynamic ranges of the QPSK and 16-QAM. If the required FER is set to 10%, a transition from QPSK to 16-QAM occurs at approximately 3 dB of Ec/No. This result is used in the system-level simulations for CQI selection and ACK/NACK decision.

10.3.3 HSUPA Link-Level Simulation

One radio frame for E-DPDCH contains five sub-frames, and each sub-frame has three slots. Therefore, one radio frame consists of 15 slots. The HSUPA system defines eight different slot formats that control the transmission rate by adjusting the spread factor (SF) from 2 to 256. Table 10.3 shows the slot formats [5]. Using the slot format shown in Table 10.3, several transmission format combinations (TFCs) are possible. In particular, E-DCH Category 6

Table 10.3   E-DPDCH Slot Formats

Slot Format	Channel Bit Rate (kbps)	SF
0	15	256
1	30	128
2	60	64
3	120	32
4	240	16
5	480	8
6	960	4
7	1920	2

allows up to four multi-code transmission for both 2-ms and 10-ms TTI [4]. Table 10.4 shows the TFC for Category 6. TFCs 0 to 6 use a single code transmission. TFC 7 transmits two SF4 channels, TFC 8 transmits two SF2 channels, and TFC 9 transmits two SF2 channels and two SF4 channels. In link-level simulation, the link performance can be analyzed for each

Table 10.4   Transmission Format Combinations (TFCs) for Category 6

Index	Bit Rate (kbps)	Combination	Bits/Frame	Bits/Sub-frame	Bits/Slot
0	15	SF256	150	30	10
1	30	SF128	300	60	20
2	60	SF64	600	120	40
3	120	SF32	1200	240	80
4	240	SF16	2400	480	160
5	480	SF8	4800	960	320
6	960	SF4	9600	1920	640
7	1920	2×SF4	19200	3840	1280
8	3840	2×SF2	38400	7680	2560
9	5760	2×SF2 + 2×SF4	57600	11520	3840

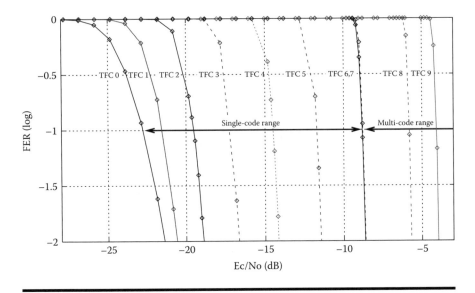

Figure 10.5 **FER performance of the HSUPA system under a Rayleigh fading channel.**

TFC in Table 10.4 using the parameters for frame (10 ms) or sub-frame (2 ms).

Figure 10.5 depicts the FER performance of the HSUPA system under a Rayleigh fading channel with 10-ms TTI. TFC 6 has a single code transmission with spreading factor 4 and TFC 7 has a two-code transmission with spreading factor 2. Both TFCs have nearly identical FER performance under a Rayleigh fading channel. However, TFC 7 transmits twice more data than TFC 6. Therefore, TFC 7 is selected rather than TFC 6 in system-level operation.

10.4 System-Level Simulation

10.4.1 System-Level Simulator Structure

The entire structure and major function blocks for the HSDPA system-level simulator is described here. Figure 10.6 represents the structure of a system-level simulator for the HSDPA system. Unlike the link-level simulator, the system-level simulator focuses on the system-level performance among multiple transmitters and multiple receivers in a multi-cell environment. The functional blocks of the system-level simulator can be grouped into three parts: initialization, transmitters (Node B), and receivers (UE).

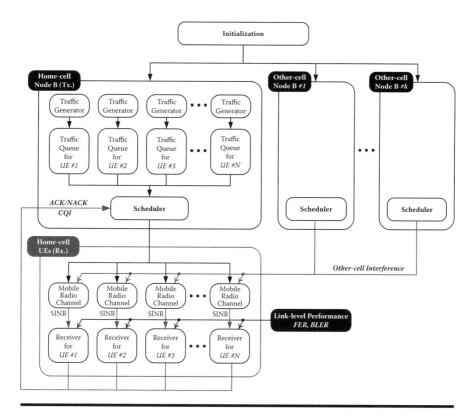

Figure 10.6   System-level simulator structure for the HSDPA system.

10.4.1.1 Initialization

10.4.1.1.1 Topology

The geographical positions of Node Bs and UEs are fixed throughout the simulation. Node Bs have deterministic locations while the positions of UEs are randomly distributed. Generally, a hexagonal cell structure is considered, and one Node B in a single-cell environment, seven Node Bs in a single-tier environment, and 19 Node Bs in a two-tier environment are implemented. Figure 10.7 shows a multi-cell model representing a home-cell, the first-tier, and the second-tier cellular environments [6]. A home-cell Node B is located at the center.

Positioning UEs follow either a uniform distribution or a Gaussian distributions. A Gaussian distribution model effectively represents an environment in which more UEs are concentrated in a cell center area. On the other hand, if UEs are located according to the uniform distribution, then more users are located in a cell boundary area. Generally, the uniform distribution model yields much poorer system-level performance than the Gaussian distribution model.

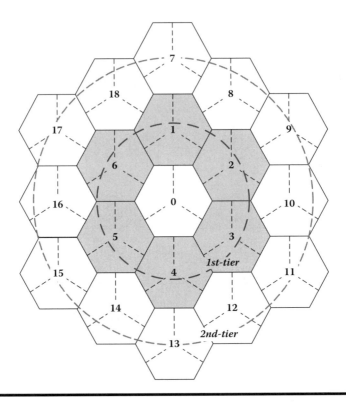

Figure 10.7 Multi-cell model.

10.4.1.1.2 Static Channel (Large-Scale Fading) Parameters

Once the locations of Node Bs and UEs are set, then static channel parameters are also determined. The static channel parameters, which represent the mobile radio channel components depending only on the geographical environment, include path loss, shadowing, antenna gain, etc.

A path-loss (PL) model widely used in urban and suburban areas can be expressed as [2,7]

$$PL[\text{dB}] = 40(1 - 4 \times 10^{-3} h_{BS}) \log_{10}(R) - 18 \log_{10}(h_{BS}) + 21 \log_{10}(f) + 80 \tag{10.1}$$

where R is the distance in kilometers from Node B to a UE, f is the carrier frequency in megahertz, and h_{BS} is the height of Node B above rooftop. For HSDPA/HSUPA with 2-GHz carrier frequency, if h_{BS} is set to 15 m above rooftop, then the path-loss formula becomes

$$PL(R)[\text{dB}] = 128.1 + 37.6 \log_{10}(R) \tag{10.2}$$

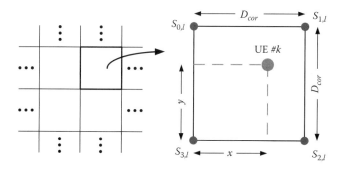

Figure 10.8   Shadowing factor grid.

The shadowing factor (SF) can be modeled using a log-normal distribution, and its standard deviation depends on the given propagation scenario. In urban or suburban macrocells, the standard deviation is set to 8 dB [7]. Because the shadow fading is caused by geographical environment, including buildings, hills, trees, and so on, the shadowing factors of the adjacent UEs are correlated. The longest distance between the UEs within which the correlation is maintained is called the *decorrelation distance*. Therefore, the shadowing factor of each UE should be generated according to the predefined decorrelation distance [7]. Figure 10.8 shows a shadowing factor grid to generate the correlated shadowing factor. The entire plane, including the home-cell shown in Figure 10.7, can be divided into multiple squares. The length and height of a single square are equal to the decorrelation distance D_{cor}, as shown in the right figure, and UE #k is located in the square. $S_{0,l}, \cdots, S_{3,l}$ at the four nodes of the square are the Gaussian distributed random variables with a standard deviation mentioned above. The $S_{n,l}$, which denotes the shadowing factor at node n and Node B l, is generated as $aZ_n + bZ_l$, where Z_n is a Gaussian random number for each node, Z_l is a Gaussian random number for each Node B, and $a^2 = b^2 = 1/2$. Then, the shadowing factor of UE #k can be calculated using the following interpolation [7]:

$$SF[\text{dB}] = \sqrt{1 - \frac{x}{D_{cor}}} \left(S_{0,l} \sqrt{\frac{y}{D_{cor}}} + S_{3,l} \sqrt{1 - \frac{y}{D_{cor}}} \right)$$

$$+ \sqrt{\frac{x}{D_{cor}}} \left(S_{1,l} \sqrt{\frac{y}{D_{cor}}} + S_{2,l} \sqrt{1 - \frac{y}{D_{cor}}} \right) \quad (10.3)$$

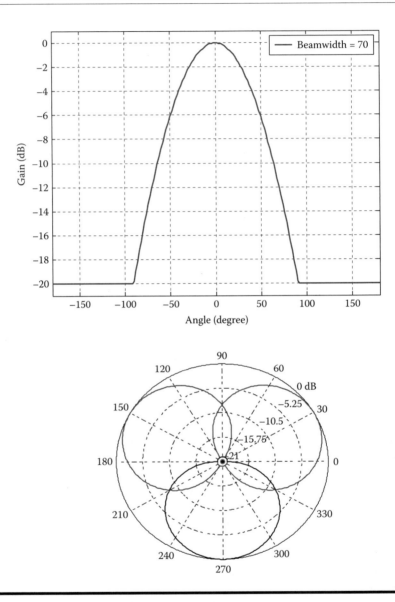

Figure 10.9   Antenna pattern for sectorization.

Figure 10.9 illustrates an antenna pattern for sectorization. The antenna gain $AG(\theta)$, shown in the upper figure of Figure 10.9, is generated as

$$AG(\theta)[\text{dBi}] = -\min\left\{12\left(\frac{\theta}{\theta_{3dB}}\right)^2, A_m\right\} \qquad (10.4)$$

where θ is the relative direction to the main robe, $-180° \leq \theta \leq 180°$, θ_{3dB} is the 3-dB beamwidth, and $A_m = 20$ dB is the maximum attenuation. The typical value of θ_{3dB} is $70°$ for a three-sector antenna system [6]. The lower figure of Figure 10.9 shows a sectorization using the antenna pattern of Equation (10.4) in a cell. The directions of main lobes are $30°$, $150°$, and $270°$.

Finally, the large-scale fading gain G_{LS} at a UE is expressed as

$$G_{LS}[dB] = -PL - SF + AG \qquad (10.5)$$

Note that the path loss and shadowing factor in Equations. (10.1) and (10.3) have positive values, and the antenna gain in Equation (10.4) has a negative value. Therefore, the large-scale fading gain G_{LS} has a negative value. G_{LS} is maintained unless the UE's position is changed.

10.4.1.2 Transmitter (Node B)

10.4.1.2.1 Traffic Generator and Traffic Queue

At Node B, there is a traffic generator and a traffic queue for each UE. The traffic types generally used for system-level simulations include full-queue, HTTP, FTP, VoIP, near-real-time video (NRTV), gaming, e-mail, and so on [6,7]. Unlike the other traffic types, full-queue traffic represents that a sufficient amount of traffic is always backlogged in the traffic queue. If full-queue traffic is applied, because there are no effects of queuing dynamics on the system performance, it is desirable to evaluate the maximum performance bound of the system.

The traffic generator produces packets containing the information about generation time and packet length which are used for measuring packet delay and throughput, respectively. The generated packets are stored in a FIFO-type traffic queue awaiting for scheduling.

10.4.1.2.2 Scheduler

The scheduler is one of the most important functions that affect system performance. The function of the scheduler is to select a UE to serve at every TTI. For the operation, the scheduler utilizes ACK/NACK and CQI feedback information from UEs and the backlogging status of each UE's traffic queue.

Because the HSDPA system adopts HARQ, a UE with NACK feedback has the highest priority for scheduling. Therefore, the scheduler always serves the UE with NACK feedback until ACK occurs or the retransmission counter expires. Generally, the number of retransmissions is limited to three or four. Once the retransmission counter value exceeds the retransmission limit, the packet is discarded and the counter is reset.

If there are no UEs with NACK feedback, then the scheduler orders and selects the UEs according to its own scheduling criterion. Well-known scheduling algorithms include Max C/I (or best SINR) and Proportionally Fair (PF) algorithms. The Max C/I scheduler selects one UE with the best SINR value. In the HSDPA system, UEs report CQI values instead of the measured SINR values. The reported CQI values can be considered as a quantized version of the SINR values. Therefore, the Max C/I scheduler simply chooses one UE with the highest CQI value among all UEs in a cell. Because the Max C/I selects the UE with the best channel quality, it can be used for evaluating the maximum limit of system throughput. However, due to a fairness problem, the Max C/I scheduler is not a feasible solution for the practical HSDPA system. The PF scheduler can be a feasible solution for solving the fairness problem while minimizing throughput degradation. The PF scheduler can be implemented using the normalized SINR algorithm, which selects one UE with the largest value of the instantaneous SINR normalized by the long-term average of its own SINR. Therefore, the PF for the HSDPA system can be implemented using the normalized CQI instead of the normalized SINR.

10.4.1.3 Receiver (UE)

10.4.1.3.1 SINR Measurement

Every UE measures the SINR value at each TTI. The desired signal is transmitted from the home-cell Node B and the interference comes from the other cells. The received SINR value of UE k, $z(k)$, can be measured as

$$z(k) = \frac{G_{LS}^{bc} G_{SS}^{bc} P_{tx}^{bc}}{N_0 W + LF \cdot \sum_{\text{other-cells}} G_{LS}^{oc} G_{SS}^{oc} P_{tx}^{oc}}, \tag{10.6}$$

where the superscripts bc and oc represent the home-cell and other-cell, respectively, G_{LS} is the large-scale fading gain, G_{SS} is the small-scale fading gain, P_{tx} is the transmit power, N_0 is the noise spectral density, W is the channel bandwidth, and LF is the loading factor of the other-cells. For more accurate simulations, other-cells need to be modeled reflecting the transmissions and reception of UEs with the corresponding Node Bs. However, it is too complex to simulate these operations. Therefore, it is assumed that UEs are located in the home-cell only, and the transmit power of the other-cell NodeB, P_{tx}^{oc} is constant. To evaluate the effect of other-cell interference level, the loading factor (LF) is used for adjusting the level. The LF is a real number in $[0, 1]$.

Once the position of the UE is determined, the large-scale fading gains, G_{LS}^{bc} and G_{LS}^{oc}, are also determined according to Equation (10.5). The noise power term is also constant. Therefore, only the small-scale fading gains, G_{SS}^{hc} and G_{SS}^{oc}, are generated at every measurement time. The small-scale fading gain depends on the given type of mobile radio channel. G_{SS} is 1 for the AWGN channel. For the independent Rayleigh fading channel, $\sqrt{G_{SS}}$ is a Rayleigh-distributed random variable. If the time-correlated Rayleigh fading channel is required, then $\sqrt{G_{SS}}$ is generated using Jakes' model. Other fading models, including Rician channel and Nakagami-m channel, also can be used.

10.4.1.3.2 CQI Determination

Each UE determines the proper CQI value based on the measured SINR value and reports it to the home-cell Node B at every TTI. The UE should select the highest CQI value with which the frame can be successfully received while satisfying a required FER of 0.1 [3]. For example, if the current measured SINR is 0 dB in Figure 10.4, then the UE selects CQI 13 because it is the highest CQI value among the CQI values achieving the required FER of 0.1.

10.4.1.3.3 HARQ Process and ACK/NACK Decision

The UE that received a frame decides ACK or NACK using the measured SINR. The estimated FER value is obtained from the link-level simulation result using the current SINR. Then, the system-level simulator generates a random number in (0, 1) using the uniform distribution and decides NACK if the random number is smaller than the given FER. ACK is decided in the other case. The ACK/NACK information is reported to Node B. If Node B receives NACK, then it retransmits the previous frame. There are two types of HARQ processes: chase-combining (CC) and incremental redundancy (IR). For the CC-type HARQ, Node B retransmits the same information and parity bits. Therefore, the CQI value of the retransmitted data is the same as the original (previous) transmission, and SINR gain can be obtained through retransmissions at the receiver. On the other hand, the IR-type HARQ uses different coding for each retransmission. Therefore, the distinct FER curve for each retransmission format is required to evaluate the system performance of the IR-type HARQ.

10.4.1.4 System-Level Simulator for the HSUPA System

Unlike the HSDPA system, multiple UEs act as transmitters in uplink and a home-cell Node B acts as a receiver as shown in Figure 10.10. Therefore, a traffic generator and traffic queue are located in each UE. The UEs transmit their data according to the absolute or relative grant information from an uplink scheduler in Node B. The uplink scheduler adjusts each UE's

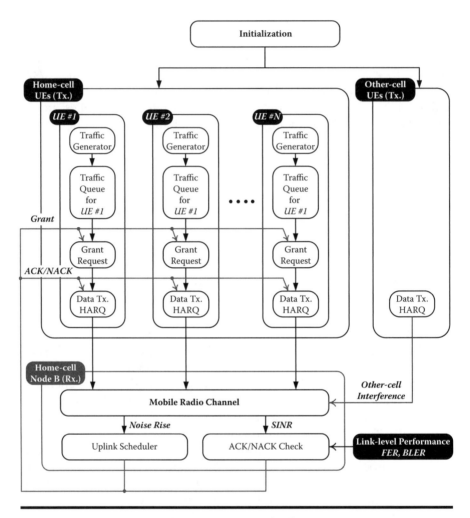

Figure 10.10 System-level simulator structure for the HSUPA system.

transmission format using the grant message for maintaining the overall noise rise in the home-cell below the allowable noise rise. Node B is also responsible for determining ACK/NACK for HARQ.

In the HSUPA system-level simulator, a number of UEs in other-cells act as interferers. To reduce computational complexity, it is required to approximate a single interference source instead of simulating a number of individual interferers in other-cells. According to the central limit theorem, the sum of a number of interferences can be approximated by the Gaussian distribution. The mean and variance of the Gaussian distribution can be varied according to the other-cell interference status.

10.4.2 Performance Metrics of System-Level Simulations

Throughput and delay are the most important performance metrics for system-level simulation. According to the different measurement perspectives, various metrics can be defined as follows:

■ *System throughput* represents the total amount of successfully transmitted data during the simulation time. Let b_u denote the successfully transmitted data of user u in bits; then the system throughput R_{sys} can be expressed as

$$R_{sys} = \frac{\sum_{\forall u}^{N_{user}} b_u}{\tau_{sim}} \qquad (10.7)$$

where N_{user} represents the total number of users and τ_{sim} is the total simulation time, respectively.

■ *Spectral efficiency* is scaled version of the system throughput expressed as

$$R_{eff} = \frac{R_{sys}}{BW} \qquad (10.8)$$

where BW is the system bandwidth in hertz.

■ *Average user throughput* is defined as

$$R_{user} = \frac{R_{sys}}{N_{user}} \qquad (10.9)$$

■ *Average packet call throughput* is measured for some specific traffic models having packet call, such as VoIP traffic model [6,7]. Packet call throughput is the total bits per packet call divided by total packet duration. Then the average packet call throughput can be expressed as

$$R_{call} = \sum_{\forall u}^{N_{user}} \frac{1}{N_{call,u}} \sum_{\forall i}^{N_{call,u}} \frac{b_{u,i}}{\delta_{u,i}} \qquad (10.10)$$

where $N_{call,u}$ denotes the number of packet calls for user u, $b_{u,i}$ represents the successfully transmitted bits in the ith packet call for user u, and $\delta_{u,i}$ is the duration of ith packet call for user u.

■ *Average user packet call throughput* is defined as

$$R_{call,user} = \frac{R_{call}}{N_{user}} \qquad (10.11)$$

■ *Average packet delay* is the average time duration for successful transmission of each packet. According to the length of a packet, it can be fragmented into several segments for transmission. Then, the packet delay is measured from the start of the first segment until the every segment is delivered to the receiver. If some of the segments are discarded during the HARQ process, then the entire packet transmission is failed and the packet is also discarded. Therefore, the packet delay only considers the successfully transmitted packets.

In addition to the itemized performance metrics above, there can be various performance metrics, including the received SINR distribution, distribution for the number of retransmission, CQI/TFC distribution, service probability, delay variation, and so on.

10.4.3 Numerical Examples of System-Level Simulations

10.4.3.1 HSDPA System

Figure 10.11 shows the system throughput for HSDPA when the full-queue traffic model, Max C/I scheduler, and Rayleigh fading channel are used for system-level simulations. The UE Category 10 shown in Table 10.2 is assumed, and link-level simulation results in Figure 10.4 are used for CQI determination and ACK/NACK decision. Because a home-cell has three sectors, the total cell throughput is three times the average sector throughput. The full-queue traffic model implies that every UE's traffic queue is always backlogged with a sufficient amount of traffic. This traffic model is generally used in a bid to analyze the maximum throughput performance of the system because the queuing dynamics according to the activity of the traffic pattern are not considered.

Figure 10.11 also shows the system throughput for different MIMO configurations that the HSDPA supports. The "No Diversity" in the figure represents the normal antenna configuration, which is one transmit antenna in the Node B and one receive antenna at the UE. HSDPA supports two types of transmit diversity schemes: space-time transmit diversity (STTD) and closed-loop transmit diversity (CLTD). Because CLTD uses the feedback information from the UE, more SNR gain can be achieved, as shown in the figure.

High system throughput performance is achieved through the wireless scheduler, which provides a multi-user diversity (MUD) gain. Figure 10.12 shows examples of the MUD gain achieved by the Max C/I scheduler. This figure presents the received SINR distributions for all users and for the selected users by the scheduler. Because the Max C/I scheduler selects the users with the best channel quality at each time, the SINR values of the

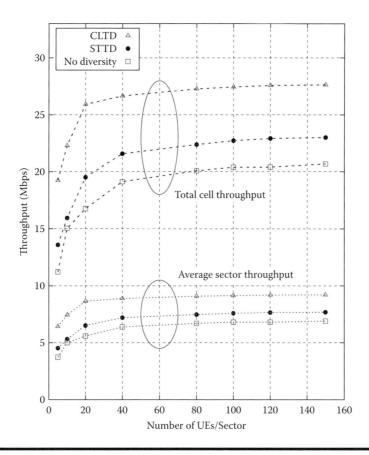

Figure 10.11   System throughput for the HSDPA, Rayleigh fading channel.

scheduled users become higher, as shown in the figure. Moreover, as the number of users increases, the MUD gain also increases.

The Max C/I provides the maximum MUD gain to maximize the system throughput while it does not consider the fairness among the users. It is possible to observe the fairness of each scheduling algorithm in Figure 10.13. This figure shows the service probability according to the distance from the Node B. The service probability is defined as the probability that a UE at a certain location is scheduled. Because the selection criterion of the Max C/I scheduler is based on the SINR value, as the UE is located far from the Node B, the service probability also decreases drastically. However, the PF scheduler is based on the normalized SINR instead of the SINR itself. Therefore, the service probability is nearly even regardless of the distances, as shown in the figure. While the PF scheduler provides a fair service, the MUD gain is degraded.

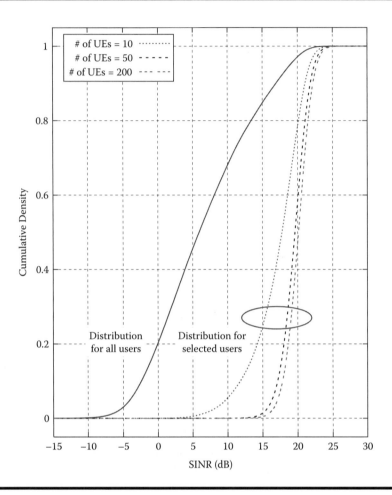

Figure 10.12 SINR distributions before and after scheduling, Rayleigh fading channel.

10.4.3.2 HSUPA System

The TFC formats shown in Table 10.4 are used to evaluate the performance of the HSUPA system. Figure 10.14 shows the system throughput performance of HSUPA with full-queue traffic in Rayleigh fading channel. The code-domain uplink scheduling is performed with the maximum allowable noise rise of 5 dB. As the number of UEs increases, the system throughput decreases. This is a different characteristic from the HSDPA system. Because the HSUPA is an uplink system, the aggregated interference at Node B (receiver) as the number of UEs (transmitters) increases. The total amount of the interference, which is measured as the *noise rise*, is controlled to be less than the target value, which is 5 dB in this case. The scheduler orders

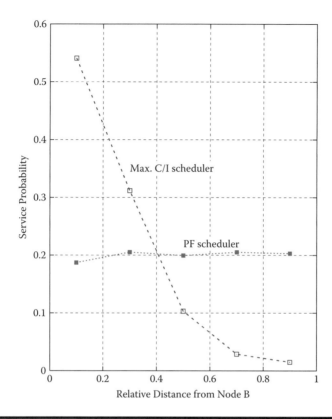

Figure 10.13 Service probabilities for two different schedulers, Rayleigh fading channel.

the UEs to lower their TFC through the absolute/relative grant process in order to support as many UEs as possible while maintaining the noise rise requirement. Consequently, the system throughput decreases as the number of UEs increases because the transmit power of each UE decreases and the total interference at Node B increases.

Figure 10.15 illustrates the system throughput of the HSUPA system when the VoIP traffic model is used. Comparing to Figure 10.14, the system throughput increases as the number of UEs increases. The VoIP traffic is generated periodically with an on/off pattern and a rate of approximately 10 kbps. Due to the relatively low rate and activity of the VoIP traffic, it is possible to maintain a low level of interference. Therefore, the system throughput increases as the number of UEs increases. Because the uplink is sensitive to the interference among the UEs, the HSUPA system is appropriate to serve medium/low rate data services, including VoIP and streaming video traffic.

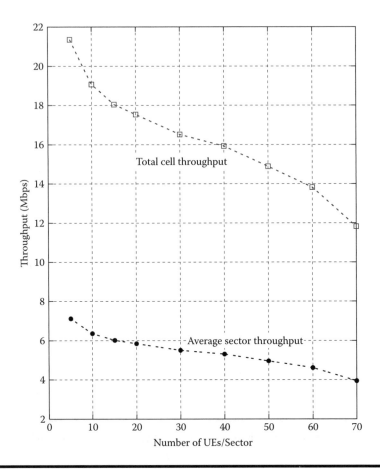

Figure 10.14 System throughput for the HSUPA, full queue traffic.

10.5 Concluding Remarks

The performance of the HSDPA/HSUPA system is evaluated through link- and the system-level simulators. The link-level simulation provides the performance of a single link between a transmitter and a receiver. The system-level simulator evaluates the performance of the system with multiple transmitters and multiple receivers in a multi-cell environment by considering the radio resource management (RRM) schemes, such as scheduling, rate control, and power control.

The HSDPA/HSUPA system achieves MUD gain through an opportunistic scheduler and yields high throughput. Through the simulator, it is possible to analyze the achievable MUD gain for the Max C/I and PF scheduling algorithms. Moreover, the fairness of each scheduler also can be evaluated.

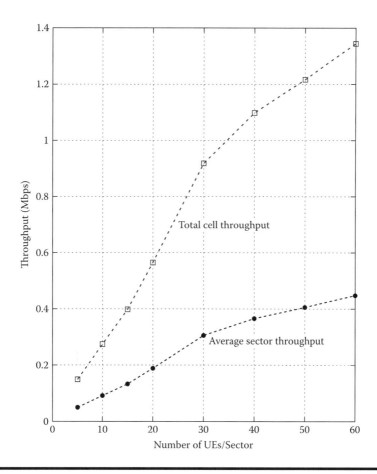

Figure 10.15   System throughput for the HSUPA, VoIP traffic.

The HSUPA system is very sensitive to co-channel interference. Therefore, the uplink scheduler should control the noise rise in the home-cell using the absolute or relative grant message. The HSUPA system provides high throughput through fast uplink scheduling at Node B, compared to the conventional WCDMA uplink.

References

[1] 3GPP. *Multiplexing and Channel Coding (FDD)*, 3GPP TS 25.212, v. 6.7.0. December 2005.

[2] International Telecommunication Union. *Guidelines for evaluation of radio transmission technologies for IMT-2000*, ITU Std. Rec.ITU-R M.1225, 1997.

[3] 3GPP. *Physical Layer Procedures (FDD)*, 3GPP TS 25.214, v. 6.3.0. September 2004.

[4] 3GPP. *UE Radio Access Capabilities*, 3GPP TS 25.306, v. 6.7.0. December 2005.

[5] 3GPP. *Physical Channel Mapping of Transport Channels onto Physical Channels (FDD)*, 3GPP TS25.211, v. 6.7.0. December 2005.

[6] *1xEV-DV Evaluation Methodology—Addendum (V6)*, WG5 Evaluation Ad-hoc Group Std. July 2001.

[7] IEEE 802.16 Broadband Wireless Access Working Group. *IEEE 802.16m Evaluation Methodology Document (EMD)*. July 2008.

Chapter 11

MIMO HSDPA Throughput Measurement Results

Christian Mehlführer, Sebastian Caban, and Markus Rupp

Contents

11.1 Introduction

In this chapter, physical layer throughput measurement results of multiple-input, multiple-output (MIMO) High-Speed Downlink Packet Access (HSDPA) [1] are presented. The results were obtained in two extensive measurement campaigns: The first campaign was carried out in an alpine valley in Austria. Here, the propagation channel had a very small mean root mean square (RMS) delay spread[1] of about one chip (260 ns) due to the fact that scattering objects existed only in the immediate vicinity of the receiver. The second campaign was carried out in the inner city of Vienna, Austria. Here, the propagation conditions were non-line-of-sight with a rather large mean RMS delay spread of about 3.8 chips (1 μs).

In the measurements we considered fast link adaptation [2] and fast hybrid automated repeat request [3], two of the key features of the HSDPA physical layer. We restricted ourselves to the single-user case due to the hardware effort required for multi-user measurements. Another reason for choosing only the single-user case for the measurements is that multi-user scheduling for HSDPA is still a topic in research [4–6]. We furthermore restricted our investigations to slow fading; that is, the channel was assumed constant during the transmission of several sub-frames. This restriction is required by our measurement procedure explained in detail in the section entitled "Measurement Setup and Procedure."

11.2 MIMO HSDPA

In this section, a mathematical description of the (MIMO) HSDPA physical layer is introduced. This description is then used to derive the equalizer at the receiver and to describe the precoding at the transmitter. Furthermore, a throughput-maximizing feedback calculation method is explained.

11.2.1 System Model

Assume the transmission of N_s independently coded and modulated data chip streams, each of length $L_c = L_b + L_f - 1$ chips; L_b and L_f correspond to the channel and the equalizer length, respectively. A block diagram of such a transmission system is shown in Figure 11.1. We define the stacked transmit chip vector $\mathbf{s}_k$ of length $N_s L_c$ at time instant k as:

$$\mathbf{s}_k = \left[\mathbf{s}_k^{(1)T}, \ldots, \mathbf{s}_k^{(N_s)T} \right]^T \tag{11.1}$$

---

[1] The mean RMS delay spread was calculated by averaging the RMS delay spreads of all channel impulse responses measured in a specific scenario.

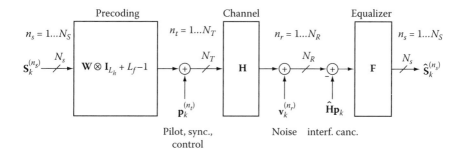

Figure 11.1   Generalized system model of the HSDPA physical layer.

The N_s chip streams are weighted by the $N_T \times N_s$ dimensional precoding matrix

$$\mathbf{W} = \begin{bmatrix} w^{(1,1)} & \cdots & w^{(1,N_s)} \\ \vdots & \ddots & \vdots \\ w^{(N_T,1)} & \cdots & w^{(N_T,N_s)} \end{bmatrix} \tag{11.2}$$

forming the data chip streams of the N_T transmit antennas. At each transmit antenna, pilot, synchronization, and control channels accumulated in

$$\mathbf{p}_k = \left[\mathbf{p}_k^{(1)T}, \ldots, \mathbf{p}_k^{(N_T)T} \right]^T \tag{11.3}$$

are added. Using the $L_c \times L_c$ dimensional identity matrix $\mathbf{I}_{L_c}$, the transmit signal vector $\mathbf{a}_k$ of length $N_T L_c$ at time instant k is given by:

$$\mathbf{a}_k = (\mathbf{W} \otimes \mathbf{I}_{L_c})\mathbf{s}_k + \mathbf{p}_k \tag{11.4}$$

The frequency-selective link between the n_t-th transmit and the n_r-th receive antenna is modeled by the $L_f \times L_c$ dimensional band matrix:

$$\mathbf{H}^{(n_r,n_t)} = \begin{bmatrix} b_0^{(n_r,n_t)} & \cdots & b_{L_b-1}^{(n_r,n_t)} & 0 \\ \ddots & & & \ddots \\ 0 & b_0^{(n_r,n_t)} & \cdots & b_{L_b-1}^{(n_r,n_t)} \end{bmatrix} ; \quad \begin{array}{c} 1 \le n_r \le N_R \\ 1 \le n_t \le N_T \end{array} \tag{11.5}$$

where the $b_i^{(n_r,n_t)}$ ($i = 0, \ldots, L_b - 1$) represent the channel impulse response between the n_t-th transmit and the n_r-th receive antenna. The entire frequency-selective MIMO channel is modeled by a block matrix $\mathbf{H}$ consisting of $N_R \times N_T$ band matrices defined in Equation (11.5):

$$\mathbf{H} = \begin{bmatrix} \mathbf{H}^{(1,1)} & \cdots & \mathbf{H}^{(1,N_T)} \\ \vdots & \ddots & \vdots \\ \mathbf{H}^{(N_R,1)} & \cdots & \mathbf{H}^{(N_R,N_T)} \end{bmatrix} \quad (11.6)$$

At the receiver, additive noise, denoted by $\mathbf{v}_k$, deteriorates the desired signal:

$$\mathbf{b}_k = \mathbf{H}\mathbf{a}_k + \mathbf{v}_k = \mathbf{H}(\mathbf{W} \otimes \mathbf{I}_{L_c})\mathbf{s}_k + \mathbf{H}\mathbf{p}_k + \mathbf{v}_k \quad (11.7)$$

The signal $\mathbf{b}_k$ is then processed in an equalizer $\mathbf{F}$ to obtain an estimate of the transmitted chip stream:

$$\hat{\mathbf{s}}_k = \left[\hat{s}_{k-\tau}^{(1)}, \ldots, \hat{s}_{k-\tau}^{(N_s)}\right]^T = \mathbf{F}\mathbf{b}_k = \mathbf{F}\mathbf{H}(\mathbf{W} \otimes \mathbf{I}_{L_c})\mathbf{s}_k + \mathbf{F}\mathbf{H}\mathbf{p}_k + \mathbf{F}\mathbf{v}_k \quad (11.8)$$

The equalizer matrix $\mathbf{F}$ defined by

$$\mathbf{F} = \left[\mathbf{f}^{(1)}, \ldots, \mathbf{f}^{(N_s)}\right]^T \quad (11.9)$$

consists of N_s vectors, each having a length of $N_R L_f$. The equalizer and the further processing of the chip stream $\hat{\mathbf{s}}_k$ are explained in the next section.

11.2.2 Receiver

At the receiver, we first perform synchronization and iterative channel estimation according to [7]. After that, the interference of the deterministic signals, that is, the pilot and synchronization channels, is cancelled. Therefore, in Equation (11.8) the term $\mathbf{F}\mathbf{H}\mathbf{p}_k$ is reduced by $\mathbf{F}\hat{\mathbf{H}}\mathbf{p}_k$ and only interference caused by the channel estimation error $(\mathbf{H} - \hat{\mathbf{H}})$ and the data channels remains [8]. Without this interference cancellation, the post equalization signal to interference and noise ratio (SINR) would saturate at about 20 dB. Channel quality indicator (CQI) values requiring higher SINR can thus not be selected, leading to a saturation of the throughput [9–11]. Alternatively to interference cancellation, interference-aware equalization is possible and has been shown to achieve high performance [12,13].

The equalizer coefficients calculated in the Linear Minimum Mean Square Error (LMMSE) sense can be shown [14–16] to be equal to

$$\mathbf{f}^{(n_s)} = \left(\mathbf{H}\left(\mathbf{W}\mathbf{W}^H \otimes \mathbf{I}_{L_c}\right)\mathbf{H}^H + \sigma_v^2 \mathbf{I}_{N_R L_c}\right)^{-1} \cdot \mathbf{H}(\mathbf{W} \otimes \mathbf{I}_{L_c})\mathbf{e}_{\tau + (n_s - 1)L_c} \quad (11.10)$$

for the n_s-th data stream. The vector $\mathbf{e}_k$ denotes a unit vector with a single "1" at cursor position k and "0" at all other positions. The calculation of the equalizer coefficients can be implemented efficiently using Fast Fourier Transform (FFT)-based algorithms as in [17,18], or the conjugated gradient algorithm [19]. This receiver therefore represents a low-complexity HSDPA receiver that is feasible for real-time implementation in a chip [20]. The output of the equalizer $\hat{\mathbf{s}}_k$ is descrambled, despread, soft-demapped, and soft-decoded in a Turbo decoder using eight iterations to obtain the data bits. For the sake of completeness, we note that more complex MIMO receivers (e.g., the LMMSE-Maximum A-Posteriori (MAP)) are known to show only about 1 dB better performance [21] than the standard LMMSE equalizer.

11.2.3 Quantized Precoding

The precoding matrix defined in Equation (11.2) is strongly quantized and chosen from a predefined codebook in HSDPA systems [1]. For single antenna transmissions in which obviously no spatial precoding can be performed, the precoding matrix $\mathbf{W}$ is reduced to a scalar equal to "1":

$$\mathbf{W}^{(\text{SISO})} = 1 \tag{11.11}$$

For multiple antenna transmissions, the precoding matrices are composed of the scalars

$$w_0 = \frac{1}{\sqrt{2}} \tag{11.12}$$

and

$$w_1, w_2 \in \left\{ \frac{1+j}{2}, \frac{1-j}{2}, \frac{-1+j}{2}, \frac{-1-j}{2} \right\}. \tag{11.13}$$

The Transmit Antenna Array (TxAA) transmission mode utilizes two antennas to transmit a single stream. In this mode, the precoding matrix is defined as:

$$\mathbf{W}^{(\text{TxAA})} = \begin{bmatrix} w_0 \\ w_1 \end{bmatrix} \tag{11.14}$$

This means that the signal at the first antenna is always weighted by the same scalar constant w_0, whereas the signal at the second antenna is weighted by w_1, which is chosen in order to maximize the received post equalization SINR [8]. In TxAA, the number of possible precoding matrices is equal to four, corresponding to an amount of 2 bit feedback.

In the case of Double Transmit Antenna Array (D-TxAA) transmission, the precoding matrix is given by:

$$\mathbf{W}^{(\text{D-TxAA})} = \begin{bmatrix} w_0 & w_0 \\ w_1 & -w_1 \end{bmatrix} \tag{11.15}$$

Note that this precoding matrix is a unitary matrix; that is, the precoding vector of the second stream is always chosen orthogonal to the one of the first stream. Although D-TxAA defines four precoding matrices, only the first two of them cause different SINRs at the receiver. In the other two cases, the SINRs of the first and the second stream are exchanged. Because the data rates of both streams can be adjusted individually, the third and fourth precoding matrices are redundant. Note also that if the user experiences low channel quality in D-TxAA, only a single stream is transmitted; that is, the precoding matrix in Equation (11.14) is applied to the data streams at the transmitter. Thus, in TxAA a single stream is always transmitted, and in D-TxAA either single-stream or double-stream transmission—whichever leads to a higher throughput—is performed. Details about the feedback calculation are provided in the next section.

11.2.4 Feedback Calculation

In MIMO HSDPA, two feedback values, the CQI and the precoding control indicator (PCI), must be calculated by the user equipment (UE) and fed back to the base station. In contrast to the receive power maximization suggested in [1], we perform a throughput-maximizing joint calculation of the CQI and the PCI [8]. This method has the advantage that it inherently provides a decision as to whether single-stream or double-stream transmission should be used.

The feedback calculation starts with the estimation of the post equalization SINR for *every* possible PCI value for single-stream and—where applicable—dual-stream transmission. All SINR values are reduced by a 1-dB margin (to account for SINR estimation errors) and then mapped to CQI values using Table 11.1 and Table 11.2 for single-stream and double-stream mode, respectively. The SINR-to-CQI mapping tables were obtained by simulating the block error ratio (BLER) performance of a category 16 UE

Table 11.1 SINR-CQI Mapping Table for Single-Stream Mode of a Category 16 UE

CQI	1	2	3	4	5	6	7	8
SINR	−3.5 dB	−2.6 dB	−1.5 dB	−0.3 dB	0.5 dB	1.7 dB	2.5 dB	3.5 dB
CQI	9	10	11	12	13	14	15	16
SINR	4.4 dB	5.5 dB	6.5 dB	7.5 dB	8.5 dB	9.5 dB	10.7 dB	11.5 dB
CQI	17	18	19	20	21	22	23	24
SINR	12.6 dB	13.4 dB	14.7 dB	15.7 dB	16.6 dB	17.5 dB	18.6 dB	19.6 dB
CQI	25	26	27	28	29	30		
SINR	20.6 dB	21.4 dB	22.6 dB	23.5 dB	24.0 dB	24.8 dB		

Table 11.2 SINR-CQI Mapping Table for Double-Stream Mode of a Category 16 UE

CQI	0	1	2	3	4	5	6	7
SINR	10.5 dB	10.5 dB	11.2 dB	12.7 dB	14.3 dB	15.7 dB	17.2 dB	18.8 dB

CQI	8	9	10	11	12	13	14	
SINR	20.4 dB	21.9 dB	23.4 dB	25.3 dB	26.0 dB	26.8 dB	28.3 dB	

for all CQI values in an additive white gaussian noise (AWGN) channel. The SINR values in the tables are equal to the AWGN signal-to-noise ratio (SNRs) at 10% BLER, as defined by the HSDPA standard [1, Section 6A.2] as the maximum BLER that should not be exceeded.

The CQI values obtained from the mapping tables correspond to transport block sizes (TBSs) defined in [1, Table 7D, p. 50] and [1, Table 7I, p. 54] for single-stream and double-stream mode, respectively. In double-stream mode, every PCI value yields two TBS values. Thus, by selecting the PCI value corresponding to the maximum overall TBS, the throughput is maximized. Note that this method inherently also provides the decision between single-stream and double-stream transmission.

In the case of Category 16 UE, the maximum TBS in single-stream mode is 25,558 bits that are transmitted in one sub-frame of length 2 ms. Thus, the maximum data rate in single-stream mode is 12.779 Mbps. In dual-stream mode, the maximum TBS of one stream is equal to 27,952 bits, allowing for a maximum overall data rate of 27.952 Mbps.

11.3 Measurement Setup and Procedure

In the following we report on our MIMO HSDPA measurement setup in an alpine and an urban scenario.[2] We furthermore explain in detail the measurement procedure utilizing our testbed [22, 23].

11.3.1 Measurement Setup

Two scenarios were as follows:

- *Alpine scenario.* The base-station antenna (KATHREIN 800 10543 [24], ±45° polarization, half-power beam width 58°/6.2°, down tilt 6°) was placed 5.7 km away from the RX unit, which was located inside a

---

[2] For both measured scenarios, detailed transmitter and receiver positions can be downloaded for Google Earth at http://www.nt.tuwien.ac.at/fileadmin/data/testbed/Vienna-and-Carinthia-TX-RX-GPS.kmz.

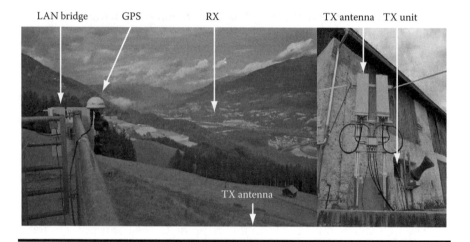

Figure 11.2 **The alpine scenario and the base-station transmit antennas, of which the left one was utilized in the measurements.**

house in a village on the opposite side of the "Drautal"-valley as shown in Figure 11.2. At the RX unit, we utilized standard Linksys WiFi-Router rod antennas. The results presented in the section "Measurement Results" were obtained in a setup in the alpine scenario in which the receive antennas were placed indoors in non-line-of-sight to the transmitter. This setup is characterized by a short mean RMS delay spread of about 1 chip (260 ns) and one major propagation path because the receive signal was mainly propagating through one window facing the transmit antennas. In addition to this setup, we also investigated RX antenna positions in different rooms where the TX antennas can and cannot be seen from the window. We also placed the RX unit outside, in the middle of a field, with direct line-of-sight to the transmitter. In all measured setups, the results obtained did not change significantly, apart from a variation in the average path loss that only shifts the throughput curves in Figure 11.6 to the left and right, respectively.

■ *Urban Scenario.* The same base-station antenna (KATHREIN 800 10543 [24], ±45° polarization, half-power beam width 58°/6.2°, down tilt 6°) was placed on the roof of a big building in the center of Vienna, Austria, 430 m away from the RX unit that was placed inside an office room (see Figure 11.3). At the RX unit, we utilized four low-cost printed monopole antennas [25] that are based on the generalized Koch pre-fractal curve. Due to their low cost and small size, such antennas are very realistic and could be built into a mobile handset or a laptop computer. In all measurements carried out in the urban scenario, the direct path from the TX antennas to the RX antennas

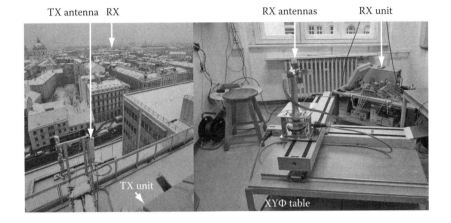

TX antenna RX   RX antennas   RX unit

Figure 11.3   The urban scenario and the XYΦ positioning table with the receive antennas.

was blocked by the building in which the RX unit was located. This scenario is characterized by a rather long mean RMS delay spread of about 3.8 chips (1 μs).

In both scenarios, the transmit antennas were placed right next to existing base stations making the measurement results obtained very realistic and representative for a mobile communication system.

To exploit polarization diversity and to obtain good separation between the two spatial data streams, differently polarized antenna elements were employed at the transmitter and at the receiver.

Figure 11.4 shows the basic measurement setup used for all our measurements as well as the important system parameters. Prior to a measurement, the following steps were carried out (the numbers in the square brackets are typical values):

■ The TX and RX units are set up and a control radio link (basically a wireless local area network (LAN) bridge) between the TX and RX units is established. [3 days]

■ All required transmit-data-blocks are pre-generated in Matlab and then stored (in the form of 14-bit complex baseband data samples) on parallel solid-state hard disks to achieve an access time in the order of 1 ms. [2 hours]

■ The rubidium+GPS stabilized "clocks" of the TX and RX units are reset to zero with a relative accuracy of ±20 ns. The handshaking required for this reset is carried out via the control radio link so that there is no need for a cable connection. To meet the accuracy required

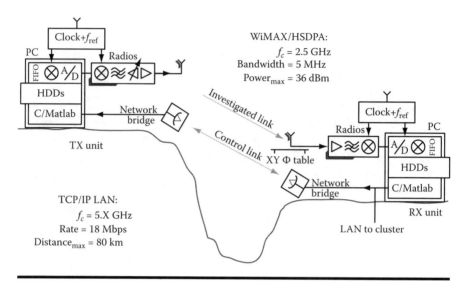

Figure 11.4   Measurement setup.

by typical measurements, this procedure takes less than half an hour from a cold start. The rubidium clocks also provide perfect frequency synchronization between transmitter and receiver. [30 min]

11.3.2 *Measurement Procedure*

Now we describe the transmission of an HSDPA block using our testbed (see also Figure 11.5). The required time for each operation is provided in square brackets at the end of each step.

■ First, the RX unit (the master) requests the transmission of a "previous block" via the control link. (HSDPA requires the receiver to feed back information calculated from the previously received block in order to transmit a channel adapted signal.) [3 ms]

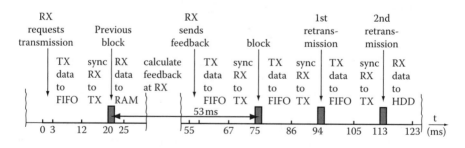

Figure 11.5   Timing of the transmission of a "single" data block.

■ Then the transmitter copies the selected block (that is, pre-generated transmit data samples) from the solid-state hard disks to the FIFOs of the transmit hardware. [9 ms]

■ Next, via the control link, the TX unit tells the RX unit the exact time that the transmission will take place—that is, current time plus 4 ms. (The delay of the control link is typically less than 4 ms.) Another 4 ms are required for the handshaking between PC and external synchronization hardware. [8 ms]

■ Consequently, at exactly the same time, the transmission and the reception of a data block are triggered by the TX and the RX hardware, respectively.

■ In real-time, the complex valued transmit data samples are interpolated to 200 MSamples/s, digitally up-converted to 70 MHz, converted to the analog domain (14 bit), analog up-converted to 2.5 GHz, attenuated (digitally adjustable), amplified, and then transmitted. At the receiver, exactly the reverse procedure takes place. At the end, the already down-sampled received complex baseband data samples are stored in the internal memory of the RX unit (not on the hard disks). [5 ms]

■ These received samples are now immediately evaluated by the CPU of the RX computer, that is, synchronization, channel estimation, and feedback calculation are carried out in Matlab [8]. No further receiver processing is performed at this point. [$N_R \times 26$ ms]

■ Now the RX unit requests the transmission of the actual channel-adapted data block via the control link (its index is determined by the feedback information calculated from the previous data block). Because we want to investigate the HSDPA performance when Hybrid Automated Repeat Request (HARQ) retransmissions are employed, the RX unit also requests the transmission of the two possible retransmissions, regardless of whether or not they are required. This is necessary because the data evaluation and determination of the required number of retransmissions are carried out later. [3 ms]

■ Transmission of the three blocks now takes place in real-time (as described above). [57 ms]

■ Finally, all four received blocks are stored on RAID hard disks for later/immediate off-line evaluation[3] in a cluster. [10 ms]

In our measurement, a typical value for the feedback delay between the "previous block" used for channel sounding and the actual transmission of the first channel-adapted data block is 53 ms, as shown in Figure 11.5. The

---

[3] We usually evaluate the first blocks of a measurement immediately in order to quickly discover possible flaws in the measurement setup (for example, power amplifiers still turned off or wrong buttons on the equipment pushed by accident).

actual time value depends on the MIMO HSDPA scheme under investigation. In a real HSDPA network, this feedback delay is on the order of a few milliseconds (at a maximum, 7.5 slots corresponding to 5 ms). Therefore, to obtain representative results, the measurement procedure requires the channel to stay constant during the feedback delay. Later off-line testing revealed that this is the case; that is, estimation of the feedback based on the "previous" blocks and the channel-adapted blocks did not show any significant difference. Note that the delays between the channel-adapted data block and the first and second retransmissions are about the same as in a real HSDPA network; thus, no further requirements on the measurement procedure are necessary.

11.3.3 Inferring the Average Performance

In the previous section, we explained in detail how a "single" HSDPA data block is transmitted. To infer the average throughput performance (as shown, for example, in Figure 11.6) of a specific scenario, we carry out the fully automatized following steps shown below:

- We repeat the procedure above for all different schemes under investigation (the curves in Figure 11.6).
- We repeat all the above for different transmit power levels (the x-axis in Figure 11.6). To achieve this, we attenuate the transmit signal prior to the power amplifier.
- We repeat all the above for different receive antenna positions (110 for the alpine scenario and 484 for the urban scenario). These realizations are created by moving the RX antennas with a fully automated XYΦ positioning table as shown in Figure 11.3. In order to minimize large scale fading effects, we measure only within a small area of 3×3 wavelengths. The measured antenna positions are uniformly distributed. By this systematic sampling approach, the distance between all positions measured is maximized and their correlations minimized.

Measuring all this typically takes from an hour up to a day. The evaluation of the data blocks (up to one terabyte of baseband data samples) is carried out using a self-programmed PC cluster software that parallelizes the off-line evaluation of the received data on a position-by-position basis on all currently available (typically about 20) employee PCs of our workgroup.

- Once calculated, we collect all results from the cluster in order to average the throughput observed over the positions measured. Several tests are carried out in order to validate the results; for example, we check for measurement outliers[4] or interference that should not exist.

---

[4] That does not mean that we discard them automatically. We check for their existence in order to detect possible flaws in the experiment setup.

- By measuring the output powers of all transmitters with a spectrum analyzer (Rohde and Schwarz FSQ26, relative accuracy $\pm 0.1\,$dB), we ensured that our individual power adjustment done in the digital domain resulted in identical output powers. Nevertheless, we observed that the average receive powers originating from the individual transmit antennas differ by about 1 to 2 dB. To compensate for this effect when comparing HSDPA schemes with different numbers of transmit antennas, we therefore performed throughput averaging over the corresponding transmit/receive antennas. For example, if the 2×2 D-TxAA system is compared to the single input single output (SISO) transmission, the SISO throughput is obtained by averaging over the four individual SISO links (TX1→RX1, TX1→RX2, TX2→RX1, and TX2→RX2).

- Finally, we estimate the accuracy of the measurement by means of bootstrapping methods [26]. In all throughput graphs (Figures 11.6–11.7), the dots represent the average throughputs, the corresponding vertical lines the 95% confidence interval, and the corresponding horizontal lines the 2.5% and 97.5% percentiles. Note that the RX antenna positions remained unchanged between measuring different schemes at different transmit power levels. This, on one hand, leads to smooth curves. On the other hand, the *relative* positions of the curves are more accurate than the confidence intervals for the *absolute* positions might suggest.

11.4 Achievable Mutual Information

This section defines a so-called "achievable mutual information," which is used as a performance bound for the actually measured data throughput. The calculation of the bound is based on the mutual information between the transmit and the receive signals, that is, the estimated frequency response. The achievable data throughput is a function of the wireless channel and the allowed precoding vectors. Thus, it incorporates the restrictions given by the transmission standard (quantized, frequency-flat precoding) but not the restrictions given by the receiver employed.

Consider the time-dispersive channel of length L_b chips between the n_t-th transmit and the n_r-th receive antenna:

$$\mathbf{h}^{(n_r, n_t)} = \begin{bmatrix} b_0^{(n_r, n_t)} & \cdots & b_{L_b-1}^{(n_r, n_t)} \end{bmatrix}^T \tag{11.16}$$

This channel can be equivalently described in the frequency domain as $\mathbf{g}^{(n_r, n_t)} = \mathcal{F}\mathfrak{F}\{\mathbf{h}^{(n_r, n_t)}\}$ using the N_{FFT} point Fourier transform $\mathfrak{F}\{.\}$. Thus, the Fourier transform separates the frequency-selective channel into N_{FFT}

frequency flat channels. The MIMO channel matrix of the m-th frequency bin can then be written as:

$$\mathbf{G}_m = \begin{bmatrix} \left(\mathbf{g}^{(1,1)}\right)_m & \cdots & \left(\mathbf{g}^{(1,N_T)}\right)_m \\ \vdots & \ddots & \vdots \\ \left(\mathbf{g}^{(N_R,1)}\right)_m & \cdots & \left(\mathbf{g}^{(N_R,N_T)}\right)_m \end{bmatrix} ; \quad m = 1, \ldots, N_{\text{FFT}} \qquad (11.17)$$

Using the well-known expressions for the MIMO capacity (see, for example, [27,28]), we obtain the achievable mutual information:

$$D_{\text{achievable}} = \max_{\mathbf{W} \in \mathcal{W}} \sum_{m=1}^{N_{\text{FFT}}} \frac{f_s}{N_{\text{FFT}}} \log_2 \det \left(\mathbf{I}_{N_R} + \frac{1}{\sigma_v^2} \mathbf{G}_m \mathbf{W} \mathbf{W}^H \mathbf{G}_m^H \right) \qquad (11.18)$$

Here, f_s corresponds to the chip rate of HSDPA (3.84 MHz) and σ_v^2 to the variance of the noise $\mathbf{v}_k$. The maximization is performed over the set of all possible precoding matrices is $\mathcal{W}$. Note that Equation (11.18) represents neither the mutual information nor the channel capacity, because only quantized and frequency flat precoding is utilized in HSDPA. Therefore, $D_{\text{achievable}}$ is referred to as "achievable mutual information." Also note that Equation (11.18) only gives the achievable mutual information for a specific channel realization at a specific receive antenna position and a given transmit power level. To obtain a mean achievable mutual information, we perform averaging over all measured receive antenna positions. Also, because the channel is not known perfectly, we use the estimated channel coefficients at the largest transmit power (thus having the smallest possible channel estimation error) to calculate the achievable mutual information. At lower transmit powers, the different channel SNRs are obtained by scaling the channel coefficients accordingly.

11.5 Measurement Results

In this section we present the throughput measurement results (the solid lines in Figures 11.6 and 11.7) and compare them to the achievable mutual information (the dashed lines in Figures 11.6 and 11.7). All throughput curves are plotted versus transmit power. Two additional axes show the average received SISO SNR and average received SISO signal power. The reason we plot the throughput versus transmit power is the following: All MIMO schemes in HSDPA utilize adaptive precoding at the transmitter that effectively increases the received power and thus also the SNR, while the total transmit power is the same as in the SISO transmission. If the throughput is plotted versus SNR rather than versus transmit power, the curves will be shifted with respect to each other. For example, in case of TxAA, this

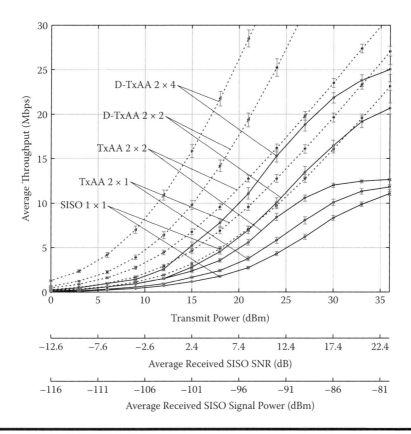

Figure 11.6 Throughput results of a category 16 UE in the alpine scenario (measurement ID "2008-09-16"). Averaging was performed over 110 receiver positions. The solid lines represent the measured throughput, the dashed lines the achievable mutual information.

shift would be about 2 dB compared to SISO. The additional x-axes (average received SISO SNR and average received SISO signal power) are thus only shown for reference reasons to indicate the approximate SNR and receive power ranges.

11.5.1 Alpine Scenario

Figure 11.6 shows the measured throughput and the achievable mutual information of the 1×1 SISO, 2×1 TxAA, 2×2 TxAA, 2×2 D-TxAA, and 2×4 D-TxAA transmission systems in the alpine scenario. Although all these schemes are standardized, only the 1×1 SISO transmission is currently (2009) implemented in HSDPA networks.

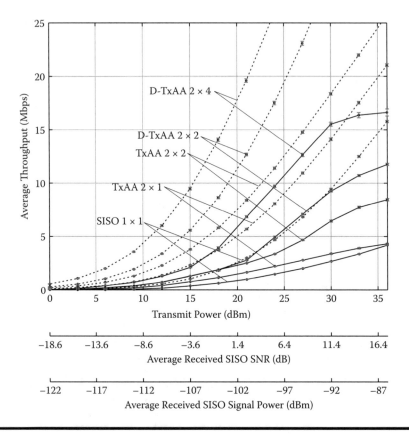

Figure 11.7 Throughput results of a category 16 UE in the urban scenario (measurement ID "2008-12-12"). Averaging was performed over 484 receiver positions. The solid lines represent the measured throughput, the dashed lines the achievable mutual information.

Figure 11.6 shows that the measured throughput of the 2×1 TxAA system is significantly (about 3 dB) better than the throughput of the SISO system. The 2×2 TxAA system performs another 3 dB better than the 2×1 TxAA system. The very good performance of the TxAA system in this scenario is explained by the rather small delay spread of the channel in which the frequency-flat precoding works very well. The 2×2 D-TxAA system performs better than the 2×2 TxAA system above transmit powers of 15 dBm. At certain receive antenna positions, D-TxAA thus employs double-stream transmissions outperforming the pure single-stream transmissions of TxAA. At all other receive antenna positions, D-TxAA switches back to single-stream operation. Compared to 2×2 D-TxAA, the 2×4 D-TxAA system achieves about 4 dB gain due to the doubled number of receive antennas.

The measured and the achievable mutual informations are compared at a transmit power of about 10 to 25 dBm or a throughput of about 5 Mbps to avoid saturation effects of the measured throughput (because outside this range either no more smaller or larger CQI values are available). At large transmit powers, when the single-stream transmission modes saturate, such a comparison would be misleading because the throughput could be easily increased by providing additional modulation and coding schemes (as it is done, for example, in UE categories 13, 14, 17, and 18 that support 64-QAM modulation [1]). At a throughput of 5 Mbps, the measured SISO system loses about 7 dB and the 2 × 2 D-TxAA system about 5 dB compared to their corresponding achievable mutual informations.

11.5.2 Urban Scenario

In Figure 11.7, the results of the standard compliant schemes in the urban scenario are shown. Here, at low SNR the 2 × 1 TxAA system only performs a little bit better than the SISO system and achieves only the same performance as SISO at large SNR. The reason for this is the rather large mean RMS delay spread of 3.8 chips (1 µs) causing the precoding to be far from optimal. Optimal precoding must be frequency dependent, for example, using a water-filling solution. The large delay spread also causes inter-code interference that can only be partially removed by the LMMSE equalizer, leading to a saturation of the throughput at larger transmit powers and SNRs. When using two or four receive antennas, the interference situation gets better because the equalizer can suppress the post-equalization interference more effectively. Thus, the two and four receive antenna schemes show greatly increased throughput in the urban scenario.

Due to the increased interference in the urban scenario, the loss between measured throughput and the achievable mutual informations is larger than in the alpine scenario. With increasing order of the MIMO system, the loss to the achievable mutual information decreases.

11.5.3 Discussion of the Throughput Loss

Although the results of the previous sections show a significant performance increase of the different MIMO schemes when compared to SISO transmission, all measured throughput curves are about 6 to 9 dB away from the achievable mutual information. The following effects contribute (next to maybe others) to this loss:

■ The rate-matched Turbo code utilized in HSDPA is good but not optimal. AWGN simulations show that at higher code rates, it loses up to 2 dB when decoded by a MAP decoder. [approx. 2 dB]

■ The LMMSE equalizer representing a low-complexity and cost-effective solution is also not optimal. Better receivers such as the LMMSE-MAP have the potential to improve the performance by about 1 dB [21]. [approx. 1 dB]

■ In the urban scenario, a larger throughput loss than in the alpine scenario was measured because of the larger delay spread and, consequently, the larger inter-code interference. For example, in the alpine scenario, the SISO system loses about 7 dB to the achievable mutual information, whereas the loss in the urban scenario is about 9 dB [>2 dB in the urban scenario].

11.6 Summary

The physical layer MIMO HSDPA throughput measurement results shown in this chapter were obtained in two extensive measurement campaigns carried out in an alpine valley in Austria and in the inner city of Vienna, Austria. These scenarios differ significantly in the delay spread of the channel and consequently in the resulting intra-cell interference. In both scenarios, the use of multiple antennas considerably increases the physical layer throughput. The 2×2 D-TxAA system increases the physical layer throughput by more than a factor of two compared to the SISO system. Absolute values of the mean measured throughput at transmit powers of 20 and 30 dBm are summarized in Table 11.3.

As a performance bound for the measured throughput, the so-called "achievable" mutual information is defined. Its calculation is based on mutual information of the channel that includes the quantized precoding employed at the transmitter. Comparing the measured throughputs to their corresponding achievable mutual informations shows that the measured

Table 11.3 Mean Measured HSDPA Throughput in the Alpine and Urban Scenarios at Transmit Powers $P_{TX} = 20$ dBm and $P_{TX} = 30$ dBm

	Alpine Scenario 260 ns Mean RMS Delay Spread		Urban Scenario 1 µs Mean RMS Delay Spread	
	$P_{TX} = 20$ dBm	$P_{TX} = 30$ dBm	$P_{TX} = 20$ dBm	$P_{TX} = 30$ dBm
1×1 SISO	2.4 Mbps	8.3 Mbps	0.9 Mbps	2.7 Mbps
2×1 TxAA	3.3 Mbps	10.1 Mbps	1.0 Mbps	2.8 Mbps
2×2 TxAA	4.9 Mbps	12.0 Mbps	1.7 Mbps	4.7 Mbps
2×2 D-TxAA	6.1 Mbps	16.5 Mbps	1.7 Mbps	7.1 Mbps
2×4 D-TxAA	10.0 Mbps	21.9 Mbps	3.3 Mbps	12.6 Mbps

throughput is far from optimal, losing between 5 and 9 dB in SNR, depending on the MIMO scheme employed. The main reasons for this are found to be coding loss, and equalizer loss, as well as losses due to residual inter-code interference. Thus, a large potential for future optimizations exists.

11.7 Acknowledgments

This work has been funded by the Christian Doppler Laboratory for Wireless Technologies for Sustainable Mobility. The authors thank Constantine Kakoyiannis (National Technical University of Athens, Greece) for providing us with the printed monopole RX antennas utilized in our measurements. The TX antennas were provided by KATHREIN-Werke KG. Also, the authors thank José Antonio García Naya, Michal Šimko, Walter Schüttengruber, and Georg Maier for supporting us with setting up the testbed.

References

[1] 3GPP, Technical specification group radio access network; physical layer procedures (FDD) (Tech. Spec. 25.214 V7.7.0), November 2007. [Online]. Available: http://www.3gpp.org/ftp/Specs/html-info/25214.htm

[2] M. Nakamura, Y. Awad, and S. Vadgama, Adaptive control of link adaptation for high speed downlink packet access (HSDPA) in W-CDMA, in *Proc. 5th Int. Symp. on Wireless Personal Multimedia Commun. 2002*, 2: 382–386, October 2002. [Online]. Available: http://ieeexplore.ieee.org/stamp/stamp.jsp?arnumber=1088198

[3] A. Das, F. Khan, A. Sampath, and H.-J. Su, Performance of hybrid ARQ for high speed downlink packet access in UMTS, in *Proc. 54th IEEE Vehicular Technol. Conf. 2001 (VTC2001-Fall)*, 4: 2133–2137, 2001. [Online]. Available: http://ieeexplore.ieee.org/stamp/stamp.jsp?arnumber=957121

[4] H. Chao, Z. Liang, Y. Wang, and L. Gui, A dynamic resource allocation method for HSDPA in WCDMA system, in *Proc. 5th IEE Int. Conf. 3G Mobile Commun. Technol. 2004 (3G 2004)*, 2004, pp. 569–573. [Online]. Available: http://ieeexplore.ieee.org/stamp/stamp.jsp?tp=&arnumber=1434541

[5] R. Naja, J.-P. Claude, and S. Tohme, Adaptive multi-user fair packet scheduling in HSDPA network, in *Proc. Int. Conf. Innovations in Information Technol. 2008 (IIT 2008)*, December 2008, pp. 406–410. [Online]. Available: http://ieeexplore.ieee.org/stamp/stamp.jsp?tp=&arnumber=4781652

[6] R. Kwan, M. Aydin, C. Leung, and J. Zhang, Multiuser scheduling in HSDPA using simulated annealing, in *Proc. Int. Wireless Commun. Mobile Computing Conf. 2008 (IWCMC 2008)*, August 2008, pp. 236–241. [Online]. Available: http://ieeexplore.ieee.org/stamp/stamp.jsp?tp=&arnumber=4599941

[7] C. Mehlführer and M. Rupp, Novel tap-wise LMMSE channel estimation for MIMO W-CDMA, in *Proc. 51st IEEE Global Telecommun. Conf. 2008*

(GLOBECOM 2008), New Orleans, LA, November 2008. [Online]. Available: http://publik.tuwien.ac.at/files/PubDat_169129.pdf

[8] C. Mehlführer, S. Caban, M. Wrulich, and M. Rupp, Joint through-put optimized CQI and precoding weight calculation for MIMO HS-DPA, in *Conf. Record 42nd Asilomar Conf. Signals, Systems and Computers*, Pacific Grove, CA, October 2008, pp. 1320–1325. [Online]. Available: http://publik.tuwien.ac.at/files/PubDat_167015.pdf

[9] M. Harteneck, M. Boloorian, S. Georgoulis, and R. Tanner, Practical as-pects of an HSDPA 14 Mbps terminal, in *Conf. Record 38th Asilomar Conf. Signals, Systems and Computers, 2004*, Pacific Grove, CA, 1: 799–803, November 2004. [Online]. Available: http://ieeexplore.ieee.org/stamp/stamp.jsp?arnumber=1399246

[10] M. Harteneck, M. Boloorian, S. Georgoulis, and R. Tanner, Through-put measurements of HSDPA 14 Mbit/s terminal, *Electronics Lett.*, 41(7): 425–427, March 2005. [Online]. Available: http://ieeexplore.ieee.org/stamp/stamp.jsp?arnumber=1421242

[11] C. Mehlführer, S. Caban, and M. Rupp, Measurement based evaluation of low complexity receivers for D-TxAA HSDPA, in *Proc. 16th Eur. Signal Processing Conf. (EUSIPCO 2008)*, Lausanne, Switzerland, August 2008. [Online]. Available: http://publik.tuwien.ac.at/files/PubDat_166132.pdf

[12] M. Wrulich, C. Mehlführer, and M. Rupp, Interference aware MMSE equaliza-tion for MIMO TxAA, in *Proc. 3rd Int. Symp. Communications, Control and Signal Processing (ISCCSP 2008)*, St. Julians, Malta, March 2008, pp. 1585–1589. [Online]. Available: http://publik.tuwien.ac.at/files/pub-et_13657.pdf

[13] C. Mehlführer, M. Wrulich, and M. Rupp, Intra-cell interference aware equal-ization for TxAA HSDPA, in *Proc. 3rd IEEE Int. Symp. Wireless Pervasive Com-puting (ISWPC 2008)*, Santorini, Greece, May 2008, pp. 406–409. [Online]. Available: http://publik.tuwien.ac.at/files/pub-et_13749.pdf

[14] S. Geirhofer, C. Mehlführer, and M. Rupp, Design and real-time measure-ment of HSDPA equalizers, in *Proc. 6th IEEE Workshop on Signal Processing Advances in Wireless Communications (SPAWC 2005)*, New York City, June 2005, pp. 166–170. [Online]. Available: http://publik.tuwien.ac.at/files/pub-et_9722.pdf

[15] L. Mailaender, Linear MIMO equalization for CDMA downlink signals with code reuse, *IEEE Trans. on Wireless Commun.*, 4(5): 2423–2434, September 2005. [Online]. Available: http://ieeexplore.ieee.org/iel5/7693/32683/01532226.pdf

[16] R. Love, K. Stewart, R. Bachu, and A. Ghosh, MMSE equalization for UMTS HSDPA, in *Proc. 58th IEEE Vehicular Technol. Conf. 2003 (VTC2003—Fall)*, 4: 2416–2420, October 2003. [Online]. Available: http://ieeexplore.ieee.org/stamp/stamp.jsp?arnumber=1285963

[17] Y. Guo, J. Zhang, D. McCain, and J. Cavallaro, Efficient MIMO equal-ization for downlink multi-code CDMA: Complexity optimization and comparative study, in *Proc. 47th IEEE Global Telecommun. Conf. 2004 (GLOBECOM 2004)*, 4: 2513–2519, November 2004. [Online]. Available: http://ieeexplore.ieee.org/stamp/stamp.jsp?arnumber=1378459

[18] Y. Guo, J. Zhang, D. McCain, and J. R. Cavallaro, An efficient circulant MIMO equalizer for CDMA downlink: Algorithm and VLSI architecture, *EURASIP Journal on Applied Signal Processing*, Vol. 2006, Article ID 57134, 2006. [Online]. Available: http://www.hindawi.com/GetPDF.aspx?doi=10.1155/ASP/2006/57134

[19] G.H. Golub and C.F. van Loan, Eds., *Matrix Computations*, 3rd ed. The Johns Hopkins University Press, 1996.

[20] D. Garrett, G. Woodward, L. Davis, G. Knagge, and C. Nicol, A 28.8 Mb/s 4x4 MIMO 3G high-speed downlink packet access receiver with normalized least mean square equalization, in *Digest of Technical Papers IEEE Int. Solid-State Circuits Conf. 2004 (ISSCC 2004)*, 1: 420–536, February 2004. [Online]. Available: http://ieeexplore.ieee.org/stamp/stamp.jsp?arnumber=1332773

[21] J. Ylioinas, K. Hooli, K. Kiiskila, and M. Juntti, Interference suppression in MIMO HSDPA communication, in *Proc. 6th Nordic Signal Processing Symp. 2004 (NORSIG 2004)*, 2004, pp. 228–231. [Online]. Available: http://ieeexplore.ieee.org/stamp/stamp.jsp?arnumber=1344565

[22] S. Caban, C. Mehlführer, G. Lechner, and M. Rupp, Testbedding MIMO HSDPA and WiMAX, in *Proc. 70th IEEE Vehicular Technol. Conf. (VTC2009—Fall)*, Anchorage, AK, September 2009. [Online]. Available: http://publik.tuwien.ac.at/files/PubDat_176574.pdf

[23] S. Caban, C. Mehlführer, R. Langwieser, A. L. Scholtz, and M. Rupp, Vienna MIMO testbed, *EURASIP J. Appl. Signal Processing, Special Issue on Implementation Aspects and Testbeds for MIMO Systems*, Vol. 2006, Article ID 54868, 2006. [Online]. Available: http://publik.tuwien.ac.at/files/pub-et_10929.pdf

[24] Kathrein, Technical Specification Kathrein Antenna Type No. 800 10543. [Online]. Available: http://www.kathrein.de/de/mcs/produkte/download/9363438.pdf

[25] C. Kakoyiannis, S. Troubouki, and P. Constantinou, Design and implementation of printed multi-element antennas on wireless sensor nodes, in *Proc. 3rd Int. Symp. Wireless Pervasive Computing 2008 (ISWPC 2008)*, Santorini, Greece, May 2008, pp. 224–228. [Online]. Available: http://ieeexplore.ieee.org/stamp/stamp.jsp?arnumber=4556202

[26] B. Efron and D.V. Hinkley, *An Introduction to the Bootstrap (CRC Monographs on Statistics & Applied Probability 57)*, 1st ed. Chapman & Hall, 1994.

[27] I.E. Telatar, Capacity of multi-antenna Gaussian channels, Eur. Trans. Telecommun. 1999, Technical Memorandum, Bell Laboratories, Lucent Technologies, 10(6): 585–595, October 1998. [Online]. Available: http://mars.bell-labs.com/papers/proof/proof.pdf

[28] G.J. Foschini and M.J. Gans, On limits of wireless communication in a fading environment when using multiple antennas, *Wireless Personal Commun.* 6(3): 311–335, 1998. [Online]. Available: http://www.springerlink.com/content/h1n7866218781520/fulltext.pdf

Chapter 12

HSDPA Indoor Planning

Tero Isotalo, Panu Lähdekorpi,
and Jukka Lempiäinen

Contents

12.1 Introduction

This chapter introduces principles of indoor coverage and capacity planning, concentrating on requirements created by High-Speed Downlink Access (HSDPA) service. The basic principles and elements for planning Universal Mobile Telecommunication System (UMTS) indoor network using wideband code division multiple access (WCDMA) are presented, and two different basic configurations to provide HSDPA service inside buildings are further investigated. In a dedicated indoor system, the service is provided by a dedicated indoor base station that is connected to an antenna system inside the building. The other approach uses a WCDMA repeater that amplifies the received signal from the outdoor UMTS network to the indoor antenna system. The superiority of dedicated indoor systems, as well as outdoor-to-indoor repeaters compared to a traditional macrocellular approach is clearly shown in the literature (e.g., [9,10]). The Release 99 (R99) specification defines the basis of the UMTS system, and the Release 5 (R5) introduces the HSDPA bringing improvements mostly to the physical layer and radio resource control [3,7]. The fundamentals of HSDPA are not discussed here but can be studied in the literature (e.g., [4,7]). This chapter introduces the basics of WCDMA indoor planning, emphasizing the requirements that HSDPA has brought out.

The first section starts with a description of the indoor radio propagation environment and the challenges it causes to WCDMA systems. Then, "Strategies and Configurations for Providing Indoor Coverage" continues by introducing different strategies for providing dedicated cellular HSDPA service to indoor locations from the network functionality point of view, and presents different system configurations. Then the "Equipment" section presents the essential technical properties of the equipment used in implementing the UMTS indoor network. In "HSDPA Indoor Network Planning," the planning process for HSDPA indoor system is assessed. Example link budgets for different UMTS/HSDPA indoor configurations are presented, the capacity of HSDPA in indoor radio channel is analyzed, and different aspects of optimizing the performance of indoor HSDPA service are discussed.

12.2 Indoor Environment and Propagation

12.2.1 Indoor Propagation Channel

The radio propagation channel describes what kind of changes the transmitted signal is undergoing while propagating through the environment. The propagation environment types can be roughly categorized into macro- and microcellular outdoor environments and picocellular indoor environment. The basic parameters that characterize the propagation environment are

- Delay profile or delay spread
- Frequency response
- Coherence bandwidth
- Angular spread
- Doppler spread
- Propagation slope
- Location variability or slow-fading standard deviation

Indoor delay profiles clearly differ from outdoor environments. Delay spread in different outdoor environments can vary between 500 and 3000 ns [13], whereas in an indoor environment, the variation is between 10 and 500 ns [5,13]. In addition, the number of multipath components is typically higher indoors. Channel frequency response describes how different frequencies fade in the radio channel, and the coherence bandwidth defines the frequency range where fading correlates. Delay spread and coherence bandwidth have a direct relation:

$$\Delta f_c = \frac{1}{2\pi S} \tag{12.1}$$

where Δf_c is the coherence bandwidth and S is the delay spread of the channel. The coherence bandwidth in outdoor environments varies between 50 kHz and 1.6 MHz, whereas indoors it can vary from 300 kHz up to even larger than 16 MHz [13,20]. The angular spread in outdoor environments is a couple of tens of degrees in the horizontal, and below 10° in the vertical direction, whereas indoors it is typical to have an angular spread of 360° in both the vertical and horizontal direction. The Doppler spread is caused by movement of the transmitter, receiver, or propagation environment. In indoor environments, the Doppler spread is typically rather narrow due to low mobile speed. Location variability, also known as slow fading standard deviation, has values between 4 and 8 dB outdoors, whereas indoors values of up to 10 dB have been reported [20]. Due to high wall and floor penetration loss, the typical propagation slope is clearly higher indoors. Finally, it can be concluded that the indoor propagation channel is quite different from the outdoor propagation channel.

12.2.2 Repeater Donor Link Propagation Channel

When considering utilization of outdoor-to-indoor repeater implementation, the donor antenna of the repeater is typically located on the roof of the building. Furthermore, the type of link between the repeater and the outdoor macro base station is considered a fixed point-to-point radio link. Obstacles in the repeater link would cause additional attenuation in the received power. For point-to-point radio links, the Fresnel zone should be left empty in order to avoid the additional attenuations in the link [20]. For a 2-GHz carrier frequency and 500-m link length, for example, the maximum radius of the Fresnel ellipsoid is approximately 4 m [20]. Based on the assumption of the non-obstructed link, the radio channel in the repeater donor side can be modeled using Friis' equation for free space radio propagation.

12.2.3 UMTS System Performance in an Indoor Environment

The bandwidth of a system, B, and the coherence bandwidth of the channel define whether a system is wideband or narrowband. In a narrowband system, the coherence bandwidth is larger than the system bandwidth; thus the whole system bandwidth fades simultaneously and the channel is flat fading. In a wideband system, the coherence bandwidth is clearly smaller than the system bandwidth; thus the channel is frequency selective (Table 12.1). So, in a wideband system, the average changes in the channel over the system bandwidth are clearly smaller compared to a narrowband system. According to Table 12.1 and Δf_c values shown above UMTS can be identified as wideband in all outdoor environments, but is changing toward narrowband in indoor environments, which may cause deterioration of system behavior.

A RAKE receiver in UMTS/WCDMA can mitigate multipath fading with maximal ratio combining. A typical receiver has a time resolution of one chip time ($1/3.84$ Mcps $= 0.26$ μs), which equals 78 meters in distance in air interface. Thus the receiver can separate multipath components that have more than one chip time separation. In indoor environments, the multipath components may have significantly shorter separations, which makes combining impossible, and therefore may degrade system performance.

Table 12.1   Definition of Narrowband and Wideband Systems.

Condition	System
$B \ll \Delta f_c$	Narrowband
$B \gg \Delta f_c$	Wideband

Source: From M.K. Simon and M.S. Alouini. Digital Communications over Fading Channels: A Unified Approach to Performance Analysis. Wiley InterScience, Sep 2000.

In WCDMA, different downlink channels from one base station are separated by synchronized orthogonal Walsh codes [6]. If the codes are fully orthogonal in the reception end, then the transmissions do not interfere with each other. However, in practice, the orthogonality is degraded due to multipath fading. The orthogonality varies as a function of multipath profile (delay spread) and distance, having values between 0 and 1, where 1 means perfect orthogonality. Longer delay spread and distance degrade the orthogonality, so indoor systems are expected to have better performance in terms of orthogonality [16]. Reported orthogonality values in dedicated indoor systems vary between 0.68 and 0.85; and when indoor coverage is provided by a macro cell, the orthogonality varies between 0.34 and 0.55 [10,21]. However, based on [16], larger variations may also occur. The results clearly indicate that dedicated indoor systems should provide better performance compared to indoor coverage from outdoor cells. The impact of changes in code orthogonality on HSDPA link- and system-level performance have been studied (e.g., [6,7]).

12.2.4 Indoor Propagation Areas

The different characteristics of indoor environments compared to outdoor environments are caused by very densely spaced obstacles in the environment. Furthermore, antennas are typically placed close to objects. Indoor environments can be categorized in many different ways, based on the usage, traffic, shape, material, etc. [5,8,20,21]. The different indoor environments have different characteristics: the propagation slope, number of scatterers and reflections, probability of line-of-sight, etc. In particular, the attenuation between floors and walls, and the number of windows can have a significant impact on signal propagation. In practice, indoor areas can be divided into a few basic categories:

■ Dense areas (e.g., multiple small offices connected with narrow corridors)
■ Corridors (e.g., long and wide corridors in buildings)

■ Open areas (e.g., entry halls, auditoriums, airports, railway stations)
■ Special areas (e.g., elevators, fire escapes, basements)

Different indoor areas have different propagation characteristics. An extensive list of references is provided in [5]; but because there is lots of variation, depending on the individual building, field measurements are recommended if accurate and reliable propagation characteristics of a certain building are needed.

12.3 Strategies and Configurations for Providing Indoor Coverage

The most basic way of providing indoor coverage is to rely on outdoor cellular networks. However, building penetration attenuates the signal typically by 15 to 20 dB. This requires large thresholds in link budget. Thus, either indoor coverage is not available on the cell edge, or large cell overlapping is needed, which increases other cell interference levels. For the inner parts of larger buildings, the attenuation may be clearly higher, and thus lack of coverage is probable. In the case of power-controlled R99 connections, indoor users with high path loss cause excess downlink interference to outdoor users, whereas with HSDPA, indoor users are expected to have poor throughput. Because it is probable to have the users requiring high throughput located indoors, improving indoor coverage may be worthwhile for network operators.

There are two basic approaches to improve indoor coverage: a dedicated indoor system and an outdoor-to-indoor repeater. When considering the cell configuration, there are two strategies: using a single cell for one building or having multiple cells in one building. In addition, the antenna configuration can use a single antenna at a cell, or a distributed antenna system (DAS) where the signal is split to several antennas or radiating cables. In the future, different optical solutions are likely to replace lossy coaxial cables, where antennas would be replaced by remote RF (radio frequency) heads, including optical interfaces, amplifiers, and antennas.

12.3.1 Single-Cell Strategy

In single-cell strategy (Figure 12.1(a)), a building is covered using one indoor cell or one outdoor-to-indoor repeater. Heavy exterior walls are blocking the signals from outdoor cells, there are no handover regions inside buildings, and also the other cell interference remains at low level. Varying numbers of antennas can be connected to the base station. A single antenna may be built into the base station, whereas distributed antenna

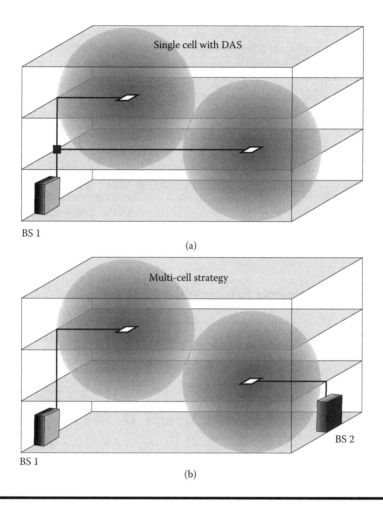

Figure 12.1 Principal difference between (a) single-cell strategy implemented using multiple antennas, and (b) multi-cell strategy.

systems may have tens of antennas. Although a DAS with several antennas can be used in the single-cell strategy to cover even larger buildings, the cell size cannot be endlessly increased, due to increasing antenna line losses. Also, the total available capacity is limited to one cell. Thus, the single-cell strategy is well suited for small buildings, or medium-sized buildings with low amounts of traffic.

12.3.2 Multi-Cell Strategy

For larger buildings with high numbers of users, the multi-cell strategy (Figure 12.1(b)) can be considered. In a multi-cell strategy, the indoor coverage is formed by several adjacent indoor cells, which creates handover regions inside the building. Therefore, the antenna locations must be more carefully selected due to the handovers taking place when a user moves between the cells. Due to slow fading variations in the received signal level, some overlapping between the cells is needed to guarantee continuous coverage. On the other hand, other cell interference must be efficiently controlled. Therefore, in a multi-cell strategy, antenna selection and placement must be done more carefully than in the single-cell strategy.

Because the cost of the multi-cell indoor network is high, the network can be deployed in parts. An outdoor-to-indoor repeater can be used as a first solution to ensure coverage. Later, it can be upgraded to a dedicated indoor system utilizing an indoor base station. As the capacity need grows, a multi-cell indoor system can be deployed. If this is taken into account at the beginning, the indoor antenna installations can remain untouched throughout the network evolution.

12.3.3 Picocells and Femtocells

So-called pico or femto base stations are base stations with small power and in-built antenna, with an option to connect external antennas. The physical size of the base station is small enough to be mounted on an indoor wall similar to antennas. Femto base stations are even planned to be userdeployable, in a similar manner to that of wireless access points.

12.3.4 Distributed Antenna System (DAS)

The most used strategy for planning indoor is the distributed antenna system, where several antennas are connected to a single base station. The idea of DAS is to divide the signal from the base station into several branches, and to connect a discrete antenna or radiating cable to each branch. Typically, coaxial cables with signal dividers (splitters and tappers) are used, but also, for example, optical connection between Node B and active RF-heads can be used. Due to a steep propagation slope in indoor environments, the longitudinal loss in cables is lower than in an air interface. Thus, it is beneficial to split the signal into multiple antennas. DAS also brings users closer to the antenna, which shortens the propagation path, thus improving orthogonality due to less multipath fading and a higher probability of line-of-sight connection. Usually in DAS installation, diversity reception or low noise amplifiers (LNAs) are not used due to high cost [13]. The basic idea of DAS is illustrated in Figure 12.1(a) and an example layout is shown in Figure 12.3(a). In the sense of interference and functionality, a single-cell configuration outperforms a multi-cell configuration, because all antennas are connected to one Node B. Thus, the interference from other cells remains rather small, and handovers do not occur when users are moving in the building.

12.3.5 Outdoor-to-Indoor Repeater

If indoor base stations are not the preferred solution, and if sufficient outdoor networks are available, outdoor-to-indoor repeaters may be considered. A WCDMA repeater can be used to amplify the WCDMA signal between a nearby macrocellular outdoor base station and an indoor antenna system. When the receiving antenna of the repeater is placed on the rooftop of the building, and the received and amplified signal is guided to the indoor antennas using a cable, the signal does not need to penetrate through the exterior walls of the building. Therefore, the building penetration loss can be bypassed. Finally, the amplified WCDMA signal can be received by the indoor users without using an expensive indoor base station. This outdoor-to-indoor repeating principle can be utilized in practice to provide low-cost HSDPA service in a building located near the edge of an outdoor macro cell. Because additional signal amplification takes place at the repeater, data throughput can be increased due to an improved signal-to-interference ratio. The link between the repeater and the macrocellular base station is called the "donor link." Furthermore, the link between the indoor antenna and the indoor user is called "service link," and antennas are called "donor antennas" and "service antennas," correspondingly. In outdoor-to-indoor repeater implementation, the donor antenna connects the repeater to the macrocellular base station, while the serving indoor antenna system is connected to the repeater to provide service for indoor users. The outdoor-to-indoor repeating principle, together with an example repeater configuration, is illustrated in Figure 12.2.

The WCDMA repeater discussed in this chapter is an analog, transparent, nonregenerative repeater. This means that the repeater device does not utilize any signaling and control information from the network, but only

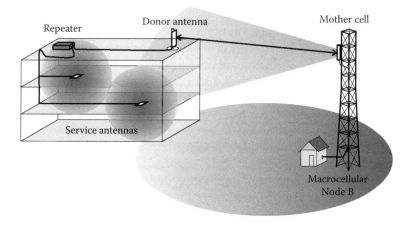

Figure 12.2 **The principle of outdoor-to-indoor repeating.**

repeats the entire desired carrier frequency band. Similarly, the network is not aware of the repeater's presence. The transparency of the repeater provides flexibility and cost efficiency. To deploy a repeater, no changes or reconfigurations are required on the network side. In addition, the cost of the repeater is low due to reduced device complexity, especially compared to a base station. Furthermore, the structural complexity of the radio network remains at the original level if a repeater is deployed instead of an indoor base station. For each new indoor base station, a connection to the radio network controller is required, which increases the structural complexity of the network. In the case of a repeater, this connection is not needed, and the existing capacity can be boosted with the existing radio resources. Because with a repeater the radio access interface is shared between the repeater users and the users of the mother cell (i.e., the cell to which the repeater is connected), repeater installation must be done with care to avoid decreasing the performance of the mother cell. The most important repeater-related parameters are repeater amplifier gain and repeater antenna deployment strategy. These planning-related parameters are further discussed later in this chapter.

12.4 Equipment

12.4.1 Antennas and Antenna Line Elements

12.4.1.1 Antenna Types

A typical indoor antenna is a small, low-gain antenna. Reasons for this are the wide angular spread in the indoor environment, coupling loss limitations when users are close to the antenna, and the need for invisible antenna installation. There are three basic types of indoor antennas: omnidirectional antennas, directional antennas, and radiating cables.

Omnidirectional antennas are typically used in an open area, such as a lobby or auditorium. The radiation pattern of an omnidirectional antenna is isotropic in the horizontal plane, and the vertical pattern is close to the ideal dipole antenna pattern; thus, the gain is approximately 2 dBi.

Directional indoor antennas are good for covering narrower areas, such as corridors. The radiation pattern is clearly wider than in directional macro cell antennas. Both the horizontal and vertical −3 dB beam width are around 70° to 90°, and the antenna gain is around 6 or 7 dBi.

Radiating cables (also called leaky feeders or coaxial antennas) can be used to provide smooth coverage for long distances. They are traditionally used in tunnels but can be used, for example, in elevator shafts, staircases, and corridors. Radiating cable is a special coaxial cable that has holes or a groove in the outer conductor and operates as a small antenna, leaking signal out from the cable.

Typical antennas for outdoor micro- and macrocellular networks can be also used at the donor side The antennas have about a 65° to 90° horizontal beam width for three-sectored, and 33° for six-sectored installations. The vertical beam width is typically between 6° and 12°. The antenna gain values are between 12 and 18 dBi. In some cases, even narrower antennas with higher gain can be considered.

12.4.1.2 Splitters and Tappers

Splitters and tappers are power dividers, and are needed in distributed antenna systems. A splitter divides the input power into equal parts, and typically has two to four ports, thus having an attenuation of 3 to 6 dB. A tapper divides the power into unequal parts, and typically has two ports, with attenuations of, for example, 1.0 and 7.0 dB, or 0.1 and 15 dB [9, 11].

12.4.1.3 Cables

Feeder cables in cellular networks are coaxial cables of between $\frac{1}{4}$ and $2\frac{1}{4}$ inch, and the attenuations at 2 GHz vary from 21 to 3.5 dB/100 m, respectively. Typical choices for micro- and macrocellular outdoor installations is $\frac{7}{8}$-inch cable with attenuation of about 6 dB/100 m; whereas in indoor installations, $\frac{1}{2}$-inch cable with attenuation of about 11 dB/100 m is a typical choice [11, 19].

More information about antenna line equipment, antenna patterns, specifications, etc. can be obtained from the manufacturers.

12.4.2 Base Station Equipment

The base station in UMTS is called Node B. Node B equipment has few indoor radio planning-related parameters, such as total transmission power and noise figure, and several radio resource control-related parameters, such as maximum power per user, pilot channel and other common channel powers, power allocated for HSDPA users, etc. Typically, macrocellular base stations have a maximum total transmission power of +43 to +50 dBm, and indoor base stations below +38 dBm. Path loss estimation is based on the primary common pilot channel (P-CPICH). Thus, the power setting of the pilot channel is an important parameter from the system performance point of view. The typical pilot channel power for a macro base station is +33 dBm, and for an indoor base station +27 to +30 dBm. In the case of a pico base station, power can be smaller than +27 dBm, and with femto base station, the total power can go down to +15 dBm.

The HSDPA power settings can be configured in various ways. Usually, R99 connections and HSDPA connections share the frequency. Thus, either a fixed part of the total power is reserved for HSDPA, or all the unused power can be allocated to HSDPA. In all configurations, a fixed amount of power must be reserved for common channels, for example, the pilot channel. If a dedicated carrier for HSDPA is used, all the power remaining from the common channels can be utilized for the HSDPA users. The actual power allocation depends on each implementation, wherein the network vendor can set some limits. The impact of power allocation on HSDPA capacity is discussed in detail in [7].

12.4.3 WCDMA Repeaters

A WCDMA repeater is a device used to amplify and retransmit the received WCDMA signal on the same carrier frequency. Here, only analog and transparent repeaters are discussed; they operate at the physical layer and are unable to reduce the noise from the received signal.

A WCDMA repeater has two connectors for connecting the donor and service antennas. Naturally, power must be supplied to the amplifier from a power source. The configuration of the repeater setting is typically made by a computer using a local or remote connection.

12.4.3.1 Parameters

The most important repeater parameters are carrier frequency or channel number selection, and repeater amplifier gain for the uplink and downlink. The repeater amplifier gain affects the reception level of the signals in the uplink and downlink receiver, and is important from a network planning point of view. The repeater amplifier gain can typically have values between 50 dB and 90 dB.

12.4.3.2 Antenna Isolation

To guarantee successful repeater operation, the donor and service antennas must maintain sufficient isolation to prevent self-oscillation. In a self-oscillating state, the repeater starts to transmit the signal received and amplified by itself. This leads to uncontrollable repeater operation, and the mother cell will be blocked due to excess interference from the repeater. Based on [1], antenna isolation should maintain a level of 15 dB larger than the selected repeater amplifier gain. The antenna isolation is equal to the total attenuation measured between the donor and service antenna ports in the repeater. The isolation can be improved by adding attenuation between the donor and service antennas, by, for example, taking advantage of obstacles in the propagation environment, or by increasing the distance between the donor and service antennas of the repeater.

12.4.3.3 Automatic Gain Control

WCDMA repeaters typically include an automatic gain control system (AGC) to keep the output power of the repeater at a decent level in order to avoid amplifier circuit oscillation due to antenna isolation problems. Thus, when setting the repeater amplifier gain, it is important to check whether or not the AGC has been activated. If AGC is active, the set repeater amplifier gain may not actually be the one used by the repeater.

12.4.3.4 Noise

Noise produced in repeater circuits plays a crucial role in repeater planning because the noise is amplified by the repeater along with the desired signal. Thus, the required repeater amplifier gain is directly proportional to the amount of noise received at the target locations (base station receiver and mobile station receiver). This is discussed in more detail later. The noise is composed of thermal noise and of impairments of the repeater device. The addition of noise caused by the repeater device impairments can be modeled using a noise figure. The noise figure of the repeater is typically 2 dB to 4 dB.

12.4.3.5 Power

A repeater amplifier has a maximum characteristic output power for the uplink and downlink. This sets the maximum average power level measured at the output ports of the repeater. According to [1], a typical value for the maximum average output power is +30 dBm in the downlink direction. With strong input signals at the repeater, the maximum average power level may start to limit the output signal level.

12.5 HSDPA Indoor Network Planning

The planning process of a WCDMA radio network is divided into dimensioning, detailed planning, and optimization. The detailed planning is further divided into configuration planning, combined coverage and capacity planning called topology planning, code planning, and parameter planning [13]. The basic guidelines used in an outdoor network planning process can be adapted to the planning of indoor networks. The main differences lie in the configuration planning and topology planning phases. In the configuration planning phase, the antenna line elements are defined and the link budget calculations are performed. The topology planning of an outdoor network is based on coverage predictions and system simulations. Due to lower reliability in propagation prediction indoors, system simulations are providing only a rough indication of system coverage and performance, while the detailed planning is more empirical and experiential.

The target of indoor planning is to design the indoor network in such a way that adequate signal quality can be ensured in the targeted indoor location. In planning a dedicated indoor system, the main challenge is to find good parameters in the base station to provide proper coverage and capacity. However, for the outdoor-to-indoor repeater solution, repeater parameters also have a direct impact on the mother cell. Thus, the surrounding macrocellular network must be considered when deploying the repeater.

The actual placement of the antennas needs careful planning. Sufficient signal levels must be provided in all important areas to ensure the required service coverage. A sufficient signal level is defined by planning thresholds, which are based on link budget calculations, system simulations, and radio interface measurements. Controlling the leakage of outdoor signal inside the building, as well as leakage of indoor signal outside the building is important in handling inter-cell interference, as well as smooth handovers. Because HSDPA does not support soft handover (SHO), large cell overlapping is inadvisable. In probable handover areas to outdoor network, such as main entrances, indoor coverage can be slightly extended to the outdoors in a controlled way to ensure successful handovers. Inside the building, the coverage provided by the indoor system should be planned in such a way that handovers to outdoor networks do not occur. Controlling the indoor coverate is especially challenging near windows, which are areas prone to signal leakage between outdoor and indoor networks. Furthermore, the indoor network should not cause excess interference to the outdoor network operating in the same frequency band. Tinted outermost windows in new buildings help in isolating outdoor and indoor networks.

Indoor planning tools with propagation models can be used in dimensioning the network, but field measurements with test transmitters are recommended to verify the characteristics of the environment.

12.5.1 Planning of Dedicated UMTS Indoor System with DAS

The basic target of planning a DAS is to provide constant coverage everywhere in the building. Typically this is best achieved by having equal effective isotropic radiated power (EIRP) at each antenna. Because it is usually impossible to keep equal cable loss for each antenna, splitters and tappers must be smartly used to keep the antenna EIRP values in balance. In the example configuration in Figure 12.3, the maximum difference between the antenna line losses to different antennas is 1.6 dB, although cable losses vary significantly. The selection of the antenna depends on the indoor area type. For open area, an omnidirectional antenna is a typical selection. On the other hand, a directional antenna may be a better fit for a corridor environment [13]. Due to its complicated installation, 1/2-inch cable is a typical selection, although the longitudinal loss is rather high [21]. Because users may be located very close to antennas in an indoor environment, it is important to take care of coupling loss (the loss between a Node B antenna connector and the mobile station antenna). If the coupling loss is

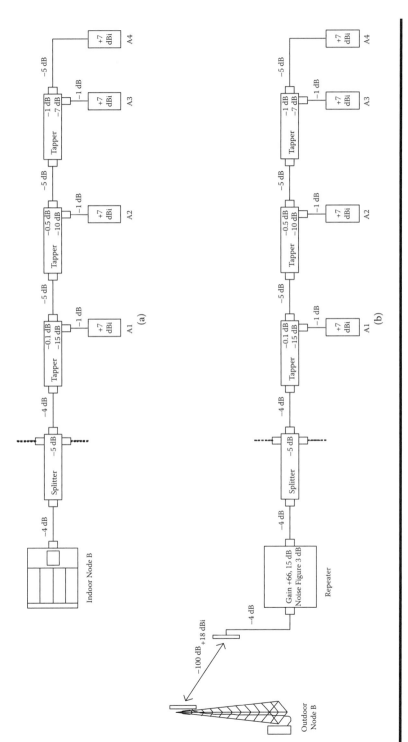

Figure 12.3 **Example DAS configuration using (a) dedicated indoor system and (b) outdoor-to-indoor repeater.**

too small, a mobile near an antenna may cause excess interference in the uplink direction due to too high minimum transmission power. A typical minimum value for coupling loss is approximately 50 dB, calculated based on a minimum mobile transmission power of −50 dBm and a receiver sensitivity of −100 dBm. Practical issues concerning the basics of indoor DAS design are well covered in literature (see, for example, [21]).

12.5.1.1 Indoor Propagation Prediction

To have reliable calculations for indoor propagation (in, for example, a planning tool), propagation models for indoor environment are needed. The indoor environment is challenging for propagation prediction. Even so, there are several propagation models available. The models are divided into empirical models (e.g., COST-231 multi-wall model, Ericsson model) and deterministic models (e.g., ray tracing) [14, 20]. The simplest empirical models only model the propagation slope for different indoor environments; thus, the accuracy is very poor. Empirical wall and floor factor models have a distance-dependent free space loss component, where losses caused by wall and floor penetration are added. Empirical models for the indoors utilize ray tracing techniques, which have good theoretical accuracy but require high computation power. However, they also require very accurate information about the building, and moved furniture or an opened door may have a significant impact on the results, and thus errors are expected in practice. There are different commercial tools available where the building layout can be uploaded. After defining the materials and setting the antenna position and EIRP values in the tool, coverage estimation is calculated. Currently, indoor propagation estimation is primarily based on empirical models, where low accuracy requires field measurements to verify the results.

12.5.2 *Planning of Outdoor-to-Indoor Repeater System*

The key parameter that defines the repeater performance is the repeater amplifier gain. The required repeater amplifier gain value to achieve a certain received DL signal strength at the repeater output port naturally depends on the donor link properties of the outdoor-to-indoor repeater, that is, the losses and gains in the donor link path. The main contributor in donor link loss is the free space attenuation due to distance between the repeater donor antenna and the mother base station antenna. In addition, the antenna line components of the donor base station and the donor antenna line of the repeater contribute to the total loss. Furthermore, to achieve a certain signal strength (EIRP) at the indoor antenna, the losses caused by the serving indoor antenna system (either single antenna or DAS) must also be considered. The boundary values limiting the acceptable repeater amplifier gain are maximum allowed noise rise at UL receiver, repeater antenna isolation, and maximum repeater output power.

Because a repeater is merely an amplifier, noise is amplified together with the desired signal at the repeater. Due to amplification of noise at repeater, the receivers affected by the repeater will experience increased noise levels, when the repeater is turned on (even with no traffic in the network). Increased received noise power at the base station directly leads to decreased base station sensitivity. In the uplink direction, the impact of repeater noise is more significant due to low coupling loss on the donor side (line-of-sight connection and two highly directive BS antennas used). In the downlink, the mobile station receiver typically has lower sensitivity, and the mobile antenna can be described as an omnidirectional antenna. Moreover, in the uplink direction, the receiver at the base station is shared by all mobile stations in the cell, and thereby the repeater directly affects the performance of the entire macro cell. Due to the noise-related phenomena described above, the total repeater performance is typically a compromise between increased repeater service area and decreased mother cell performance in the uplink. Hence, both the uplink and downlink must be considered in outdoor-to-indoor repeater planning. The mathematical presentation of the noise and sensitivity behavior in analog repeater systems can be found in [18].

A narrow beam repeater donor antenna is typically preferred to direct and gather the signal energy toward and from the mother cell antenna. Due to the narrow beam, low angular spread and thus line-of-sight between the repeater donor antenna and the mother cell antenna are typically required in order to be able to pick up the energy of the desired signal. When selecting the donor link configuration (i.e., donor antenna type and direction), emphasis should be placed on establishing high cell isolation. The cell isolation here means that the difference in the received signal levels from the first and second best cells at the donor antenna location is high enough to avoid amplifying an excess amount of other cell interference using the repeater. The impacts of donor cell isolation on system performance are discussed in more detail in [9]. If the two macro cells are repeated with almost equal level, all users in the repeater service area will be in soft handover, thereby wasting resources from the participating cells. Usually it can be assumed that the donor link connection has line-of-sight to the mother cell. Thus, to increase cell isolation, accurate antenna directioning and narrow beam antennas are preferred to separate the desired mother cell signal from the other interfering cells.

12.5.3 Handover Functionality

The fundamental difference in the functionality between a dedicated indoor system and an outdoor-to-indoor repeater is the handover operation when users are moving from outdoors to indoors, and vice versa. The difference can be assessed by considering a situation where a user of the network is approaching the planned indoor location (i.e., building) with an active HSDPA packet data connection. In the case of a dedicated indoor system, the user must be handed over from the current outdoor macro cell to the new indoor cell while entering the building. Here, the importance of proper indoor planning is emphasized. Users must be seamlessly and successfully handed over to the indoor cell. If the handover area between the outdoor macro cell and the indoor cell is poorly planned, unnecessary handovers from and to the indoor cell may occur.

The handover operation with outdoor-to-indoor repeater is different compared to the dedicated indoor system. Because the indoor coverage produced by the repeater originates from the same logical cell, that is, from the outdoor mother cell (assuming that the building is located in a cell dominance area rather than cell edge), no handovers are required. The mobile station only experiences a sudden increase in the received signal strength when the mobile station moves into the building and becomes served by the indoor antennas. Thus, the utilization of outdoor-to-indoor repeaters also reduces the amount of required signaling load in the cell. However, the correct repeater planning is emphasized here, in particular to ensure the proper operation of the surrounding macrocellular network. The difference in handover functionality of the two indoor implementation approaches is visualized in Figure 12.4.

12.5.4 HSDPA Indoor Link Budget

The fundamental tools of coverage planning are link budget (also called power budget) and propagation model. The outputs of a link budget calculation are the maximum allowed path losses in the uplink and downlink directions. With a proper propagation model, the maximum coverage in decibels can be converted to maximum distance in meters when the propagation environment is known. Example link budgets for outdoor macrocellular WCDMA/UMTS Release 99 configuration can be found from the literature (e.g., [6,12,13]). Therefore, the link budget discussion here focuses on the HSDPA indoor configuration.

In Table 12.4, link budget for dedicated indoor systems using UMTS R99 384-kbps service, Table 12.3, UMTS R5 HSDPA service, and also Table 2.4, UMTS HSDPA R5 outdoor-to-indoor repeater configuration are shown[1]. In HSDPA link budgets, the uplink service is planned to provide sufficient capacity for HSDPA uplink feedback information. The link budgets are only

---

[1] The example link budget can be downloaded in Excel format from http://www.cs. tut.fi/tlt/RNG/IndoorLinkBudget.xls.

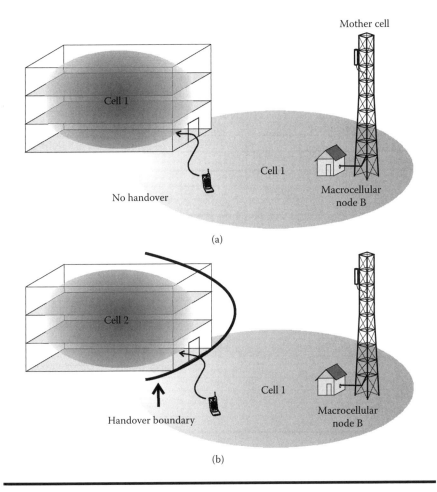

Figure 12.4 Difference in handover functionality of (a) outdoor-to-indoor repeater and (b) dedicated indoor system.

examples, and the actual link budget must always be done carefully to match each network implementation. All the details of the link budget calculations are not shown here, but can be found in literature (e.g., [6,12,13]). The link budget is divided to general parameters, service profile, receiving end, and transmitting end. For the outdoor-to-indoor repeater configuration, repeater parameters are introduced.

12.5.4.1 General Parameters and Service Profile

The general parameters include the fundamentals of the radio interface, and are the same for each configuration. In the service profile, the network load is the upper limit before load control and admission control functions start to limit connections. Here the load is fixed to 50%, but to get more accurate

Table 12.2   Link Budget for UMTS R99 Indoor, 384-kbps Service

Parameter	Unit	Value	
General parameters		*Uplink*	Downlink
Chip rate	cps	$3.84 \cdot 10^6$	
Noise bandwidth ($\alpha = 1.22$)	Hz	$4.68 \cdot 10^6$	
Temperature	K	293	
Boltzmann constant	J/K	$1.38 \cdot 10^{-23}$	
Frequency	Mhz	1950	2150
Service profile	Unit	Uplink	Downlink
Load	%	50	50
Required bit rate (phy layer)	kbps	480	480
Spreading factor		4	8
Receiving end		NodeB	Mobile
Noise figure	dB	4	8
Noise power	dBm	−103.2	−99.2
Interference margin	dB	3.0	3.0
Total interference level	dBm	−100.2	−96.2
Required Eb/N0	dB	4	7
Processing gain	dB	6.0	9.0
Antenna diversity gain	dB	0	0
SHO diversity gain	dB	1	2
Power control headroom	dB	3	0
Required C/I	dB	0.0	−4.0
Receiver sensitivity	dBm	−100.2	−100.3
RX antenna gain	dBi	7	0
LNA gain	dB	0	0
DAS antenna line losses	dB	30	0
Required signal power	dBm	−77.2	−100.3
Transmitting end		Mobile	NodeB
Indoor Node B total power	W		8
Indoor Node B total power	dBm		39.0
TX Power / connection	W	0.13	0.8
TX Power / connection	dBm	21.0	29.0
Antenna gain	dBi	0	7
DAS antenna line losses	dB	0	30
Peak EIRP	dBm	21.0	6.0
Maximum path loss	dB	98.2	106.3

Table 12.3   Link Budget for UMTS HSDPA R5 Indoor

Parameter	Unit	Value	
General parameters		*Uplink*	*Downlink*
Chip rate	cps	$3.84.10^6$	
Noise bandwidth ($\alpha = 1.22$)	Hz	$4.68.10^6$	
Temperature	K	293	
Boltzmann constant	J/K	$1.38.10^{-23}$	
Frequency	Mhz	1950	2150
Service profile	Unit	Uplink	Downlink
Load	%	50	50
Number of HS-DSCH codes			10
Required bit rate (phy layer)	kbps	240	2000
Spreading factor		8	16
Receiving end		NodeB	Mobile
Noise figure	dB	4	8
Noise power	dBm	−103.2	−99.2
Interference margin	dB	3.0	3.0
Total interference level	dBm	−100.2	−96.2
Required Eb/N0 (UL) / SINR (DL) [1]	dB	4	13
Processing gain	dB	9.0	12.0
Antenna diversity gain	dB	0	0
SHO diversity gain	dB	1	0
Power control headroom	dB	3	0
Required C/I	dB	−3.0	1.0
Receiver sensitivity	dBm	−103.3	−95.3
RX antenna gain	dBi	7	0
LNA gain	dB	0	0
DAS antenna line losses	dB	30	0
Required signal power	dBm	−80.3	−95.3
Transmitting end		Mobile	NodeB
Indoor Node B total power	W		8
Indoor Node B total power	dBm		39.0
HS-DSCH Power	W	0.13	5
HS-DSCH Power	dBm	24.0	37.0
Antenna gain	dBi	0	7
DAS antenna line losses	dB	0	30
Peak EIRP	dBm	24.0	14.0
Maximum path loss	dB	104.3	109.3

Table 12.4 Link Budget for UMTS HSDPA Outdoor-to-Indoor Repeater

Parameter	Unit	Value	
General parameters		*Uplink*	*Downlink*
Chip rate	cps	$3.84.10^6$	
Noise bandwidth ($\alpha = 1.22$)	Hz	$4.68.10^6$	
Temperature	K	293	
Boltzmann constant	J/K	$1.38.10^{-23}$	
Carrier frequency	Mhz	1950	2150
Service profile	Unit	Uplink	Downlink
Load	%	50	50
Number of HS-DSCH codes			10
Required bit rate (phy layer)	kbps	240	2000
Spreading factor		8	16
Repeater parameters	Unit	Uplink	Downlink
Repeater noise figure	dB	3	3
Repeater donor link loss (FSL)	dB	100	100
Repeater donor antenna gain	dBi	18	18
Repeater donor antenna line loss	dB	4	4
Repeater amplifier gain	dB	66.2	66.2
Repeater system gain	dB	−19.9	−19.9
Receiving end		NodeB	Mobile
Noise figure	dB	4	8
Effective noise figure with repeater	dB	5.0	8.0
Noise power	dBm	−102.2	−99.2
Interference margin	dB	3.0	3.0
Total interference level	dBm	−99.2	−96.2
Required Eb/N0 (UL) / SINR (DL) [1]	dB	4	13
Processing gain	dB	9.0	12.0
Antenna diversity gain	dB	0	0
SHO diversity gain	dB	1	0
Power control headroom	dB	3	0
Required C/I	dB	−3.0	1.0
Receiver sensitivity	dBm	−102.2	−95.3
Outdoor Node B antenna line loss	dB	0	
Outdoor Node B antenna gain	dBi	15	
Repeater system gain	dB	−19.85	
RX antenna gain	dBi	7	0
Outdoor Node B LNA gain	dB	0	0
DAS antenna line losses	dB	30	0
Required signal power	dBm	−74.4	−95.3

Table 12.4   Link Budget for UMTS HSDPA Outdoor-to-Indoor Repeater (Continued)

Parameter	Unit	Value	
General parameters		Uplink	Downlink
Transmitting end		Mobile	NodeB
Outdoor Node B total power	W		20
Outdoor Node B total power	dBm		43.0
Tx Power / HS-DSCH Power	W	0.13	8
Tx Power / HS-DSCH Power	dBm	24.0	39.0
Outdoor Node B antenna line loss	dB		0
Outdoor Node B antenna gain	dBi		15
Repeater system gain	dB		−19.9
Antenna gain	dBi	0	7
DAS antenna line losses	dB	0	30
Peak EIRP	dBm	24.0	11.2
Maximum path loss	dB	98.4	106.4

results, the actual load of the HSDPA cell should be calculated based on High-Speed Downlink Shared Channel (HS-DSCH) signal-to-interference-plus-noise ratio (SINR) (see Equation (12.2)), inter-cell interference, and network geometry factor [7]. Either a physical layer bit rate or a spreading factor is needed to calculate processing gain (PG). For R99, the used bit rates are the maximum available. In HSDPA, the bit rate changes every 2 ms due to adaptive modulation and coding. Therefore, the bit rate in the link budgets is an adjustable average target value for which the maximum path loss is calculated.

12.5.4.2 Receiving and Transmitting Ends

In the receiving end, the noise figures are typical values for UE and Node B. The noise power equals the thermal noise on the system noise bandwidth plus receiver noise figure. After calculating the interference margin from the cell load level, the total interference level at the receiver can be calculated.

The required signal level divided by the noise plus interference level is called E_b/N_0, which depends on several variables [13]; and therefore, only estimates can be given. Because for HSDPA connections the bit rate is changing every 2 ms, a single E_b/N_0 requirement cannot be given. However, the HSDPA bit rate with a certain number of codes can be connected to a certain SINR. The value in the link budget is based on the simulations in [7], with ten codes. In indoor systems, diversity reception is not typically used. Also, the HSDPA connection does not support SHO or power control; thus, for HSDPA, the required carrier-to-interference ratio (C/I) is simply SINR − PG. Receiver sensitivity is equal to total interference level plus required C/I.

The impact of antenna line elements is added to the receiver sensitivity to obtain the required signal power at the receiver antenna. The Node B antenna line parameters are based on the example DAS shown in Figure 12.3(a). At the UE, the antenna is typically approximated as isotropic, and the antenna line losses are close to zero. The power settings are typical for indoor Node B. The maximum transmission power for R99 UE is +21 dBm; and for HSDPA UE, it is +24 dBm [2]. After calculating the EIRP at the transmitter antenna, the maximum path loss can be calculated.

12.5.4.3 Repeater

Some changes exist in the link budget for the outdoor-to-indoor repeater configuration, when compared to the link budget for the dedicated indoor system. The selected values for the donor antenna line loss and repeater noise figure correspond to typical ones in the situations where a repeater is deployed to a macrocellular network. The values for donor link loss and donor antenna type are very case specific, but here a donor link loss of 100 dB and an antenna of 18 dBi gain has been selected[2]. Furthermore, macro cell antennas that have relatively narrow horizontal beam width (30° half-power beam width) typically have a gain value near 18 dBi.

To take into account the increased noise level at the base station receiver, a modified noise level value has been added to the link budget for the outdoor-to-indoor repeater case. The effective noise figure of the base station receiver is calculated based on the repeater link properties (gains and losses) according to the approach presented in [18]. The modified noise level value has a direct impact on the sensitivity level of the receiver and thus on the required signal power in the uplink direction of the link budget. A practical guideline for selecting the repeater amplifier gain is a value such that the noise level in the base station receiver has increased 1 dB from the original level after deploying the repeater. An increase of 1 dB can be considered acceptable. The repeater system gain is equal to the combined gain from the components in the repeater donor antenna line, donor link loss, and repeater amplifier gain.

12.5.4.4 Maximum Path Loss

The maximum allowed path loss is the difference between the required signal power and the EIRP calculated separately for both directions. It is worth noting that the planning margins, such as body loss, slow fading, etc., are not included in the link budget, but must be taken into account in the actual planning. Based on the calculated link budgets, very similar downlink maximum path loss is observed with outdoor-to-indoor repeater

---

[2] Free space attenuation of 100 dB is equal to a distance of about 1100 m.

implementation and with dedicated indoor systems. These maximum path loss values also correspond quite well to the maximum path loss for R99 indoor packet data service (Table 12.4), and the difference with respect to the indoor HSDPA system is approximately 3 dB in downlink and 6 dB in uplink direction. Thus, note how nearly equal coverage can be achieved using the outdoor-to-indoor repeater or dedicated indoor system. However, it should be emphasized that the behavior of the two systems in case of increasing load is different. While a dedicated indoor system provides additional radio resources for indoor users, in the outdoor-to-indoor repeater approach, the radio resources available for the indoor users are shared between the indoor users and the outdoor mother cell users. Thus, the repeater only helps the network utilize the free existing radio resources by amplifying the signal rather than providing new resources. Because the availability of radio resources of the dedicated indoor system are independent of the surrounding cells, clearly higher maximum capacity can be achieved using the dedicated indoor system—when compared to the repeater approach.

12.5.5 HSDPA Coverage and Capacity

Due to adaptive behavior, the coverage and capacity in HSDPA are even more strongly connected together than in traditional WCDMA planning. The link budget calculation (previous subsection) gives some indication of coverage requirements. A deeper understanding of HSDPA indoor performance requires system simulations and field measurements. In the radio interface, the instantaneous SINR of HS-DSCH is fundamentally defining the achievable HSDPA throughput. Thus, planning should be targeted at maximizing the SINR in the network.

The instantaneous SINR of HS-DSCH channel can be calculated:

$$SINR = SF_{16} \frac{P_{HS\text{-}DSCH}}{(1-\alpha)P_{own} + P_{other} + P_{noise}} \tag{12.2}$$

where SF_{16} is spreading factor of 16, $P_{HS\text{-}DSCH}$ is the total HSDPA power in own cell, α is the code orthogonality, P_{own} is the total power from own cell, P_{other} is the total power from all neighboring cells, and P_{noise} is the thermal noise power [7]. All the powers are received powers; thus, the corresponding link losses must be taken into account if the calculation is based on transmitted power.

In addition to the parameters in Equation (12.2), the achieved throughput with certain SINR depends on the propagation environment, system capabilities, and system parameters. The maximum number of HS-DSCH codes that a receiver at mobile can handle is 5, 10, or 15, depending on the mobile category [2,3] and the maximum available throughput is 3.6, 7.2, or 10.8 Mbps, respectively. The number of errors and retransmissions,

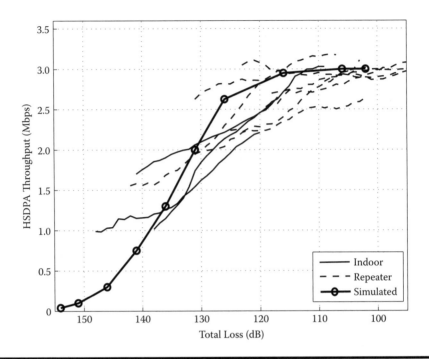

Figure 12.5 Measured [9] and simulated [7] single-user throughput values as a function of total path loss. HSDPA with five codes, 5-W power allocation.

as well as handling of retransmitted data, also affect the achieved throughput. The practical throughput can, of course, be further limited by other system-related issues, such as bottlenecks in different interfaces or lack of processing capacity in system elements.

In Figure 12.5, HSDPA throughput as a function of total path loss for single user with five codes and 5 W power allocation is plotted. The total path loss consists of antenna line losses and air interface path loss. The simulation-based values adopted from [7] are averaged between 3 W and 7 W power allocation, for Pedestrian-A channel (outdoor). The measured values are gathered from [9], where the path loss estimation is based on received signal code power (RSCP) measurements that are made with a commercial HSDPA UE (category 5/6 [2]) connected to a field measurement tool [17]; thus, some errors are expected due to calibration error and low time resolution. The measurements were made with a single user in a dedicated indoor system and outdoor-to-indoor repeater configurations, both connected to a distributed antenna system in different indoor environments.

The simulated curve in Figure 12.5 shows the expected throughput, and the measured values can be compared with reasonable accuracy. The graph shows that both the dedicated indoor system and outdoor-to-indoor repeater are able to operate as expected based on the simulations. It is also worth noting that it is possible to get high throughput with 16QAM modulation via repeater link (QPSK modulation with five codes can provide maximum 1.8 Mbps of throughput). Thus, outdoor-to-indoor repeating is a feasible solution for improving HSDPA indoor coverage.

Measurements for a UMTS indoor system with different antenna configurations in different indoor environments are presented, for example, in [8,9]. The results show that it is beneficial to increase the antenna density—that is, to decrease the average antenna spacing in DAS—by splitting the signal into several antennas. Higher antenna density provides smoother signal coverage and better signal level, finally leading to better HSDPA performance. Increasing the antenna density from 1 to 3 antennas at 50 m measurement route improved the pilot coverage in the range of 1 and 5 dB, and HSDPA throughput in the range of 5% and 30%. A radiating cable configuration was also tested, and it was observed to provide very smooth coverage close to the cable, but the signal level compared to discrete antennas was low even close to the cable. Keeping the signal level above coverage threshold requires a dense network of radiating cables installed close to user locations. Thus, discrete antennas outperform radiating cable solutions in terms of coverage and implementation complexity.

Furthermore, pilot coverage planning thresholds for indoor environments to achieve certain average throughput are presented in, for example, [9]. The throughput values are applicable for HSDPA with five codes. With typical P-CPICH and HSDPA power settings for indoor environments, a pilot coverage threshold of −50 dBm was measured, to provide average throughput of 3 Mbps; a threshold of −80 dBm gave an average throughput of 2.5 Mbps; and for threshold below −80 dBm, the throughput was significantly decreased. Therefore, a pilot coverage threshold of −80 dBm can be recommended for indoor planning. Figure 12.5 can be further used to estimate throughput thresholds for the total path loss.

In Figure 12.6, a measurement example of HSDPA with ten codes in an indoor environment is shown. The RSCP of a pilot channel and HSDPA throughput are plotted as a function of time. The measurement starts below an antenna, and ends when the HSDPA connection breaks due to low coverage. The measurement was taken in a long office corridor. Close to the antenna (about 40 seconds from the beginning), the received signal level was high enough to achieve maximum throughput almost constantly. After that point in time, maximum throughput cannot be achieved, but yet remains at a very good level. The first part (75 seconds from the beginning) of the measurement route is partly line-of-sight, and the quick drop after that is partly caused by going behind a corner to a non-line-of-sight environment. The example measurement emphasizes the importance of good coverage if high throughput values are required.

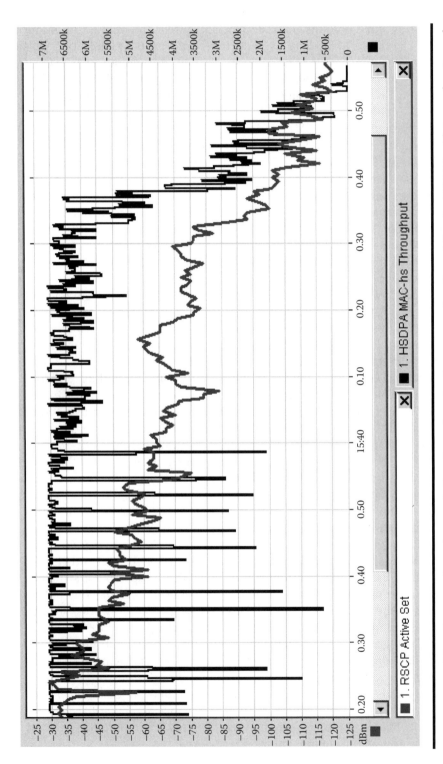

Figure 12.6 Achieved HSDPA throughput and pilot channel received power as a function of time for HSDPA with ten codes. Example for single user, measured from below the antenna to the cell edge in an indoor environment.

12.5.6 Quality of Planning

The quality of planning defines the capability of the network to function properly when it is utilized with full capacity, as well as the capability to provide sufficient coverage over the entire network area. As discussed previously, with WCDMA and HSDPA, coverage and capacity are tied together. Thus, the available capacity is defined by comparing the desired signal power levels to the noise and interference power levels, e.g. as in Equation (12.2) for HSDPA. Therefore, a common planning target of all cellular WCDMA-based systems is to minimize the interference between the cells.

The effect of surrounding macro cells must be considered to prevent interference from penetrate inside the building. Correspondingly, high-quality indoor planning guarantees that no interference is caused to the surrounding macro cells from the indoor antennas. The high quality can be maintained by proper transmission power and indoor antenna configuration for the dedicated indoor system. In case of an outdoor-to-indoor repeater, interference in the indoor location is also produced by the macrocellular outdoor network when the signals are amplified by the repeater. However, orthogonality between the codes of the mother cell users should be relatively good due to line-of-sight propagation from the macro cell to the repeater. Also, careful coverage planning is important for efficient indoor HSDPA operation. Sufficient coverage must be provided over the entire network area, which should always be verified by field measurements. Furthermore, the quality of coverage can be improved by optimizing the antenna configuration, for example, by increasing the antenna density in DAS to enhance HSDPA throughput.

References

[1] 3GPP Technical Recommendation 25.956 V6.0.0 , Release 6. Universal Terrestrial Radio Access (UTRA); repeater planning guidelines and system analysis.

[2] 3GPP Technical Specification, TS 25.306, Version 5.9.0, Release 5. UE Radio Access Capabilities Definition.

[3] 3GPP Technical Specification, TS 25.308, Version 5.7.0, Release 5. High Speed Downlink Packet Access (HSDPA); Overall description; Stage 2.

[4] E. Dahlman, S. Parkvall, J. Sköld, and P. Beming. *3G Evolution: HSPA and LTE for Mobile Broadband*. Academic Press, 2007.

[5] H. Hashemi. The indoor radio propagation channel. In *Proceedings of the IEEE*, 81: 943–968, September 1993.

[6] H. Holma and A. Toskala, Editors. *WCDMA for UMTS*. John Wiley & Sons Ltd., 2000.

[7] H. Holma and A. Toskala, Editors. *HSDPA / HSUPA for UMTS*. John Wiley & Sons Ltd., 2006.

[8] T. Isotalo, J. Lempiäinen, and J. Niemelä. Indoor planning for high speed downlink packet access in WCDMA cellular network. *Wireless Personal Communications Journal*, 52(1): 89–104, 2010.

[9] T. Isotalo, P. Lähdekorpi, and J. Lempiäinen. Improving HSDPA indoor coverage and throughput by repeater and dedicated indoor system. *EURASIP Journal on Wireless Commun. Networking*, April 2009. In press.

[10] K. Hiltunen, B. Olin, and M. Lundevall. Using dedicated in-building systems to improve HSDPA indoor coverage and capacity. In *IEEE 61st Vehicular Technology Conference, VTC 2005-Spring*, pp. 2379–2383, 2005.

[11] Kathrein. http://www.kathrein.de, 2009.

[12] J. Laiho, A. Wacker, and T. Novosad, Editors. *Radio Network Planning and Optimisation for UMTS*. John Wiley & Sons Ltd., 2002.

[13] J. Lempiäinen and M. Manninen, Editors. *UMTS Radio Network Planning, Optimization and QoS Management*. Kluwer Academic Publishers, 2003.

[14] M.J. Nawrocki, M. Dohler, and A.H. Aghvami, Editors. *Understanding UMTS Radio Network Modelling, Planning and Automated Optimisation*. John Wiley & Sons, Ltd., 2008.

[15] M.K. Simon and M.S. Alouini. Digital Communications over Fading Channels: A Unified Approach to Performance Analysis. Wiley InterScience, Sep 2000.

[16] K. Pedersen and P. Mogensen. The downlink orthogonality factors influence on WCDMA system performance. In *IEEE 56th Vehicular Technol. Conf. VTC 2002-Fall*, pp. 2061–2065, 2002.

[17] Anite plc. Anite Nemo Outdoor Product description. http://www.anite.com/nemo/, March 2009.

[18] R. Anderson, B. Arend, and K. Baker. Power Controlled Repeaters for Indoor CDMA Networks. In White paper, Qualcomm Inc., 2003.

[19] Radio Frequency Systems (RFS). http://www.rfsworld.com, 2009.

[20] S.R. Saunders. *Antennas and Propagation for Wireless Communication Systems*. John Wiley & Sons, Ltd., 1999.

[21] M. Tolstrup. *Indoor Radio Planning*. John Wiley & Sons, Ltd., 2008.

Chapter 13

Dimensioning HSPA Networks: Principles, Methodology, and Applications

---

Salah E. Elayoubi

Contents

13.1 Introduction

Operators are now taking another step in the development of their broadband networks by launching 3G+ and providing customers with faster mobile data services. 3G+, or 3.5G, is a natural evolution of the current (Release 99) 3G networks and promises much faster data speeds, based on HSDPA (High-Speed Downlink Packet Access) in the forward link and HSUPA (High-Speed Uplink Packet Access) in the reverse link.

In this context, capacity calculations are needed in order to dimension the system and determine whether an upgrading of the existing UMTS (Universal Mobile Telecommunications System) base stations is necessary, and, if necessary, determine if it is better to add a new carrier or share the existing carrier between 3G and 3G+. This chapter aims to respond to these questions. We present a capacity analysis of a network carrying both real-time and elastic traffic over classical 3G dedicated channels (DCHs), or on HSUPA Enhanced DCH (E-DCH) and HSDPA High-Speed Downlink Shared Channels (HS-DSCHs).

The method developed in this chapter is based on the analytical model of [1] that is adapted to the real network using HSDPA and HSUPA link-level simulations.

The methodology of capacity assessment is as follows:

- For a given offered traffic, estimate the average load of the network.
- Using the estimated load, calculate SINR (signal-to-interference-plus-noise ratio) and admission control constraints as in [1].
- Deduce quality-of-service (QoS) for DCH calls as well as the resources they consume, based on multi-Erlang analysis. As DCH traffic has guaranteed throughput, the QoS is expressed by blocking, that is, the probability that the call, upon arrival, does not have the required DCH.
- Based on the estimated load and the DCH resources, obtain the HSPA throughput values from link budget calculations.
- Using the calculated throughputs, apply processor sharing (PS) in order to obtain the HSPA QoS. The latter is expressed in terms of average throughput, or in terms of probability of having less than a target throughput.

The remainder of this chapter is organized as follows:

- In "DL R99 Model" we calculate the capacity of a Release 99 UMTS network.
- In "Capacity Analysis" we consider a UMTS carrier that has been upgraded to HSDPA and calculate its capacity.

- The same analysis is done for the uplink in "Shared R99/HSDPA Model" and "UL R99 Model."
- "R99/HSUPA Model" illustrates the analysis by numerical results.
- The "Conclusion" summarizes the chapter.

13.2 DL R99 Model

13.2.1 Radio Model

In the downlink of WCDMA, the SINR achieved for user u situated at distance r_0 from the target cell base station is given by

$$SINR_u^D(r_0) = \frac{S^D P_{u,0}/q_{u,0}^D}{I_{inter,u}^D + I_{intra,u}^D + N_0}$$

where S^D is the spreading factor; $P_{u,0}$ is the power received by user u from the target cell base station; $I_{inter,u}^D = \sum_{l \neq 0} \frac{P_{tot,l}}{q_{u,l}^D}$ is the inter-cell interference, with $P_{tot,l}$ the total power transmitted by the base station l and $q_{u,l}^D$ the path-loss between user u and base station l; and $I_{intra,u}^D = \alpha \frac{P_{tot,0} - P_{u,0}}{q_{u,0}^D}$ is the intra-cell interference originating from the common channels and from other users, with α the orthogonality factor and N_0 the thermal noise.

The total power of a typical base station in the network can be written as $\bar{\chi}^D P_{max}$, where $\bar{\chi}^D$ is the average load and P_{max} is the maximal transmission power. Let us note that the 3G mean load is defined as the ratio between the used and the total power. This average load is not known but can be determined iteratively, as will be shown later. The inter-cell interference can be expressed in terms of the well-known F-factor [1] illustrated in Figure 13.1.

The SINR can be rewritten in the following way:

$$SINR_u^D(r_0) = \frac{S^D \times P_{u,0}}{\alpha(P_{tot} - P_{u,0}) + \bar{\chi}^D P_{max} F_u + N_0 q_{u,0}^D}$$

where P_{tot} is the total power emitted by the target cell. Hence, after defining the β-factor of user u,

$$\beta_u = \frac{SINR_u^D}{S^D + \alpha.SINR_u^D}$$

we obtain

$$\beta_u = \frac{P_{u,0}}{\alpha.P_{tot} + \bar{\chi}^D P_{max} F_u + N_0 q_{u,0}^D}$$

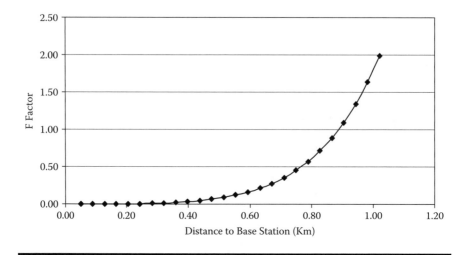

Figure 13.1   F-factor versus distance to base station.

If there are M^D UMTS users in the cell, the average total power used from the base station in the target cell is

$$P_{tot} = \frac{P_{Com} + \sum_{u=1}^{M^D} \left(\bar{\chi}^D P_{max} F_u + N_0 q_{u,0}^D\right) \beta_u}{1 - \alpha \sum_{u=1}^{M^D} \beta_u}$$

Here we denote the power associated with the common channels with P_{Com}.

Differentiating n regions in the cell as described in Figure 13.2 and considering C classes corresponding to the nature of the DCH bearer (e.g., voice calls with 12.2 kbps, and packet-switched calls carried over channels of 64, 128, or 384 kbps), this expression of the power becomes

$$P_{tot} = \frac{P_{Com} + \sum_{i=1}^{n} \left(\bar{\chi}^D P_{max} F_i + N_0 q_i^D\right) \left(\sum_{c=1}^{C} \beta_c M_{i,c}^D\right)}{1 - \alpha \sum_{i=1}^{n} \left(\sum_{c=1}^{C} \beta_c M_{i,c}^D\right)}$$

where $M_{i,c}^D$ is the number of class-c UMTS users in region i, for $i = 1, \ldots, n$; F_i and q_i^D are, respectively, the average F-factor and path-loss for region i.

Figure 13.2   Cell decomposition.

The constraint on the maximal transmission power gives the constraint on the number of users:

$$\sum_{i=1}^{n} \left(\alpha P_{\max} + \bar{\chi}^{D} P_{\max} F_i + N_0 q_i^D \right) \left(\sum_{c=1}^{C} \beta_c M_{i,c}^D \right) \leq P_{\max} - P_{Com}$$

Let A_D be the space of states $\vec{M}^D = (M_{i,c}^D)$, $i = 1, \ldots, n$, $c = 1, \ldots, C$, where this condition is verified.

13.3 Capacity Analysis

The resolution of this system is possible using the multi-Erlang approach [4]. In fact, real-time classes will be characterized by the average call duration T_{rt}, while non-real-time classes will be characterized by the size of the file to be downloaded Z_{nrt}, which, combined with the throughput of the bearer D_{nrt} (= 12.2, 64, 128, 384), gives an average transmission time of

$$T_{nrt} = \frac{D_{nrt}}{Z_{nrt}}.$$

However, the cell is decomposed into n zones; each class is then split into n sub-classes. If the overall available capacity is $P_{\max} - P_{Com}$, a call of class-c

in zone i will consume a part of the capacity equal to $(\alpha P_{max} + \bar{\chi}^D P_{max} F_i + N_0 q_i^D)\beta_c$.

Let the arrival rate for the class c of traffic be λ_c and its average duration be T_c (we will have values for AMR 12.2, PS 64, PS 128, and PS 384 classes).

Once the traffic characteristics are obtained, the capacity that a call from each class will use must be determined. To do this, each class c is split into three sub-classes, leading to a new class (c, i), $i = 1, \ldots, 3$ with arrival rate $\lambda_{c,i} = \lambda_c/3$ and average duration $T_{c,i} = T_c$. Each sub-class will represent the behavior of calls in different portions of the cell (near the base station, cell center, and cell edge). If the capacity is normalized to 1, the consumed capacity of a call of class (c, i) will be equal to

$$C_i = \frac{(\alpha P_{max} + \bar{\chi}^D P_{max} F_i + N_0 q_i^D)\beta_c}{P_{max} - P_{Com}}$$

The offered traffic of class (c, i) will be equal to

$$E_{c,i} = \lambda_{c,i} T_c$$

A multi-Erlang approach can thus be used, and the blocking probabilities for each service in each zone can be obtained.

13.3.1 Load Estimation

As can be shown from the above analysis, the performance measures, including the load of the cell, depend on the load of neighboring cells. However, this neighboring cell load is not an input of the model: It is an output that depends on the amount of traffic in each cell of the network. Indeed, a large offered traffic leads to a large number of simultaneous users and, consequently, a larger allocated power. As demonstrated in [2], the load of neighboring cells for a given offered traffic (in Erlang) can be approximated as follows:

$$\bar{\chi}^D = \begin{cases} \frac{\bar{E}\bar{\beta} N_0 \bar{q} + P_{Com}}{P_{max}(1-(\alpha+F)\bar{E}\bar{\beta})}, & if \ 0 \le \frac{\bar{E}\bar{\beta} N_0 \bar{q} + P_{Com}}{P_{max}(1-(\alpha+F)\bar{E}\bar{\beta})} \le 1 \\ 1 & otherwise \end{cases}$$

where $\bar{E}$ is the total offered traffic (in Erlang), $\bar{\beta}$ is the average required β-factor, $\bar{q}$ is the average path loss, and $\bar{F}$ is the average F-factor over the cell.

Once this load is known, the above-described capacity analysis can be performed and the performance measures calculated.

13.4 Shared R99/HSDPA Model

13.4.1 Radio Model

When both HSDPA and R99 calls share the same bandwidth, the power that is not used by R99 users is used for HSDPA (Figure 13.3).

The base station emits with its maximal power; so, for each UMTS user, we have

$$\beta_u = \frac{P_{u,0}}{\alpha . P_{\max} + \bar{\chi}^D P_{\max} F_u + N_0 q_{u,0}^D}$$

thus giving the power emitted by the base station to UMTS users equal to

$$P_{R99} = \sum_{i=1}^{n} \left(\alpha P_{\max} + \bar{\chi}^D P_{\max} F_i + N_0 q_i^D\right) \left(\sum_{c=1}^{C} \bar{\beta}_c M_{i,c}^D\right)$$

The admission control equation for R99 in the presence of HSDPA thus remains the same:

$$\sum_{i=1}^{n} \left(\alpha P_{\max} + \bar{\chi}^D P_{\max} F_i + N_0 q_i^D\right) \left(\sum_{c=1}^{C} \beta_c M_{i,c}^D\right) \leq P_{\max} - P_{Com}$$

Multi-Erlang analysis [4] allows obtaining, in addition to the blocking rate for R99 users, the average power $\bar{P}_{R99}$ that is used by R99 users.

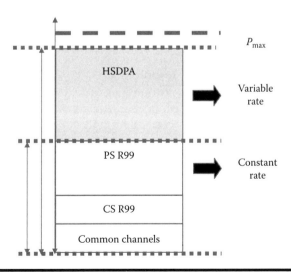

Figure 13.3   Power allocation between HSDPA and R99 users.

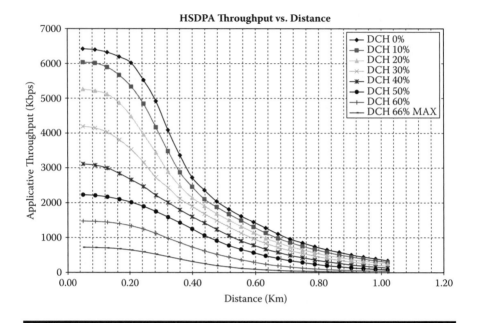

Figure 13.4   HSDPA throughput versus distance to base station, for a network load of 20% and for different values of DCH power.

Let $x = \frac{\bar{P}_{R99}}{P_{\max} - P_{Com}}$, the SINR of an HSDPA user in region j is given by

$$SINR_j^{DH}(x) = \frac{S^{DH}[(P_{\max} - P_{Com})(1 - x)]}{\bar{\chi} P_{\max} F_j + N_0 q_j^D + \alpha P_{Com} + \alpha x (P_{\max} - P_{Com})}$$

S^{DH} is the spreading factor for HSDPA.

For a given proportion of resources used by DCH traffic, link level curves [5] allow calculating the throughput of HSDPA users at each point of the cell. An example of these values is illustrated in Figure 13.4.

Let the average throughput obtained from this link budget tool in region j be equal to

$$T_j^{DH}(x, \bar{\chi}^D)$$

To assess the performance of HSDPA, we can use a processor sharing model, as in [3]. To do this, we define the load in region j by

$$\rho_j^{DH}(x, \bar{\chi}^D) = \frac{\lambda_j^{DH} Z}{T_j^{DH}(x, \bar{\chi}^D)}$$

where λ_j^{DH} is the arrival rate of HSDPA calls in region j and Z is the average file size.

The total load of the processor sharing queue is

$$\rho^{DH}(x, \bar{\chi}^D) = \sum_j \rho_j^{DH}(x, \bar{\chi}^D) = \frac{\lambda^{DH} Z}{T^{DH}(x, \bar{\chi}^D)}$$

where $T^{DH}(x, \bar{\chi}^D)$ is the harmonic mean of the peak throughput over the cell surface.

If the CAC imposes that the maximal number of HSDPA users is equal to $M_{\max}^{DH}$, then the distribution of the number of users is given by

$$\Pr(M^{DH} = i \mid x, \bar{\chi}^D) = \frac{[\rho^{DH}(x, \bar{\chi}^D)]^i}{\displaystyle\sum_{m=0}^{M_{\max}^{DH}} [\rho^{DH}(x, \bar{\chi}^D)]^m}$$

and the blocking probability for HSDPA users is

$$B^{DH}(x, \bar{\chi}^D) = \frac{[\rho^{DH}(x, \bar{\chi}^D)]^{M_{\max}^{DH}}}{\displaystyle\sum_{m=0}^{M_{\max}^{DH}} [\rho^{DH}(x, \bar{\chi}^D)]^m}$$

The probability of having at least one HSDPA user in the cell is

$$\Pr(M^{DH} > 0 \mid x, \bar{\chi}^D) = \frac{\displaystyle\sum_{m=1}^{M_{\max}^{DH}} [\rho^{DH}(x, \bar{\chi}^D)]^m}{\displaystyle\sum_{m=0}^{M_{\max}^{DH}} [\rho^{DH}(x, \bar{\chi}^D)]^m}$$

If a target throughput $T_{\min}$ is fixed for HSDPA users, the probability of having, in region j, a throughput that is less than $T_{\min}$ is equal to

$$\Pr(Throughput \ in \ j < T_{\min} \mid x, \bar{\chi}^D) = \sum_{i = \lceil T_j^{DH}(x)/T_{\min} \rceil}^{M_{\max}^{DU}} \Pr(M^{DH} = i \mid x, \bar{\chi}^D)$$

and the probability of being under this target is given by

$$\Pr(Throughput < T_{\min} \mid x, \bar{\chi}^D) = \frac{1}{n} \sum_j \sum_{i = \lceil T_j^{DH}(x)/T_{\min} \rceil}^{M_{\max}^{DU}} \Pr(M^{DH} = i \mid x, \bar{\chi}^D)$$

The total power emitted by the base station when there is at least one HSDPA user is P_{max}. As the exact number of R99 users cannot be calculated using the multi-Erlang approach, but only the proportion of occupied resources, we can only approximate the power emitted for R99 users where there are no HSDPA calls as $P_{Com} + x(P_{max} - P_{Com})$. The average power emitted by the cell is then

$$\bar{P} = [P_{max} \Pr(M^{DH} > 0 \,|\, x) + [P_{Com} + x(P_{max} - P_{Com})]$$
$$\times (1 - \Pr(M^{DH} > 0 \,|\, x))] \Pr(x)$$

13.4.2 Load Estimation

For a given network load $\bar{\chi}^D$, the average percentage of power used by R99 traffic can be estimated by

$$\bar{P}_{R99}(\bar{\chi}^D) = \frac{1}{P_{max}} \min \left[\sum_{i=1}^{n} (\alpha P_{max} + \bar{\chi}^D P_{max} F_i + N_0 q_i^D) \right.$$

$$\left. \times \left(\sum_{c=1}^{C} \bar{\beta}_c E_{c,i} \right), \, P_{max} - P_{com} \right]$$

The minimum in this equation ensures that, for large offered traffic, the estimated power emitted by the base station does not exceed P_{max}. If we neglect blocking, the Processor Sharing queue load is equal to: $\rho^{DH}(\bar{P}_{R99}(\bar{\chi}^D), \bar{\chi}^D)$. The network load is then given by

$$\bar{\chi}^D = \min \left[\frac{P_{com}}{P_{max}} + \bar{P}_{R99}(\bar{\chi}^D) \right.$$

$$\left. + \left(1 - \frac{P_{com}}{P_{max}} + \bar{P}_{R99}(\bar{\chi}^D) \right) \rho^{DH}(\bar{P}_{R99}(\bar{\chi}^D), \bar{\chi}^D), 1 \right]$$

This fixed point equation can easily be solved to obtain the network load.

13.5 UL R99 Model

13.5.1 Radio Model

In the uplink, the SNIR received from a class c mobile at the base station BS of a given cell 0 must be greater than a given constant to guarantee the reception of the signal at the BS:

$$SINR_c^U = \frac{P_c}{N_0 + I_{intra,0}^U + I_{inter,0}^U - P_c} \geq \tilde{\Delta}_c$$

for $c = 1, \ldots, C$; where $\tilde{\Delta}_c = \frac{E_c}{N_0} \times \frac{R_c}{W}$ is the required SINR for calls of class c; E_c/N_0 is the minimum allowed ratio between the bit energy and the interference plus noise density, which guarantees the target in terms of bit error probability; W/R_c is the processing gain, that is, the ratio between the chip rate and the source bit rate; N_0 is the background noise, and $I_{intra,0}$ and $I_{inter,0}$ are the total power received from other mobiles within the considered cell and all its neighbors, respectively.

The number of class-c calls in the uplink is

$$I^U_{intra,0} = \sum_{c=1}^{C} M^U_c P_c, M^U_c$$

where P_c is the constant power received by a base station for class c calls to avoid the near-far effect.

$$I^U_{inter,0} = \sum_{i=1}^{Ncell} \sum_{j} P_{i,j}$$

where $P_{i,j}$ is the power emitted by call j in cell i. We use the following analysis to calculate this interference.

The ratio of emitted power by a call in another cell j to the path loss between him and the BS 0. is

$$I^U_{inter,0} = \sum_{i=1}^{Ncell} \sum_{j} \frac{P^e_{i,j}}{q^{(0)}_{i,j}}; \quad \frac{P^e_{i,j}}{q^{(0)}_{i,j}}$$

Knowing that the transmission is power controlled in the uplink, in order to avoid the near-far effect, all calls of class c are received at their own base station with the same power P_c. On average, the inter-cell interference can then be approximated by

$$I^U_{inter,0} = E\left[\sum_{c=1}^{C} M^U_c P_c\right] \times E\left[\sum_{i=1}^{Ncell} \frac{q^{(i)}_i}{q^{(0)}_i}\right]$$

where M^U_c is the number of calls of class c in a typical cell of the system, and $q^{(j)}_i$ is the path loss between a typical position in cell i and the base station of cell j.

Let $f = E[\sum_{i=1}^{Ncell} \frac{q^{(i)}_i}{q^{(0)}_i}]$ be the interference factor in the uplink; it is obtained by integrating over the interfering cells.

$\bar{P} = E[\sum\limits_{c=1}^{C} M_c^U P_c]$ is the average received power by a base station in the system; we will show next how to calculate it.

Considering the minimal power that can achieve the target SIR, we obtain

$$\tilde{\Delta}_c = \frac{P_c}{N_0 + \sum\limits_{m=1}^{C} M_m^U P_m + \bar{P} \cdot f - P_c}$$

Defining $\Delta_c = \frac{\tilde{\Delta}_c}{1+\tilde{\Delta}_c}$, we can obtain

$$P_c = \Delta_c \left(N_0 + \bar{P} \cdot f + \sum\limits_{m=1}^{C} M_m^U P_m \right)$$

which leads to

$$\sum\limits_{m=1}^{C} M_m^U P_m = \frac{(N_0 + \bar{P} \cdot f) \sum\limits_{m=1}^{C} M_m^U \Delta_m}{1 - \sum\limits_{m=1}^{C} M_m^U \Delta_m}$$

and

$$P_c = \frac{(N_0 + \bar{P} \cdot f)\Delta_c}{1 - \sum\limits_{m=1}^{C} M_m^U \Delta_m}$$

At admission control, a constraint on the load of the cell must be respected. This is expressed as a limitation on the noise rise at the reception, defined by: $\chi^U = \frac{I_{tot} + N_0}{N_0}$, where I_{tot} is the overall power received by the base station. This leads to

$$\eta^U = \frac{\sum\limits_{m=1}^{C} M_m^U P_m + \bar{P} \cdot f + N_0}{N_0} = \frac{N_0 + \bar{P} \cdot f}{N_0 \left(1 - \sum\limits_{m=1}^{C} M_m^U \Delta_m \right)}$$

A condition on the maximal value of this noise rise is to be imposed:

$$\eta^U < \eta_{max}^U$$

leading to the condition:

$$\sum_{c=1}^{C} M_c^U \Delta_c < 1 - \frac{N_0 + \bar{P} \cdot f}{N_0 \eta_{\max}^U}$$

This equation describes the number of circuits required by calls. If we normalize the capacity to 1, each call of class c requires a number of circuits equal to $\frac{\Delta_c}{1 - \frac{N_0 + \bar{P} \cdot f}{N_0 \eta_{\max}^U}}$. The capacity of the system can then be analyzed using multi-Erlang as described for the downlink.

13.5.2 Load Estimation

In a homogeneous network, we can say that the powers received by base stations from their own users are, on average, equal:

$$\bar{P} = \sum_{m=1}^{C} M_m^U P_m$$

leading to

$$\bar{P} = E\left[\frac{(N_0 + \bar{P} \cdot f) \sum_{m=1}^{C} M_m^U \Delta_m}{1 - \sum_{m=1}^{C} M_m^U \Delta_m} \right]$$

A good approximation of $\bar{P}$ would thus be as follows:

$$\bar{P} = \frac{N_0 \bar{\Delta}(\bar{P})}{1 - (1 + f)\bar{\Delta}(\bar{P})}$$

where $\bar{\Delta}(\bar{P})$ is the approximation of $\sum_{m=1}^{C} M_m^U \Delta_m$ given the offered traffic and $\bar{P}$:

$$\bar{\Delta}(\bar{P}) = \min\left[\sum_{c=1}^{C} E_c \Delta_m, 1 - \frac{N_0 + \bar{P} \cdot f}{N_0 \eta_{\max}^U} \right]$$

A simple fixed-point solution of this equation is again possible.

13.6 R99/HSUPA Model

13.6.1 Radio Model

HSUPA aims to offer high data rates on the uplink (up to 5.76 Mbps) using key techniques implemented in HSDPA, such as fast scheduling, link adaptation, and Hybrid ARQ (HARQ). Unlike HSDPA, HSUPA does not use a shared channel for delivering the data calls. By structure, it is considered more of an add-on to the UMTS R99 standard rather than a replacement. The study of HSUPA performance will, consequently, be based on the study for the UMTS R99 uplink model.

For the admission control, the condition concerning the maximum load value won't change. This condition will then be used to allocate the available resources to the pool of the HSUPA users, knowing that there is a larger number of possible throughputs. The resulting interference configuration is described in Figure 13.5.

For R99 users, a maximal noise rise is specified taking into account only in-cell DCH users:

$$\eta_{R99}^U = \frac{I_{tot} - I_{HSUPA} + N_0}{N_0} \leq \eta_{R99}^{max}$$

where I_{HSUPA} is the power received from HSUPA users.

For HSUPA users, the admission control condition is on the total noise rise:

$$\eta_{tot}^U = \frac{I_{tot} + N_0}{N_0} \leq RoT^{max}$$

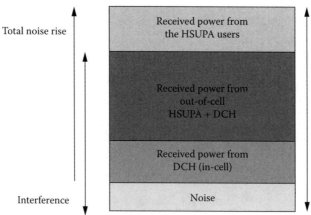

Figure 13.5 Interference budget for the uplink.

The equality holds except for coverage-limited cases where cell edge users are not able to use the overall capacity, even when they transmit with their maximal power. In this work, we limit ourselves to the RoT-limited cells, which is the case in most dense urban and urban areas.

Knowing that in the presence of HSUPA users in the cell, the maximal noise rise is attained, the SINR received at the base station for a R99 class-c user is

$$\tilde{\Delta}_c = \frac{P_c}{N_0 + I_{tot} - P_c} = \frac{P_c}{N_0 \, RoT^{\max} - P_c}$$

Defining, as before, $\Delta_c = \frac{\tilde{\Delta}_c}{1+\tilde{\Delta}_c}$, we can obtain

$$P_c = N_0 RoT^{\max} \Delta_c$$

Following the above-mentioned R99 noise rise constraint, the admission control equation for R99 users becomes

$$\eta_{R99}^U = \frac{\sum_{c=1}^{C} M_c^U P_c + \bar{P}f + N_0}{N_0} = RoT^{\max} \sum_{c=1}^{C} M_c^U \Delta_c + \frac{\bar{P}f + N_0}{N_0} \leq \eta_{R99}^{\max}$$

or

$$\sum_{c=1}^{C} M_c^U \Delta_c \leq \frac{\left(\eta_{R99}^{\max} - 1\right) N_0 - \bar{P} \cdot f}{N_0 RoT_{\max}^U}$$

This, again, can be resolved using multi-Erlang analysis. Such analysis gives, in addition to the blocking rate, the average resource utilization, that is, $E[\sum_{c=1}^{C} M_c^U \Delta_c]$. The average noise rise due to R99 users is then calculated by

$$x = \frac{E\left[\sum_{c=1}^{C} M_c^U P_c\right] + \bar{P}f + N_0}{N_0} = RoT^{\max} E\left[\sum_{c=1}^{C} M_c^U \Delta_c\right] + \frac{\bar{P}f + N_0}{N_0}$$

As for HSUPA users, their throughput will depend on three factors:

1. The power used by R99 users, expressed in a noise rise value x
2. The number of HSUPA users in the cell: M_H^U
3. The load in adjacent cells $\bar{\chi}^U$, crucial to determine the inter-cell interference

HSUPA link-level curves allow us to calculate the per-user throughput knowing these two values. An example of these throughputs is given in Figure 13.6 for one and three HSUPA users and a full HSUPA network.

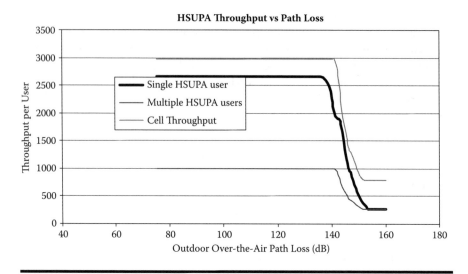

Figure 13.6   HSUPA throughput.

Let $T^{DH}(x, M_H^U, \bar{\chi}^U)$ be the overall cell throughput when the noise rise from R99 users is equal to x, the neighboring cell load is equal to $\bar{\chi}^U$ and the number of HSUPA users is equal to M_H^U. The performance of HSUPA users can then be obtained using a generalized processor sharing queue:

$$\Pr[M_H^U | x, \bar{\chi}^U] = \Pr[0 | x] \frac{(\lambda_H^U Z)^{M_H^U}}{\prod_{m=1}^{M_H^U} T^{DH}(x, m, \bar{\chi}^U)}$$

with λ_H^U the arrival rate of HSUPA calls and Z the average file size. The normalizing constant is a calculated function of the maximal allowed number of simultaneous HSUPAs in the cell M_{max}^U:

$$\Pr[0 | x, \bar{\chi}^U] = \left[1 + \sum_{M_H^U=1}^{M_{max}^U} \frac{(\lambda_H^U Z)^{M_H^U}}{\prod_{m=1}^{M_H^U} T^{DH}(x, m, \bar{\chi}^U)} \right]^{-1}$$

The HSUPA blocking rate is thus equal to $\Pr[M_{max}^U | x, \bar{\chi}^U]$, and the probability of being below a given target is

$$\Pr(Throughput < T_{min} | x, \bar{\chi}^U) = \sum_{i=M_{max}(x, T_{min})}^{M_{max}^U} \Pr(M_H^U = i | x, \bar{\chi}^U)$$

where $M_{\max}(x, \bar{\chi}^U, T_{\min})$ is the minimal number of HSUPA users, so that

$$T^{DH}\left(x, M_{\max}(x, T_{\min}), \bar{\chi}^U\right) < T_{\min}$$

13.6.2 Load Estimation

For a given offered traffic, two unknowns are to be determined: (1) the average power received by a base station from its own (R99+HSUPA) users $\bar{P}$, and (2) the load of neighboring cells $\bar{\chi}^U$ necessary for SINR calculations. The load of a cell is defined as the ratio of the measured noise rise to the maximal Rise over Thermal (RoT). These two values are thus correlated. In fact, we can approximate the total noise rise of the cell by

$$\bar{\eta}_{tot}^U = E\left[\frac{I_{tot} + N_0}{N_0}\right] = \frac{\bar{P} + \bar{P}f + N_0}{N_0} = \bar{\chi}^U RoT^{\max}$$

leading to

$$\bar{P}(\bar{\chi}^U) = \frac{N_0(\bar{\chi}^U RoT^{\max} - 1)}{(1 + f)}$$

More in depth, the load of the network is related to the offered traffic. Let us first approximate the noise rise due to R99 users:

$$\bar{x}(\bar{\chi}^U) = RoT^{\max}\bar{\Delta}(\bar{\chi}^U) + \frac{\bar{P}(\bar{\chi}^U)f + N_0}{N_0}$$

where

$$\bar{\Delta}(\bar{\chi}^U) = \min\left[\sum_{c=1}^{C} E_c\Delta m, \frac{(\eta_{R99}^{\max} - 1) N_0 - \bar{P}(\bar{\chi}^U) \cdot f}{N_0 RoT_{\max}^U}\right]$$

As for HSUPA users, they will take, if present, all the resources that are not used by R99 ones. The probability that there is at least one HSUPA user in the cell is equal to the load of the processor sharing queue. This load is given by

$$\rho_H^U(\bar{\chi}^U) = 1 - \Pr[0 \,|\, \bar{x}(\bar{\chi}^U), \bar{\chi}^U] = 1 - \left[1 + \sum_{M_H^U=1}^{M_{\max}^U} \frac{(\lambda_H^U Z)^{M_H^U}}{\prod\limits_{m=1}^{M_H^U} T^{DH}(\bar{x}(\bar{\chi}^U), m, \bar{\chi}^U)}\right]^{-1}$$

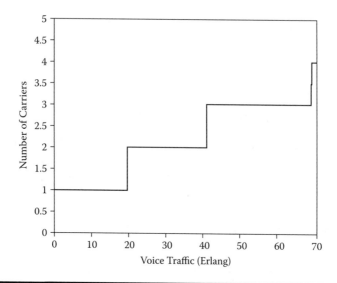

Figure 13.7 **Number of carriers necessary to carry the offered voice traffic.**

The average load of a cell in the network can thus be approximated by

$$\bar{\chi}^{U} = \frac{\bar{x}(\bar{\chi}^{U})}{RoT_{\max}} + \left(1 - \frac{\bar{x}(\bar{\chi}^{U})}{RoT_{\max}}\right) \rho_{H}^{U}(\bar{\chi}^{U})$$

This is, again, a fixed-point equation that should be solved before evaluating the performance of the target cell.

13.7 Example of Results

Figure 13.7 shows the number of dedicated carriers necessary to carry a given offered voice traffic in a pure R99 network. In the case of a shared DCH/HSPA carrier, Figure 13.8 shows the maximal HSPA offered traffic that can be supported by one carrier.

13.8 Conclusion

In this chapter we studied the capacity of 3G/3G+ networks. We developed radio models for both downlink and uplink, and used them to derive the Erlang-like capacity of the network. We studied the cases of carriers dedicated for HSPA, as well as shared carriers where HSPA users coexist with legacy UMTS ones. These capacity models allow us to know, for a

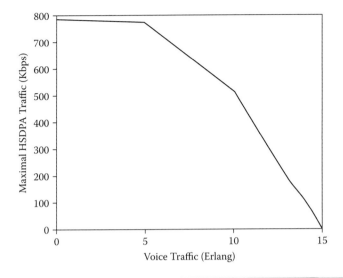

Figure 13.8 Maximal HSDPA traffic versus the voice traffic for one carrier.

given traffic pattern, the best network configuration—whether to upgrade the network by adding HSPA or add a new carrier to guarantee QoS for users.

References

[1] A. Baroudy and S.E. Elayoubi, HSUPA/HSDPA systems: Capacity and dimensioning, *Proc. IEEE Future Generation Communication and Networking 2007*, IEEE Computer Society, December 2007.

[2] B. Blaszczyszyn and M.K. Karray, An efficient analytical method for dimensioning of CDMA cellular networks serving streaming calls, *Proc. 3rd Int. Conf. Performance Evaluation Methodologies and Tools*, 2008.

[3] T. Bonald and A. Proutière, Wireless downlink data channels: User performance and cell dimensioning, *ACM Mobicom*, 2003.

[4] J.W. Roberts, A service system with heterogeneous user requirements, in G. Pujolle (Ed.), *Performance of Data Communications Systems and Their Applications*, North-Holland, Amsterdam, 1981, pp. 423–431.

[5] J.B. Landre and A. Saadani, HSDPA 14,4 Mbps mobiles—Realistic throughputs evaluation, *IEEE VTC – Spring*, 2008.

Keywords and Glossary

Keywords

16QAM
ALCAP
antenna
antenna tilt
bandwidth
bandwidth allocation
beamforming
Best-Effort scheduler
bit rate calculation algorithm
buffer management
buffering
CAC, call admission control
CCTrCH
Chase Combining
CD-EDD algorithm
cell planning tool
C/I, carrier-to-interference ratio
CMCC, Chain Mobile Communication Corporation
CNB
congestion
congestion action
congestion avoidance mechanism
cross-layer backpressure
decorrelation distance
dimensioning
EPC
equalizer
E-TFC
fading, fast fading and slow fading
feedback

packet scheduling
PAG
PCCPCH
PCPICH
PER
performance evaluation
power
precoding
propagation
protocol structure
PSCH
radio bearer allocation
RAKE receiver
reasonable peak allocation
Reed-Muller code
RNL
Round Robin scheduler, scheduling
SAG
SAP
scheduler
scheduling algorithm
scheduling principles
search algorithm
serving cell
signaled maximum flow bit rate
simulated annealing algorithm
simulation, link-level
simulation, system-level
simulation, system-level
simulator, link-level
smart antenna
soft handover
SS
SSCH
Subframe Number
sum-capacity
system-level simulation
tabu search
TF
throughput
time-space priority queuing
TPC
traffic model-based dimensioning
transport network

transport network layer (TNL)
transport network overhead
Turbo code
UTRAN architecture
UTRAN functionalities
WSS-US model

Glossary

3G: 3rd generation
3GPP: 3rd generation partnership project
AAL2/ATM: ATM adaptation layer 2/asynchronous transfer mode
ACK/NACK: acknowledge/negative acknowledge
AG: absolute (scheduling) grant
AGC: automatic gain control
AIAD: additive increase/additive decrease
AICH: acquisition indicator channel
AIMD: additive increase, multiplicative decrease
AM: acknowledged mode
AMC: adaptive modulation and coding
AMR: adaptive multirate
ARQ: automatic repeat request
ATM: asynchronous transfer mode
automated optimization:
AVI: actual value interface
AWGN: additive white Gaussian noise
BAC: buffer admission control
BE: best effort
BER: bit error rate
BLER: block error rate
BLER: block error ratio
BMA: buffer management algorithm
BS: base station
CA: capacity allocation
CACF: capacity allocation control frame
CATT: China Academy of Telecommunications Technology
CC: congestion control
CCDF: complementary cumulative distribution function
CCH: common channel
CCSA: China Communications Standards Association
CCTrCH: coded composite transport channel
CD-EDD: channel-dependent earliest due deadline
CDF: cumulative density function, cumulative distribution function

CDMA: code division multiple access
CF: control frame chip equalizer
CF: control frame
CFN: connection frame number
CLTD: closed-loop transmit diversity
CmCHPI (CmCH-PI, CmCHPi): common transport channel priority
CN: core network
CPICH: common pilot channel
CPS: common part sublayer
CQI: channel quality indicator
CRC: cyclic redundancy check
CRNC: controlling RNC, radio network controller
CS: circuit switched
Cwnd: congestion window
DAS: distributed antenna system
dB: decibel, a logarithmic unit of measurement that expresses the magnitude of a physical quantity (usually power or intensity) relative to a specified or implied reference level.
dBi: a unit measuring the gain of an antenna
dBm: standard unit for measuring power levels in relation to a 1-milliwatt reference signal
DCH: dedicated channel
DDD: dynamic delay detection
DDI: data description indicator
DF: data frame
DFD: destroyed frame detection
DFE: decision feedback equalizer
DL: downlink
DPCH: dedicated physical channel
DRNC: drift RNC
DRT: delay reference time
DS-CDMA: direct sequence CDMA
DT: discard timer
D-TSP: dynamic time-space priority
D-TxAA: dual-stream transmit antenna array
DWTPS: downlink pilot time slot
E-AGCH: E-DCH absolute grant channel
ECN: explicit congestion notification
E-DCH: enhanced-dedicated channel
E-HICH: E-DCH Hybrid ARQ indicator channel
EIRP: effective isotropic radiated power
E-PUCH: E-DCH physical uplink channel
ER scheduler: exponential rule scheduler
E-RGCH: E-DCH relative grant channel

E-RUCCH: E-DCH random access uplink control channel
Erlang capacity regions
ET gain: early termination gain
E-TSP (enhanced TSP): enhanced time-space priority
E-UCCH: E-DCH uplink control channel
EUL: enhanced uplink
EURANE: enhanced UMTS radio access network extensions for NS-2
EXP scheduler: exponential rule scheduler
F: flag
FC: flow control
FDD: frequency division duplexing
FER: frame error rate
FFT: Fast fourier transform
FFT scheduler: fast fair throughput scheduler
FIFO: first-in, first-out
FP: frame protocol, framing protocol
FSN: frame sequence number
FT scheduler: fair throughput scheduler
FWHT: fast Walsh-Hadamard transformation
GBR: guaranteed bit rate
GoS: grade-of-service
GPF scheduler: generic proportionally fair scheduler
G-RAKE receiver: generalized RAKE receiver
Gram-Schmidt procedure:
GRASP: greedy randomized adaptive search procedure
HARQ, Hybrid ARQ: hybrid automatic repeat request
HSDPA: high-speed downlink pack access
HS-DSCH (HSDSCH): high-speed downlink shared channel
HS-PDSCH: high-speed physical downlink shared channel
HS-SCCH (HSSCCH): high-speed shared control channel
HS-SICH (HSSICH): high-speed shared information channel
HSPA: high-speed packet access
IC: interference control
ICI: inter-chip interference
IE: information element
IR: incremental redundancy
ISI: inter-symbol interference
ITU: International Telecommunications Union
LAN: local area network
L-AWDF: largest average weighted delay first
LMMSE equalizer: linear minimum mean square error equalizer
LMMSE receiver: linear minimum mean square error receiver
LMMSE-ZF: linear minimum mean square error zero forcing
LNA: low noise amplifier

LPIC: linear parallel interference cancellation
LTE architecture: long-term evolution architecture
MAC layer: medium access control layer
MAC-d: dedicated MAC
MAC-e: enhanced MAC
MAP: maximum a-posteriori
Maximum C/I (Max-C/I)P: maximum carrier-to-interference ratio
MBMS: multicast broadcast multimedia services
MC-IDS: multi-carrier independence scheduling
MC-IS: multi-carrier integrated scheduling
MCS: modulation and coding scheme
MC-SS: multi-carrier separately scheduling
ME: mobile equipment
MFB: matched filter bound
MIAD: multiplicative increase/additive decrease
MIMD: multiplicative increase/multiplicative decrease
MIMO: multiple-input, multiple-output
MISO: multi-input, single-output
ML principle: maximum likelihood principle
MLPPP: multi-link point-to-point protocol
MLWDF, M-LWDF: Modified Largest Weighted Delay First rule
MMSE: minimum mean square error
MMSE-ZF: minimum mean square error zero forcing
MSWF: multi-stage Wiener filtering
MTU: maximum transfer unit
MUB: multi-user beamforming
MUD: multi-user diversity gain, multi-user detection
MUI: multi-user interference
MU-TxAA: multi-user TxAA
MU: multi-user
MUD: multi-user diversity
MUI: multi-user interface
MVU estimator: minimum variance unbiased estimator
N**:** total queue capacity
NACK: negative ACK
NBAP: Node B application part
N**:** number of MAC-d PDUs
Node-B (Node B): base station
NP**-hard (NP-hardness):** refers to a class of computationally difficult optimization problems
NRG scheduler: normalized rate guarantee scheduler
NRT: non-real-time; near-real-time
NRTV: near-real-time video
O & M: operations and management

OF: orthogonality factor
OSIC: ordered successive interference cancellation
OVSF: orthogonal variable spreading factor
PARC: per-antenna rate control
PCI: precoding control indicator
PE: polynomial expansion
PIC: parallel interference cancellation
PCI: precoding control indicator
PDF: probability density function
PDU: protocol data unit/packet data unit??
PE: polynomial expansion
PF scheduler: proportional fair scheduler, proportionally fair scheduler
PIC: parallel interference cancellation
PICH: paging indicator channel
PQ: priority queue
PS: packet switched
PS: processor sharing
PSNR: peak-signal-to-noise ratio
PSTN: public switched telephone network
QAM: quadrature amplitude modulation
QCIF: quarter common intermediate format
QoE: quality-of-experience
QoS: quality-of-service
QPSK: quadrature phase-shift keying
quality-of-service: QoS
Queue ID: Queue Identifier
R99: 3GPP Release 99
RAB: radio access bearer
RACH: random access channel
RAD scheduler: required activity detection scheduler
RAN: radio access network
RF: radio frequency
RG scheduler: rate-guarantee scheduler
RG: relative (scheduling) grant
RLC: radio link control
RLC layer: radio link control layer
RLC AM: RLC acknowledged mode
RMF: recommended modulation format
RMS: root mean square
RNC: radio network controller
RNS: radio network subsystem
RoT: rise over thermal noise
RR scheduler: Round Robin scheduler
RRC algorithms: radio resource control algorithms

RRM: radio resource management
RSN: retransmission sequence number
RT: real-time
RTBS: recommended transport-block size
RTT: round-trip time
RV: redundancy version
RX: receiver
SAW: stop-and-wait
SB scheduler: score based scheduler
SCH: synchronization channel
SDMA: spatial division multiple access
SDU: service data unit
SF: spreading factor, spread factor
Shaper: in the RNC, the Shaper shapes the arriving MAC-d PDUs according to the shaping
SHO: soft handover
SIC: successive interference cancellation
SID: size index identifier
signal-to-interference plus noise ratio: SINR
SIMO: single-input, multiple output; single-input, multi-output
SINR: signal-to-interference plus noise ratio
SISO: single-input, single-output
SNR: signal-to-noise ratio
SPI: scheduling priority indicator
SR: selective repeat
SR-ARQ: selective repeat automatic repeat request
SRNC (Serving RNC): serving radio network controller
STTD: space-time transmit diversity
SU: single user
SUMF: single-user matched filter
TBS: transport block size
TCI CF: TNL congestion indication control frame
TCP: transmission control protocol
TDD: time division duplexing
TDMA: time division multiple access
TD-SCDMA: time division duplex-synchronous code division multiple access
TD-SCDMA DCH: time division duplex-synchronous code division multiple access dedicated channel
TD-SCDMA HSDPA/HSUPA: time division duplex-synchronous code division multiple access high-speed downlink/uplink packet access
TFC: transmission format combination
TFRC: transport format and resource combination
TNL: transport network layer

TSN: transmission sequence number
TSP: time-space priority
TTI: transmission time interval
Tx window: transmission window
TxAA: transmit antenna array
UBS: user buffer size
UDP/IP: user datagram protocol/internet protocol
UE: user equipment
UL: uplink
UMTS: universal mobile telecommunications systems
UP: user plane
UpTPS: uplink pilot time slot
USIM: UMTS subscriber identity module
UTRAN: UMTS terrestrial radio access network
Uu interface: HSUPA air interface
Uu scheduler: HSUPA air interface scheduler
VF: version flag
VoIP: Voice over IP, Voice over Internet Protocol
WCDMA: wideband code division multiple access
WFQ scheduler: weighted fair queuing scheduler
WHT: Walsh-Hadamard transformation
ZF: zero forcing

Index

16QAM, 118, 207
  bit rates, 214
  HSDPA indoor planning, 406
  HSDPA link-level simulation, 336
  TD-SCDMA HSDPA, 5
    HS-DSCH channel processing, 12, 13
    HS-SCCH channel processing, 16
  TD-SCDMA HSUPA
    E-DCH, 28
    HARQ, 24
  UMTS-FDD mode system with HSDPA support, 120–121

A

AAL2/ATM (ATM adaptation layer 2/asynchronous transfer mode), 305, 329
Absolute grant (AG), 248, 315, 355
Absolute grant enhanced-dedicated channel (E-AGCH)
  congestion control, 315
  TD-SCDMA HSUPA, 23, 25–26
    association and timing, 27
    channel processing, 31–33
Access Link Control Application Part (ALCAP), 244
ACK/NACK (acknowledge/negative acknowledge)
  HSDPA, 209
  HSDPA link-level simulation, 338
  Iub/Iur architecture and protocols, 300
  performance evaluation, system-level simulation, 345, 347

performance evaluation, system-level simulator for HSUPA, 348
TD-SCDMA HSDPA
  HS-DSCH link adaptation, 18
  HS-SCCH and HS-SICH, 11
  HS-SICH channel processing, 14
TD-SCDMA HSUPA, E-HICH, 33
UMTS slot structures and timing, 120
UTRAN protocol structure, 250
Acknowledged mode (AM), 225
Acknowledge/negative acknowledge; See ACK/NACK (acknowledge/negative acknowledge)
Acquisition indicator channel (AICH), 279
Active multirate systems, iterative receivers based on chip equalizers, 153
Actual value interface (AVI), 34
Adaptation
  data rate, UTRAN architecture, 239
  filter, polynomial expansion receiver, 157–160
  link, 118
Adaptation layer, AAL2/ATM, 305, 329
Adaptive equalizers, 120
Adaptive Iub link management algorithm, 265–268
Adaptive modulation and coding (AMC), 176–177, 245, 336
  HSDPA link-level simulation, 336
  MIMO systems, 49
  TD-SCDMA HSDPA, 5, 6–7
Adaptive multirate (AMR) voice codec, 309, 414